T0262154

Mobile Computing: Technology and Applications

Mobile Computing: Technology and Applications

Edited by Joseph Anderson

CLANRYE
INTERNATIONAL
www.clanryeinternational.com

Clanrye International,
750 Third Avenue, 9th Floor,
New York, NY 10017, USA

Copyright © 2020 Clanrye International

This book contains information obtained from authentic and highly regarded sources. Copyright for all individual chapters remain with the respective authors as indicated. All chapters are published with permission under the Creative Commons Attribution License or equivalent. A wide variety of references are listed. Permission and sources are indicated; for detailed attributions, please refer to the permissions page and list of contributors. Reasonable efforts have been made to publish reliable data and information, but the authors, editors and publisher cannot assume any responsibility for the validity of all materials or the consequences of their use.

Trademark Notice: Registered trademark of products or corporate names are used only for explanation and identification without intent to infringe.

ISBN: 978-1-63240-921-8

Cataloging-in-Publication Data

Mobile computing : technology and applications / edited by Joseph Anderson.
    p. cm.
Includes bibliographical references and index.
ISBN 978-1-63240-921-8
1. Mobile computing. 2. Portable computers. 3. Electronic data processing--Structured techniques.
I. Anderson, Joseph.
QA76.59 .M63 2020
004--dc23

For information on all Clanrye International publications
visit our website at www.clanryeinternational.com

Contents

Permissions

List of Contributors

Index

Preface

Every book is a source of knowledge and this one is no exception. The idea that led to the conceptualization of this book was the fact that the world is advancing rapidly; which makes it crucial to document the progress in every field. I am aware that a lot of data is already available, yet, there is a lot more to learn. Hence, I accepted the responsibility of editing this book and contributing my knowledge to the community.

Mobile computing refers to the human-computer interaction which allows the transmission of data, video and voice using a computer or any other wireless device without it being connected to a fixed physical link. It involves mobile hardware, mobile software and mobile communication. Mobile hardware deals with mobile devices or components. Mobile software encompasses the requirements and characteristics of mobile applications. Mobile communication includes the use of infrastructure networks and ad hoc networks as well as communication protocols, data formats and concrete technologies. Some mobile computing devices are portable computers, cellular telephones, smart cards and wearable computers. The chief principles of mobile computing are portability, social interactivity, connectivity and individuality. This book outlines the processes and applications of mobile computing in detail. It is a compilation of chapters that discuss the most vital concepts and emerging trends in this field. A number of latest researches have been included to keep the readers up-to-date with the global concepts in this area of study.

While editing this book, I had multiple visions for it. Then I finally narrowed down to make every chapter a sole standing text explaining a particular topic, so that they can be used independently. However, the umbrella subject sinews them into a common theme. This makes the book a unique platform of knowledge.

I would like to give the major credit of this book to the experts from every corner of the world, who took the time to share their expertise with us. Also, I owe the completion of this book to the never-ending support of my family, who supported me throughout the project.

Editor

PRESENCE: Monitoring and Modelling the Performance Metrics of Mobile Cloud SaaS Web Services

Abdallah A. Z. A. Ibrahim (iD),[1] **Muhammad Umer Wasim**,[1,2] **Sebastien Varrette** (iD),[1] **and Pascal Bouvry** (iD)[1,2]

[1]*FSTC-CSC/ILIAS–Parallel Computing and Optimization Group (PCOG), University of Luxembourg, 2 Avenue de l'Université, L-4365 Esch-sur-Alzette, Luxembourg*
[2]*Interdisciplinary Centre for Security, Reliability and Trust (SnT), Luxembourg City, Luxembourg*

Correspondence should be addressed to Abdallah A. Z. A. Ibrahim; abdallah.ibrahim@uni.lu

Academic Editor: Andrea Gaglione

Service Level Agreements (SLAs) are defining the quality of the services delivered from the Cloud Services Providers (CSPs) to the cloud customers. The services are delivered on a pay-per-use model. The quality of the provided services is not guaranteed by the SLA because it is just a contract. The developments around mobile cloud computing and the advent of edge computing technologies are contributing to the diffusion of the cloud services and the multiplication of offers. Although the cloud services market is growing for the coming years, unfortunately, there is no standard mechanism which exists to verify and assure that delivered services satisfy the signed SLA agreement in an automatic way. The accurate monitoring and modelling of the provided Quality of Service (QoS) is also missing. In this context, we aim at offering an automatic framework named PRESENCE, to evaluate the QoS and SLA compliance of Web Services (WSs) offered across several CSPs. Yet unlike other approaches, PRESENCE aims at quantifying in a fair and by stealth way the performance and scalability of the delivered WS. This article focuses on the first experimental results obtained on the accurate modelisation of each individual performance metrics. Indeed, 19 generated models are provided, out of which 78.9% accurately represent the WS performance metrics for two representative SaaS web services used for the validation of the PRESENCE approach. This opens novel perspectives for assessing the SLA compliance of Cloud providers using the PRESENCE framework.

1. Introduction

As per NIST definition [1], Cloud Computing (CC) is a recent computing paradigm for *"enabling ubiquitous, convenient, on-demand network access to a shared pool of configurable computing resources (e.g., networks, servers, storage, applications, and services) that can be rapidly provisioned and released with minimal management effort or service provider interaction."* These resources are operated by a Cloud Services Provider (CSP), which typically delivers its services using one of three traditional models: Infrastructure-as-a-Service (IaaS), Platform-as-a-Service (PaaS), or Software-as-a-Service (SaaS) [2, 3]. This classification has since evolved to take into account the federation of more and more diverse computing resources. For instance, recent developments around Fog and Edge computing permitted to enlarge the scope of CC around Mobile CC, which offer new types of services and facilities to mobile users [4–7]. This leads to stronger business perspectives bringing more and more actors in the competition as CSPs.

In this article, we focus on the performance evaluation of Web Services (WSs) deployed in the the context of the SaaS model by these actors acting as CSPs. These services could be used through cloud users' mobile devices or normal computers [8, 9]. In practice, WSs are delivered to the cloud customers on a pay-per-use model while the performance and Quality of Service (QoS) of the provided services are defined using services contracts or Service Level Agreement (SLA) [10, 11]. In particular, SLAs define the conditions and characteristics of the provided WS and its costs and the penalties encountered when the expected QoS is not met [12, 13]. Unfortunately, there is no standard mechanism which exists to verify and assure that delivered services satisfy the signed SLA agreement. Accurate measures of the provided Quality of Service (QoS) is also missing most of the time, which render even more difficult the possibility to evaluate on a fair basis different CSPs. The ambition of the proposed PRESENCE framework (PeRformance Evaluation of SErvices on the Cloud) is to fill this gap by offering an automated approach to evaluate and classify in a fair and by stealth way the performance and scalability of the delivered WS across multiple CSPs.

In this context, the contributions of this paper are four-fold:

(1) The presentation of the PRESENCE framework, the reasoning behind its design and organization

(2) The definition of the different module composing the framework and based on a Multi-Agent System (MAS) acting behind the PRESENCE client, which aim at tracking and modeling the WS performance using a predefined set of common performance metrics

(3) The validation and first experimental results of this module over two representative WSs relying on several reference backend services at the heart of most WSs: Apache HTTP, Redis, Memcached, MongoDB, and PostgreSQL

(4) The cloud Web Service (WS) performance metrics models such as throughput, transfer rate and latency (read and write) for HTTP, Redis, Memcached, MongoDB, and PostgreSQL.

The appropriate performance modeling depicted in this paper is crucial for the accuracy and dynamic adaptation expected within the *stealth* module of PRESENCE, which will be the object of another article. This article is an *extended* version of our presented paper during the 32nd IEEE International Conference of Information Networks (ICOIN), 2018 [14], which received the best paper award. Compared to this initial paper, the present article details the modelling of the Web Services performance metrics and brings 19 generated models, out of which 78.9% accurately represent the WS performance metrics for the two SaaS WSs.

This paper is organized as follows: Section 2 details the background of this work and reviews related works. The PRESENCE framework is described in Section 3, together with some implementation details. We then focus on the validation of the monitoring module on several reference backend services at the heart of most WSs—details and experiment results are discussed in Section 4. Finally, Section 5 concludes the paper and provides some future directions and perspectives opened by this study.

2. Context and Motivations

As mentioned before, a SLA defines the conditions and characteristics of a given WS, their costs and the penalties encountered when the expected QoS is not met. Measuring the performances of a given WS is therefore key to evaluate

whether or not the corresponding SLA is satisfied—especially from a user point of view which can thus request penalties to the CSP. However, accurate measures of the provided Quality of Service (QoS) is missing most of the time as performance evaluation is challenging in a cloud context considering that the end-users do not have a full control of the system running the service. In this context, Stantchev in [15] provides a generic methodology for the performance evaluation of cloud computing configurations based on the Non-Functional Properties (NFP) (such as, response time) of individual services. Yet, none of the steps were clearly detailed, and the evaluation is based on a single benchmark, measuring a single metric. Lee et al. in [16] propose a comprehensive model for evaluating quality of SaaS after defining the key features of SaaS, deriving the quality attributes from the key features and defining the metrics for the quality attributes. This model serves as a guideline to SaaS provider to characterize and improve the provided QoS but obviously does not address user-based evaluation. However, we used part of the proposed ontology to classify our own performance metrics. Gao et al. [17] propose a Testing-as-a-Service (TaaS) infrastructure and report a cloud-based TaaS environment with tools (known as CTaaS) developed to meet the needs in SaaS testing, performance, and scalability evaluation. One drawback of this approach is that its deployment cannot be hidden from the CSP, which might in return maliciously increase the capacity of the allocated resources to mitigate artificially the evaluation in favor of its own offering. Wen and Dong in [18] propose a quality characteristics and standards for the security and the QoS of the SaaS services. Unfortunately, the authors did not propose any practical steps to evaluate the cloud services, but only a set of recommendations.

Beyond pure performance evaluation, and to the best of our knowledge, the literature around SLA assurance and violation monitoring is relatively sparse. Cicotti et al. in [19, 20] propose a quality of services monitoring approach and SLA violation reporter which are based on APIs queries and events. Called QoSMONaaS, this proposed approach is measuring the Key Performance Indicator (KPI) for the services provided from the CSPs to the cloud customers. Ibrahim et al. in [10, 21] provide a framework to assure SLA and evaluate the performance of the cloud applications. They use the simulation and local scenarios to test the cloud applications and services. Hammadi and Hussain in [22] propose a SLA monitoring framework by a third party. The third party assesses the QoS and assures the performance and no violation in the SLA. Nevertheless, none of the above mentioned approaches feature the dynamic adaptation of the evaluation campaign as foreseen within the PRESENCE proposal.

Finally, as regards to the CSPs ranking and classification, Wagle et al. in [23, 24] provide a ranking based on the estimation and prediction of the quality of the cloud provider services. The model of estimating the performance is based on the prediction methods such as ARIMA & ETS.

Motivated by recent scandals in the automotive sector, which demonstrate the capacity of solution providers to adapt the behaviour of their product when submitted to an evaluation campaign to improve the performance results, this article presents PRESENCE, which aims at covering a large set of real benchmarks contributing to all aspects of the performance analysis while hiding the true nature of the evaluation to the CSP. Our proposal is detailed in the next section.

3. PRESENCE: Performance Evaluation of Services on the Cloud

An overview of the PRESENCE framework is proposed in Figure 1 and is now depicted. It is basically composed of five main components:

(1) A set of *agents*, each of them responsible for a specific performance metric measuring a specific aspect of the WS QoS. Those metrics have been designed to reflect scalability and performance in a representative cloud environment. In practice, several reference benchmarks have been considered and evaluated to quantify this behaviour.

(2) The *monitoring and modeling module*, responsible for collecting the data from the agent, which is used together with an application performance model [25] to assess the performance metric model.

(3) The *stealth module*, responsible to dynamically adapt and balance the workload pattern of the combined metric agents to make the resulting traffic indistinguishable from a regular user traffic from the CSP point of view.

(4) The *virtual QoS aggregator and SLA checker module*, which takes care of evaluating the QoS and SLA compliance of the WS offered across the considered CSPs. This ranking will help decision maker to determine which provider is better to use for the analyzed WS.

(5) Finally, the *PRESENCE client* (or *Auditor*) is responsible for interacting with the selected CSPs and evaluating the QoS and SLA compliance of Web Services. It is meant to behave as a regular client of the WS and can eventually be distributed across several parallel instances even if our first implementation operates a single sequential client.

3.1. The PRESENCE Agents. The PRESENCE approach is used to collect the data which represent the behaviour of the CSP and reflect the performance of the delivered services. In this context, the PRESENCE agents are responsible for a set of performance metrics measuring a specific aspect of the WS QoS. Those metrics have been designed to reflect scalability and performance in a representative cloud environment, covering different criteria summarized in Table 1. The implementation status and coverage of these metrics within PRESENCE at the time of writing is also detailed.

Most of these metrics are measured through a set of reference benchmarks, and each agent is responsible for a specific instance of one of these benchmarks. Then a multiobjective optimization heuristic is applied to evaluate the audited WS according to the different performance

domains raised by the agents. In practice, a low-level hybrid approach combining Machine Learning (for deep and reinforcement learning) and evolutionary-based metaheuristics compensate the weaknesses of one method with the strengths of the other. More specifically, a Genetic Programming Hyper-heuristic (GPHH) approach will be used to automatically generate heuristics using building blocks extracted from the problem definition and the benchmarks domains. Such a strategy has been employed with success in [26, 27], and we are confident it could be efficiently applied to fit the context of this work. It is worth to note that the metrics marked as not yet implemented within PRESENCE at the time of writing are linked to the cost and the availability of the checked service. The current paper validates the approach against a set of classical WSs deployed in a local environment and introduces a complex modelling and evaluation for the performance metrics.

As regards the *stealth* module, PRESENCE aims at relying on a GA [28] approach to mitigate and adapt the concurrent executions of the different agents by evolving their respective parameters and thus the visible load pattern toward the CSP. An *Oracle* is checked upon each iteration of the GA and based on a statistical correlation of the resulting pattern against a reference model corresponding to a regular usage. When this oracle is unable to statistically distinguish the outgoing modeled pattern of the client from a regular client, we consider that we can apply one of the found solutions for a real evaluation campaign of the checked CSP. Finally, the virtual QoS aggregator and SLA checker rely on the CSP ranking and classification proposed by Wagle et al. in [23, 24].

3.2. Monitoring and Modeling for the Dynamic Adaptation of the Agents. The first step to ensure the dynamic adaptation of the workload linked to the evaluation process resides in the capacity to model accurately this workload based on the configuration of a given agent. Modelling the performance metric will help the other researchers to generate data representing the CSP's behaviour under a high load and under the normal usage in just a couple of minutes without any experiments. In this context, the multiple runs of the agents are stored and analyzed in a Machine Learning process. During the training part, the infrastructure model representing the CSP side which contains the SaaS services is first virtualized locally to initiate the first collection of data sample and setup the PRESENCE client (i.e., the auditor) based on a representative environment. The second phase of the training involves "real" runs permitting the modeling of each metrics. In practice, PRESENCE relies on a simulation software called *Arena* [29] to analyse the data returned from the agents and get the model for each individual performance metric. Arena is a simulation software by Rockwell Corporation. It is used in different application domains, from manufacturing to supply chain and from customer service and strategies to internal business processes. It includes three modules, respectively, called *Arena Input Analyser*, *Output Analyser*, and *Process Analyser*. Among the three, the *Input Analyser* is useful for determining an appropriate distribution for collected data. It allows the user to make a sample set of raw data (e.g., latency of Cloud-based

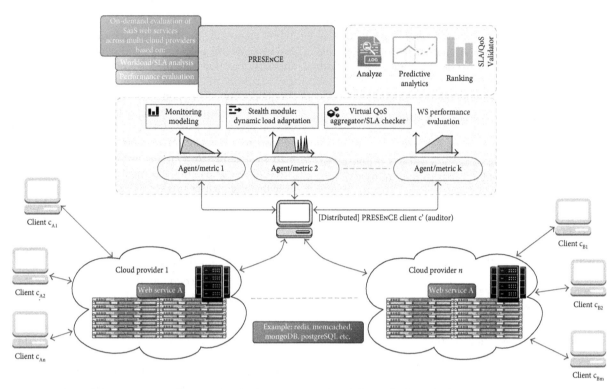

FIGURE 1: Overview of the PRESENCE framework. The figure is reproduced from Ibrahim et al. [14].

TABLE 1: Performance metrics used in PRESENCE.

Domain	Metric/Implementation status		Metric type
Scalability	Number of transactions	✓	Workload/ performance indicator
	Number of requests	✓	
	Number of operations	✓	
	Number of records	✓	
	Number of fetches	✓	
Reliability	Parallel connections (clients)	✓	Workload
	Number of pipes	✓	
	Number of threads	✓	
	Workload size	✓	
Availability	Response time	✗	Performance indicator
	Up time	✗	
	Down time	✗	
	Load balancing	✗	
Performance	Latency	✓	Performance indicator
	Throughput	✓	
	Transfer rate	✓	
	Miss/hit rate	✓	
Costs	Installing costs	✗	Quality indicator
	Running costs	✗	
Security	Authentication	✓	Security indicator
	Encryption	✓	
	Auditability	✓	

WS) and fit it into a statistical distribution. This distribution then can be incorporated directly into a model to develop and understand the corresponding system performance.

Then, assuming such a model is available for each considered metric, PRESENCE aims at adapting its client c'

(i.e., the auditor) to ensure the evaluation process is performed in a stealth way. In this paper, 19 models have been generated for each agent—they are listed in the next section. Of course, once a model is provided, we should *validate* it, that is, ensure that the model is an accurate representation of the actual metric evolution and behaves in the same way.

There are many tests that could be used to validate on the models generated. These tests are used to check the accuracy of the models by verifying on the null hypothesis. The tests are such as t-test, Wilcoxon–Mann–Whitney test, and Anova test. In PRESENCE, this is achieved by using t-test by comparing means of raw data and statistical distribution generated by the agent analysis. The use of t-test is based on the fact that the variable(s) in question (e.g., Throughput) is normally distributed. When this assumption is in doubt, the nonparametric Wilcoxon–Mann–Whitney test is used as an alternative to the t-test [30]. As we will see, 78.9% of the generated models are proved as an accurate representation of the WS performance metrics exhibited by the PRESENCE agents. In the next section, the modelling of the performance metrics are detailed besides the experiment results of PRESENCE.

4. Validation and First Experimental Results

This section presents the results obtained from the PRESENCE approach within the *monitoring and modeling* module as a prelude to the validation of the stealth module and the virtual QoS aggregator and SLA checker left for the sequel of this work.

4.1. Experimental Setup. In an attempt to validate the approach on realistic workflows, we tested PRESENCE against two traditional and core web services:

(1) A multi-*DataBase (DB) WS*, which offers both SQL (i.e., PostgreSQL 9.4) and NoSQL DBs. For the later case, we deployed two reference in-memory data structure stores, used as a database, cache, and message broker, that is, Redis 2.8.17 (redis.io) and Memcached 1.5.0 (memcached.org), as well as MongoDB 3.4, an open-source document database that provides high performance, high availability, and automatic scaling.

(2) The reference HTTP Apache server (version 2.2.22-13), which is used for testing a traditional *HTTP WS* on the cloud.

Building such a CSP environment was performed on top of two physical servers running Ubuntu 14.04 LTS (Trusty 64) connected over a 1 GbE network.

On the PRESENCE client side, 8 agents are deployed as KVM guests, that is, virtual machines running CentOS 7.3 over 4 physical servers. Each agent is running one of the benchmarking tool listed in Table 2 to evaluate the WS performance and collecting data about the CSP behaviour. Each PRESENCE agent thus measures a specific subset of performance metrics and attributes and also deals with specific kinds of cloud servers. Each measurement is consisting of an average over 100 runs collected by the PRESENCE agent to make the data statistically significant. The tools used for performance evaluation are several reference benchmarking such as Yahoo Cloud Serving (YCSB) [31], Memtire [32], Redis benchmark [33], Twitter RPC [34], Pgbench [35], HTTP load [36], and Apache AB [37]. In addition, iperf [38] (a tool for active measurements of the maximum achievable bandwidth on IP networks) is used in the closed environment for the validation of PRESENCE, as it provides an easy testimonial for the WS access capacity. The general overview of the deployed infrastructure is provided in Figure 2.

4.2. PRESENCE Agents Evaluation Results. The targeted WS of each deployed agent is precised in Table 2. The PRESENCE approach is used to collect the data which represent the behaviour of the CSP, and these data also can indicate and evaluate the performance of the services. As mentioned in the previous section, there are many metrics that can represent the performance. PRESENCE uses some of these metrics as a workload to the CSP's servers and the others as results from the experiments. For example, the number of requests, operations, records, transactions, and fetches are metrics which representing the scalability of the CSP and are used by PRESENCE to increase the workload to see the behaviour of the servers under the workload. Other metrics like parallel or concurrency connections, number of pipes, number of threads, and the workload size are representing the CSP reliability are also used by PRESENCE to increase the workload during the test. Other metrics like response time, up time, down time, transfer rate, latency (read and update), and throughput are indicating the CSP performance and availability and PRESENCE used them to

TABLE 2: Benchmarking tools used by PRESENCE agents.

Benchmark tool	Version	Targeted WS
YCSB	0.12.0	Redis, MongoDB, memcached, DynamoDB, etc.
Memtire-Bench	1.2.8	Redis, memcached
Redis-Bench	2.4.2	Redis
Twitter RPC-Perf	2.0.3-pre	Redis, memcached, Apache
PgBench	9.4.12	Postgresql
Apache AB	2.3	Apache
HTTP Load	1	Apache
Iperf	v1, v3	Iperf server

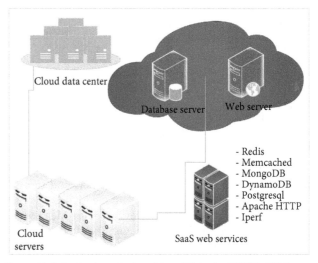

FIGURE 2: Infrastructure deployed to validate the PRESENCE framework. The figure is reproduced from Ibrahim et al. [14].

evaluate the services performance. Because PRESENCE uses many tools to evaluate and benchmark the services, it can deal with most of the metrics. But, there are two or three common metrics we will model and represent them in the results, such as latency, throughput, and transfer rate. There are other metrics that represent the security and the costs of the CSPs, and all those metrics are summarized in Table 1 in the previous section. The different parameters (both input and output) which are used for the PRESENCE validation are provided in Table 3. We now provide some of the numerous traces produced by the execution of the PRESENCE agents when checking the performance of the DB and HTTP WSs.

Figure 3 shows the Redis, Memcached, and Mongo measured WS performance under the statistically significant stress produced by the PRESENCE agent running the YCSB benchmarking tool. Figure 3(a) shows the throughput of the three backends and demonstrates that the Redis WS is the best in this metric when compared to the other two WSs where it has the highest throughput. This trend is confirmed when the latency metric is analysed in Figures 3(b) and 3(c).

Figure 4 shows the Redis and Memcached measured WS performance under the stress produced by the PRESENCE agent running the Memtier benchmarking tool. The previous trend is again confirmed in our runs; that is, the Redis-based WS performs better than the Memcached backend for

TABLE 3: Input/output metrics for each PRESENCE Agent.

PRESENCE agent	Input parameters								
	#Transactions	#Requests	#Operations	#Records	#Fetches	#Parallel clients	#Pipes	#Threads	Workload size
YCSB			✓	✓				✓	✓
Memtire-Bench		✓				✓		✓	✓
Redis-Bench		✓				✓	✓		✓
Twitter RPC-Perf		✓							✓
PgBench	✓					✓		✓	
Apache AB		✓				✓			
HTTP Load					✓	✓			
Iperf								✓	

PRESENCE agent	Output parameters								
	Throughput	Latency	Read latency	Update latency	CleanUp latency	Transfer rate	Response time	Miss	Hits
YCSB	✓		✓	✓	✓			✓	✓
Memtire-Bench	✓	✓				✓		✓	✓
Redis-Bench	✓								
Twitter RPC-Perf	✓	✓						✓	✓
PgBench	✓	✓							
Apache AB	✓						✓		
HTTP Load	✓	✓					✓		
Iperf							✓		

all metrics, for example, throughput, latency, and transfer rate. As noticed in the three plots from the Memtier agents, the latency, throughput, and transfer rate of the Memcached WS have increased suddenly in the end. Such behaviour was consistent across all runs and was linked to the memory saturation reached by the server process before being clearer. Still upon DB WS performance evaluation, Figure 5 details the performance of the PostgreSQL WS under the stress produced by the PRESENCE agent executing the PgBench benchmarking tool. The figure shows the normalized response time of the server and the normalized (standardized) number of Transactions per Second (TPS). The response time is the latency of the service when the TPS corresponds to its throughput. The performance of the WS is affected by the increased workload which is represented by the increasing number of TPSs and parallel clients. The increasing of the TPS let the response time increasing even if the TPS was going down, and after filling in the memory, the TPS decreased again and response time returned back again to a decrease. This behaviour was consistent across the runs of the PgBench agent. Finally, Figure 6 shows the average runs of the PRESENCE agent executing the HTTP Load benchmark tool when assessing the performance of the HTTP WS. We exhibit on each subfigure the behaviour of both the latency and throughput against an increasing number of fetches and parallel clients, which increases the workload of the WS.

Many more traces of all considered agent runs are available but were not displayed in this article for the sake of conciseness. Overall, and to conclude on the collected traces from the many agent runs depicted in this section, we were able to reflect several complementary aspects of the two WSs considered in the scenario of this experimental validation. Yet, as part of the contributions of this article, the generation of accurate models for these evaluations is crucial. They are detailed in the next section.

4.3. WS Performance Metrics Modeling. Outside the description of the PRESENCE framework, the main contribution of this article resides more on the *modeling* of the measured WS performance metrics from the data collected by the PRESENCE agents rather than the runs in themselves depicted in the previous section. Once these models are available, they can be used to estimate and dynamically adapt the behaviour of the PRESENCE client (which combine and schedule the execution of all considered agents in parallel) so as to hide the true nature of the evaluation by making it indistinguishable from a regular client traffic. But this corresponds to the next step of our work. In this paper, we wish to illustrate the developed model from the PRESENCE agents evaluations reported in the previous section. The main objective of the developed model is to understand the system performance behaviour relative to various assumptions and input parameters discussed in previous sections. As mentioned in Section 3.2, we rely on the simulation software called *Arena* to analyse the data returned from the agents and get the model for each individual performance metric. We have performed this analysis for each agents, and the results of the models are presented in the below tables. Of course, such a contribution is pertinent only if the generated model is validated—a process consisting in ensuring the accurate representation of the actual system and its behaviour. This is achieved by using a set of statistical t-tests by comparing the means of raw data and statistical distribution generated by the Input Analyzer of the Arena system. If the result shows that both samples are analytically similar, then the model developed from statistical distribution is an accurate representation of the actual system and behaves in the same way. For each model detailed in the tables, the outcomes of the t-tests in the form of the computed p value (against the common significance level of 0.05) is provided and demonstrate if present the accuracy of the proposed models.

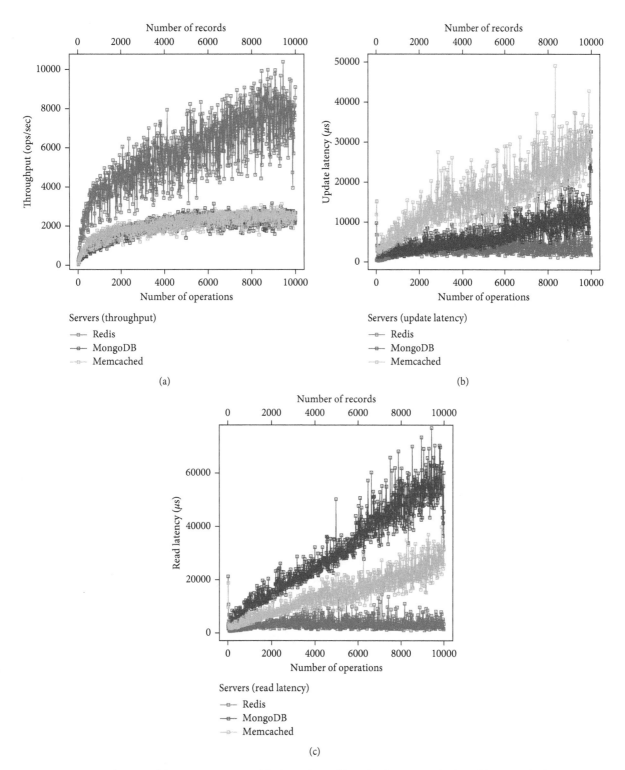

FIGURE 3: YCSB Agent Evaluation of the NoSQL DB WS. (a) Throughput, (b) update latency, and (c) read latency. The figure is reproduced from Ibrahim et al. [14].

Tables 4–6 show the model of the performance metrics such as latency (read and update) and throughput for the Redis, Memcached, and MongoDB services by using the data collected with the YCSB PRESENCE agents. In particular, Table 4 shows that the Redis throughput is Beta increasing, Redis latency (read) is Gamma increasing, and Redis latency (update) is Erlang increasing with respect to the configuration of the experimental setup discussed in the previous sections. Moreover, as p values (0.757, 0.394, and 0.503) are greater than 0.05, the null hypothesis (the two samples are the same) is accepted as compared to alternate hypothesis (the two samples are different). Hence, the models for throughput, that is, $-0.001 + 1 * \text{BETA}(3.63, 3.09)$, latency (read), that is,

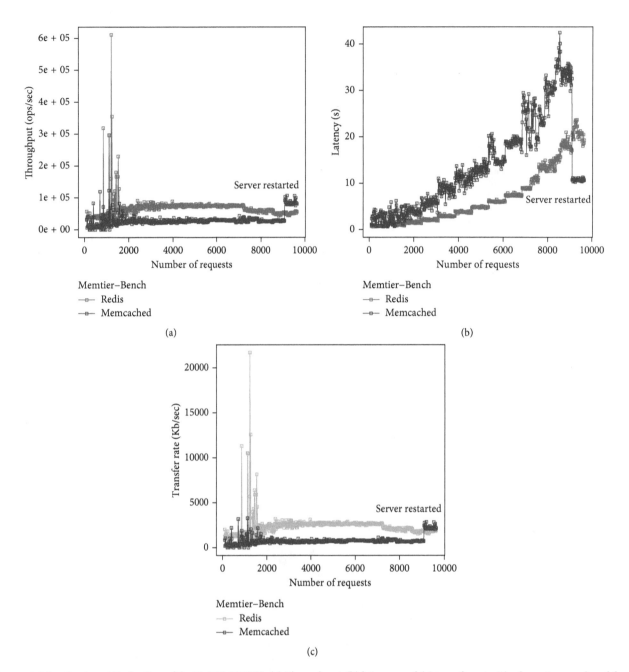

FIGURE 4: Memtier Agent Evaluation of the NoSQL DB WS. (a) Throughput, (b) latency, and (c) transfer rate. The figure is reproduced from Ibrahim et al. [14].

$-0.001 + \text{GAMM}(0.0846, 2.39)$, and latency (update), that is, $-0.001 + \text{ERLA}(0.0733, 3)$, are an accurate representation of the WS performance metrics exhibited by the PRESENCE agent in Figure 3.

Similarly, Table 5 shows that MongoDB throughput and latency (read) are Beta increasing and latency (update) is Erlang increasing with respect to the configuration setup. Moreover, as p values (0.388, 0.473, and 0.146) are greater than 0.05, the null hypothesis (the two samples are the same) is accepted as compared to alternate hypothesis (the two samples are different). Hence, the models for throughput, that is, $-0.001 + 1*\text{BETA}(3.65, 2.11)$, latency (read), that is, $-0.001 + 1*\text{BETA}(1.6, 2.48)$, and latency (update), that is,

$-0.001 + \text{ERLA}(0.0902, 2)$, are an accurate representation of the WS performance metrics exhibited by the PRESENCE agents in Figure 3.

Finally, Table 6 shows that Memcached throughput and latency (read) is Beta increasing and latency (update) is Normal increasing with respect to configuration setup. Again, as p values (0.106, 0.832, and 0.794) are greater than 0.05, the null hypothesis is accepted. Hence, the models for throughput, that is, $-0.001 + 1*\text{BETA}(4.41, 2.48)$, latency (read), that is, $-0.001 + 1*\text{BETA}(1.64, 3.12)$, and latency (update), that is, $\text{NORM}(0.311, 0.161)$, are an accurate representation of the WS performance metrics exhibited by the PRESENCE agents in Figure 3.

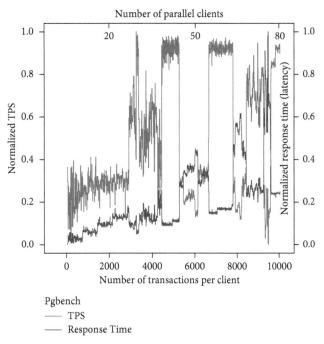

FIGURE 5: PgBench Agent on the SQL DB WS. The figure is reproduced from Ibrahim et al. [14].

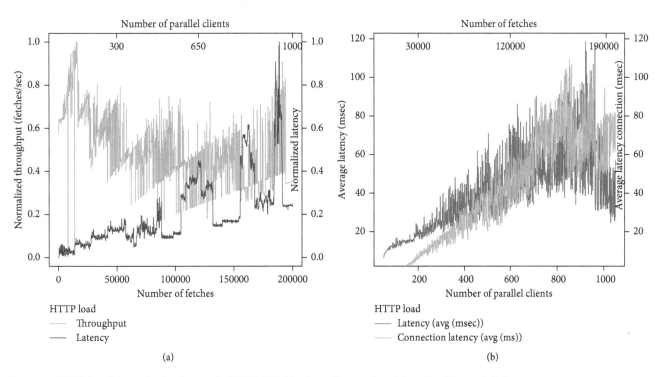

FIGURE 6: HTTP Load Agent Evaluation on the HTTP WS. The figure is reproduced from Ibrahim et al. [14].

The above analysis is repeated for the Memtier PRESENCE agent—the corresponding models are provided in Tables 7 and 8 and summarize the computed model for the WS performance metrics such as latency, throughput, and transfer rate for the Redis and Memcached services by using the data collected from PRESENCE Memtier agents. For instance, it can be seen in Table 7 that the Redis throughput is Erlang increasing with respect to time and assumptions in previous section. Moreover, as p value (0.902) is greater than 0.05, the null hypothesis is again accepted as compared to alternate hypothesis (the two samples are different). Hence, the Erlang distribution

TABLE 4: Modelling DB WS performance metrics from YCSB PRESENCE Agent: Redis.

Metric	Distribution	Model	Expression	p value (t-test)
Throughput	Beta	$-0.001 + 1 * \text{BETA}(3.63, 3.09)$ where $\text{BETA}(\beta, \alpha)$ $\beta = 3.63$ $\alpha = 3.09$ offset $= -0.001$	$f(x) = \begin{cases} (x^{\beta-1}(1-x)^{\alpha-1})/B(\beta,\alpha) & \text{for } 0 < x < 1 \\ 0 & \text{otherwise,} \end{cases}$ where β is the complete beta function given by $B(\beta,\alpha) = \int_0^1 t^{\beta-1}(1-t)^{\alpha-1}\,dt$	$0.757\,(>0.05)$
Latency read	Gamma	$-0.001 + \text{GAMM}(0.0846, 2.39)$ where $\text{GAMM}(\beta, \alpha)$ $\beta = 0.0846$ $\alpha = 2.39$ offset $= -0.001$	$f(x) = \begin{cases} (\beta^{-\alpha} x^{\alpha-1} e^{-(x/\beta)})/\Gamma(\alpha) & \text{for } x > 0 \\ 0 & \text{otherwise,} \end{cases}$ where Γ is the complete gamma function given by $\Gamma(\alpha) = \int_0^{\inf} t^{\alpha-1} e^{-1}\,dt$	$0.394\,(>0.05)$
Latency update	Erlang	$-0.001 + \text{ERLA}(0.0733, 3)$ where $\text{ERLA}(\beta, k)$ $k = 3$ $\beta = 0.0733$ offset $= -0.001$	$f(x) = \begin{cases} (\beta^{-k} x^{k-1} e^{-(x/\beta)})/(k-1)! & \text{for } x > 0 \\ 0 & \text{otherwise} \end{cases}$	$0.503\,(>0.05)$

TABLE 5: Modelling DB WS performance metrics from YCSB PRESENCE Agent: MongoDB.

Metric	Distribution	Model	Expression	p value (t-test)
Throughput	Beta	$-0.001 + 1 * \text{BETA}(3.65, 2.11)$ where $\text{BETA}(\beta, \alpha)$ $\beta = 3.65$ $\alpha = 2.11$ offset $= -0.001$	$f(x) = \begin{cases} (x^{\beta-1}(1-x)^{\alpha-1})/B(\beta,\alpha) & \text{for } 0 < x < 1 \\ 0 & \text{otherwise,} \end{cases}$ where β is the complete beta function given by $B(\beta,\alpha) = \int_0^1 t^{\beta-1}(1-t)^{\alpha-1}\,dt$	$0.388\,(>0.05)$
Latency read	Beta	$-0.001 + 1 * \text{BETA}(1.6, 2.48)$ where $\text{BETA}(\beta, \alpha)$ $\beta = 1.6$ $\alpha = 2.48$ offset $= -0.001$	$f(x) = \begin{cases} (x^{\beta-1}(1-x)^{\alpha-1})/B(\beta,\alpha) & \text{for } 0 < x < 1 \\ 0 & \text{otherwise,} \end{cases}$ where β is the complete beta function given by $B(\beta,\alpha) = \int_0^1 t^{\beta-1}(1-t)^{\alpha-1}\,dt$	$0.473\,(>0.05)$
Latency update	Erlang	$-0.001 + \text{ERLA}(0.0902, 2)$ where $\text{ERLA}(\beta, k)$ $k = 2$ $\beta = 0.0902$ offset $= -0.001$	$f(x) = \begin{cases} (\beta^{-k} x^{k-1} e^{-(x/\beta)})/(k-1)! & \text{for } x > 0 \\ 0 & \text{otherwise} \end{cases}$	$0.146\,(>0.05)$

model of Throughput, that is, $-0.001 + \text{ERLA}(0.0155, 7)$, is an accurate representation of the WS performance metrics exhibited by the PRESENCE agents in Figure 4(a). The same conclusions on the generated models can be triggered for the other metrics collected by the PRESENCE Memtier agents, that is, latency and transfer rates. Interestingly, Table 8 reports an approximation (blue curve line) for multimodel distribution of memcached throughput and transfer rate. This shows a failure of a single model to capture the system behaviour with respect to the configuration setup. However for the same setup, memcached latency is Beta increasing, and as its p value (0.625) is greater than 0.05, the null hypothesis is accepted. Hence, the model for latency, that is, $-0.001 + 1*\text{BETA}(0.99, 2.1)$, is an accurate representation

of the WS performance metrics exhibited by the PRESENCE agents in Figure 4(b).

To finish on the DB WS performance analysis using PRESENCE, Table 9 exhibits the generated model for the performance model from the Pgbench PRESENCE agents. The first row in the table shows the approximation (blue curve line) for multimodel distribution of the throughput metric, thus demonstrating the failure of a single model to capture the system behaviour with respect to the configuration setup. However for the same setup, the latency metric is log normal increasing, and as its p value (0.682) is greater than 0.05, the null hypothesis (i.e., the two samples are the same) is accepted as compared to alternate hypothesis, that is, the two samples are different. Hence, the model for latency, that is, $-0.001 + \text{LOGN}(0.212, 0.202)$, is an accurate representation

TABLE 6: Modelling DB WS performance metrics from YCSB PRESENCE Agent: Memcached.

Metric	Distribution	Model	Expression	p value (t-test)
Throughput	Beta	$-0.001 + 1 * \text{BETA}(4.41, 2.48)$ where $\text{BETA}(\beta, \alpha)$ $\beta = 4.41$ $\alpha = 2.48$ offset $= -0.001$	$f(x) = \begin{cases} (x^{\beta-1}(1-x)^{\alpha-1})/B(\beta,\alpha) & \text{for } 0 < x < 1 \\ 0 & \text{otherwise,} \end{cases}$ where β is the complete beta function given by $B(\beta,\alpha) = \int_0^1 t^{\beta-1}(1-t)^{\alpha-1}\, dt$	$0.106\,(>0.05)$
Latency read	Beta	$-0.001 + 1 * \text{BETA}(1.64, 3.12)$ where $\text{BETA}(\beta, \alpha)$ $\beta = 1.64$ $\alpha = 3.12$ offset $= -0.001$	$f(x) = \begin{cases} (x^{\beta-1}(1-x)^{\alpha-1})/B(\beta,\alpha) & \text{for } 0 < x < 1 \\ 0 & \text{otherwise,} \end{cases}$ where β is the complete beta function given by $B(\beta,\alpha) = \int_0^1 t^{\beta-1}(1-t)^{\alpha-1}\, dt$	$0.832\,(>0.05)$
Latency update	Normal	$\text{NORM}(0.311, 0.161)$ where $\text{NORM}(\text{mean}\mu, \text{stdDev}\sigma)$ $\mu = 0.311$ $\sigma = 0.161$	$f(x) = (1/\sigma\sqrt{2\pi})e^{-(x-\mu)^2/2\sigma^2}$ for all real x	$0.794\,(>0.05)$

TABLE 7: Modelling DB WS performance metrics from Memtier PRESENCE Agent: Redis.

Metric	Distribution	Model	Expression	p value (t-test)
Throughput	Erlang	$-0.001 + \text{ERLA}(0.0155, 7)$ where $\text{ERLA}(\beta, k)$ $k = 7$ $\beta = 0.0155$ offset $= -0.001$	$f(x) = \begin{cases} (\beta^{-k}x^{k-1}e^{-(x/\beta)})/(k-1)! & \text{for } x > 0 \\ 0 & \text{otherwise} \end{cases}$	$0.767\,(>0.05)$
Latency	Beta	$-0.001 + 1 * \text{BETA}(0.648, 1.72)$ where $\text{BETA}(\beta, \alpha)$ $\beta = 0.648$ $\alpha = 1.72$ offset $= -0.001$	$f(x) = \begin{cases} (x^{\beta-1}(1-x)^{\alpha-1})/B(\beta,\alpha) & \text{for } 0 < x < 1 \\ 0 & \text{otherwise,} \end{cases}$ where β is the complete beta function given by $B(\beta,\alpha) = \int_0^1 t^{\beta-1}(1-t)^{\alpha-1}\, dt$	$0.902\,(>0.05)$
Transfer rate	Erlang	$-0.001 + \text{ERLA}(0.0155, 7)$ where $\text{ERLA}(\beta, k)$ $k = 7$ $\beta = 0.0155$ offset $= -0.001$	$f(x) = \begin{cases} (\beta^{-k}x^{k-1}e^{-(x/\beta)})/(k-1)! & \text{for } x > 0 \\ 0 & \text{otherwise} \end{cases}$	$0.287\,(>0.05)$

of the WS performance metrics exhibited by the PRESENCE agents in Figure 6. As regards the HTTP WS performance analysis using PRESENCE, we decided to report in Table 10 the model generated from the performance model from the HTTP Load PRESENCE agents. Again, we can see that we fail to model the throughput metric with respect to the configuration setup discussed in the previous section. However, for the same setup, the response time is Beta increasing, and as its p value (0.165) is greater than 0.05, the null hypothesis is accepted. Hence, the model for latency, that is, $-0.001 + 1 * \text{BETA}(1.55, 3.46)$, is an accurate representation of the WS performance metrics exhibited by the PRESENCE agents in Figure 6.

Summary of the obtained Models: In this paper, 19 models were generated which represent the performance metrics for the SaaS Web Service by using the PRESENCE approach. Out of the 19 models, 15 models, that is, 78.9% of the analyzed models are proved to have accurately represent the performance metrics collected by the PRESENCE agents, such as throughput, latency, transfer rate, and response time in different contexts depending on the considered WS. The accuracy of the proposed models is assessed by the reference statistical t-tests, performed against the common significance level of 0.05.

5. Conclusion

Motivated by recent scandals in the automotive sector (which demonstrate the capacity of solution providers to adapt the behaviour of their product when submitted to an evaluation campaign to improve the performance results), this paper presents PRESENCE, an automatic framework

TABLE 8: Modelling DB WS performance metrics from Memtier PRESENCE Agent: Memcached.

Metric	Distribution	Model	Expression	p value (t-test)
Throughput	—	—		—
Latency read	Beta	$-0.001 + 1 * \text{BETA}(0.99, 2.1)$ where $\text{BETA}(\beta, \alpha)$ $\beta = 0.99$ $\alpha = 2.1$ offset $= -0.001$	$f(x) = \begin{cases} (x^{\beta-1}(1-x)^{\alpha-1})/B(\beta, \alpha) & \text{for } 0 < x < 1 \\ 0 & \text{otherwise,} \end{cases}$ where β is the complete beta function given by $B(\beta, \alpha) = \int_0^1 t^{\beta-1}(1-t)^{\alpha-1}\, dt$	0.625 (>00.05)
Latency update	—	—		—

TABLE 9: Modelling DB WS performance metrics from Pgbench PRESENCE Agent.

Metric	Distribution	Model	Expression	p value (t-test)
Throughput	—	—		—
Latency	Log normal	$-0.001 + \text{LOGN}(0.212, 0.202)$ where $\text{LOGN}(\log \text{Mean}\mu, \text{LogStd}\sigma)$ $\mu = 0.212$ $\sigma = 0.202$ offset $= -0.001$	$f(x) = \begin{cases} (x^{\beta-1}(1-x)^{\alpha-1})/B(\beta, \alpha) & \text{for } 0 < x < 1 \\ 0 & \text{otherwise,} \end{cases}$ where β is the complete beta function given by $B(\beta, \alpha) = \int_0^1 t^{\beta-1}(1-t)^{\alpha-1}\, dt$	0.682 (>0.05)

TABLE 10: Modelling HTTP WS performance metrics from HTTP load PRESENCE Agent.

Metric	Distribution	Model	Expression	p value (t-test)
Throughput	—	—		—
Latency	Log normal	$-0.001 + 1 * \text{BETA}(1.55, 3.46)$ where $\text{BETA}(\beta, \alpha)$ $\beta = 1.55$ $\alpha = 3.46$ offset $= -0.001$	$f(x) = \begin{cases} 1/(\sigma x \sqrt{2\pi})e^{-(\ln(x) - \mu)^2/2\sigma^2} & \text{for } x > 0 \\ 0 & \text{otherwise} \end{cases}$	0.165 (>0.05)

which aims at evaluating, monitoring, and benchmarking Web Services (WSs) offered across several Cloud Services Providers (CSPs) for all types of Cloud Computing (CC) and Mobile CC platforms. More precisely, PRESENCE aims at evaluating the QoS and SLA compliance of Web Services (WSs) by stealth way, that is, by rendering the performance evaluation as close as possible from a regular yet heavy usage of the considered service. Our framework is relying on a Multi-Agent System (MAS) and a carefully designed client (called the *Auditor*) responsible to interact with the set of CSPs being evaluated.

The first step to ensure the dynamic adaptation of the workload to hide the evaluation process resides in the capacity to model accurately this workload based on the configuration of the agents responsible for the performance evaluation.

In this paper, a nonexhaustive list of 22 metrics was suggested to reflect all facets of the QoS and SLA compliance of a WSs offered by a given CSP. Then, the data collected from the execution of each agent within the PRESENCE client can be then aggregated within a dedicated module and treated to exhibit a rank and classification of the involved CSPs. From the preliminary modelling of the load pattern and performance metrics of each agent, a stealth module takes care of finding through a GA the best set of parameters for each agent such that the resulting pattern of the PRESENCE client is indistinguishable from a regular usage. While the complete framework is described in the seminal paper for PRESENCE [14], the first experimental results presented in this work focus on the performance and networking metrics between cloud providers and cloud customers.

In this context, 19 generated models were provided, out of which 78.9% accurately represent the WS performance metrics for the two SaaS WSs deployed in the experimental setup. The claimed accuracy is confirmed by the outcome of reference statistical t-tests and the associated p values computed for each model against the common significance level of 0.05.

This opens novel perspectives for assessing the SLA compliance of Cloud providers using the PRESENCE framework. The future work induced by this study includes the modelling and validation of the other modules defined within PRESENCE and based on the monitoring and modelling of the performance metrics proposed in this article. Of course, the ambition remains to test our framework against a real WS while performing further experimentation on a larger set of applications and machines.

Acknowledgments

The authors would like to acknowledge the funding of the joint ILNAS-UL Programme on Digital Trust for Smart-ICT. The experiments presented in this paper were carried out using the HPC facilities of the University of Luxembourg [39] (see http://hpc.uni.lu).

References

[1] P. M. Mell and T. Grance, "SP 800–145. The NIST definition of cloud computing," Technical Report, National Institute of Standards & Technology (NIST), Gaithersburg, MD, USA, 2011.

[2] Q. Zhang, L. Cheng, and R. Boutaba, "Cloud computing: state-of-the-art and research challenges," *Journal of Internet Services and Applications*, vol. 1, no. 1, pp. 7–18, 2010.

[3] A. Botta, W. De Donato, V. Persico, and A. Pescapé, "Integration of cloud computing and internet of things: a survey," *Future Generation Computer Systems Journal*, vol. 56, pp. 684–700, 2016.

[4] N. Fernando, S. W. Loke, and W. Rahayu, "Mobile cloud computing: a survey," *Future generation computer systems Journal*, vol. 29, no. 1, pp. 84–106, 2013.

[5] M. R. Rahimi, J. Ren, C. H. Liu, A. V. Vasilakos, and N. Venkatasubramanian, "Mobile cloud computing: a survey, state of art and future directions," *Mobile Networks and Applications Journal*, vol. 19, no. 2, pp. 133–143, 2014.

[6] Y. Xu and S. Mao, "A survey of mobile cloud computing for rich media applications," *IEEE Wireless Communications*, vol. 20, no. 3, pp. 46–53, 2013.

[7] Y. Wang, R. Chen, and D.-C. Wang, "A survey of mobile cloud computing applications: perspectives and challenges," *Wireless Personal Communications Journal*, vol. 80, no. 4, pp. 1607–1623, 2015.

[8] P. R. Palos-Sanchez, F. J. Arenas-Marquez, and M. Aguayo-Camacho, "Cloud computing (SaaS) adoption as a strategic technology: results of an empirical study," *Mobile Information Systems Journal*, vol. 2017, Article ID 2536040, 20 pages, 2017.

[9] M. N. Sadiku, S. M. Musa, and O. D. Momoh, "Cloud computing: opportunities and challenges," *IEEE Potentials*, vol. 33, no. 1, pp. 34–36, 2014.

[10] A. A. Ibrahim, D. Kliazovich, and P. Bouvry, "On service level agreement assurance in cloud computing data centers," in *Proceedings of the 2016 IEEE 9th International Conference on Cloud Computing*, pp. 921–926, San Francisco, CA, USA, June-July 2016.

[11] S. A. Baset, "Cloud SLAs: present and future," *ACM SIGOPS Operating Systems Review*, vol. 46, no. 2, pp. 57–66, 2012.

[12] L. Sun, J. Singh, and O. K. Hussain, "Service level agreement (SLA) assurance for cloud services: a survey from a transactional risk perspective," in *Proceedings of the 10th International Conference on Advances in Mobile Computing & Multimedia*, pp. 263–266, Bali, Indonesia, December 2012.

[13] C. Di Martino, S. Sarkar, R. Ganesan, Z. T. Kalbarczyk, and R. K. Iyer, "Analysis and diagnosis of SLA violations in a production SaaS cloud," *IEEE Transactions on Reliability*, vol. 66, no. 1, pp. 54–75, 2017.

[14] A. A. Ibrahim, S. Varrette, and P. Bouvry, "PRESENCE: toward a novel approach for performance evaluation of mobile cloud SaaS web services," in *Proceedings of the 32nd IEEE International Conference on Information Networking (ICOIN 2018)*, Chiang Mai, Thailand, January 2018.

[15] V. Stantchev, "Performance evaluation of cloud computing offerings," in *Proceedings of the 3rd International Conference on Advanced Engineering Computing and Applications in Sciences, ADVCOMP 2009*, pp. 187–192, Sliema, Malta, October 2009.

[16] J. Y. Lee, J. W. Lee, D. W. Cheun, and S. D. Kim, "A quality model for evaluating software-as-a-service in cloud computing," in *Proceedings of the 2009 Seventh ACIS International Conference on Software Engineering Research, Management and Applications*, pp. 261–266, Haikou, China, 2009.

[17] C. J. Gao, K. Manjula, P. Roopa et al., "A cloud-based TaaS infrastructure with tools for SaaS validation, performance and scalability evaluation," in *Proceedings of the CloudCom 2012-Proceedings: 2012 4th IEEE International Conference on Cloud Computing Technology and Science*, pp. 464–471, Taipei, Taiwan, December 2012.

[18] P. X. Wen and L. Dong, "Quality model for evaluating SaaS service," in *Proceedings of the 4th International Conference on*

Emerging Intelligent Data and Web Technologies, EIDWT 2013, pp. 83–87, Xi'an, China, September 2013.

[19] G. Cicotti, S. D'Antonio, R. Cristaldi, and A. Sergio, "How to monitor QoS in cloud infrastructures: the QoSMONaaS approach," *Studies in Computational Intelligence*, vol. 446, pp. 253–262, 2013.

[20] G. Cicotti, L. Coppolino, S. D'Antonio, and L. Romano, "How to monitor QoS in cloud infrastructures: the QoSMONaaS approach," *International Journal of Computational Science and Engineering*, vol. 11, no. 1, pp. 29–45, 2015.

[21] A. A. Z. A. Ibrahim, D. Kliazovich, and P. Bouvry, "Service level agreement assurance between cloud services providers and cloud customers," in *Proceedings–2016 16th IEEE/ACM International Symposium on Cluster, Cloud, and Grid Computing, CCGrid 2016*, pp. 588–591, Cartagena, Colombia, May 2016.

[22] A. M. Hammadi and O. Hussain, "A framework for SLA assurance in cloud computing," in *Proceedings–26th IEEE International Conference on Advanced Information Networking and Applications Workshops, WAINA 2012*, pp. 393–398, Fukuoka, Japan, March 2012.

[23] S. S. Wagle, M. Guzek, and P. Bouvry, "Cloud service providers ranking based on service delivery and consumer experience," in *Proceedings of the 2015 IEEE 4th International Conference on Cloud Networking (CloudNet)*, pp. 209–212, Niagara Falls, ON, Canada, October 2015.

[24] S. S. Wagle, M. Guzek, and P. Bouvry, "Service performance pattern analysis and prediction of commercially available cloud providers," in *Proceedings of the International Conference on Cloud Computing Technology and Science, CloudCom*, pp. 26–34, Hong Kong, China, December 2017.

[25] M. Guzek, S. Varrette, V. Plugaru, J. E. Pecero, and P. Bouvry, "A holistic model of the performance and the energy-efficiency of hypervisors in an HPC environment," *Concurrency and Computation: Practice and Experience*, vol. 26, no. 15, pp. 2569–2590, 2014.

[26] M. Bader-El-Den and R. Poli, "Generating sat local-search heuristics using a gp hyper-heuristic framework," in *Artificial Evolution*, N. Monmarché, E.-G. Talbi, P. Collet, M. Schoenauer, and E. Lutton, Eds., pp. 37–49, Springer, Berlin, Heidelberg, Germany, 2008.

[27] J. H. Drake, E. Özcan, and E. K. Burke, "A case study of controlling crossover in a selection hyper-heuristic framework using the multidimensional knapsack problem," *Evolutionary Computation*, vol. 24, no. 1, pp. 113–141, 2016.

[28] A. Shrestha and A. Mahmood, "Improving genetic algorithm with fine-tuned crossover and scaled architecture," *Journal of Mathematics*, vol. 2016, Article ID 4015845, 10 pages, 2016.

[29] T. T. Allen, *Introduction to ARENA Software*, Springer, London, UK, 2011.

[30] R. C. Blair and J. J. Higgins, "A comparison of the power of Wilcoxon's rank-sum statistic to that of Student's t statistic under various nonnormal distributions," *Journal of Educational Statistics*, vol. 5, no. 4, pp. 309–335, 1980.

[31] B. F. Cooper, A. Silberstein, E. Tam, R. Ramakrishnan, and R. Sears, "Benchmarking cloud serving systems with YCSB," in *Proceedings of the 1st ACM symposium on Cloud computing SoCC' 10*, Indianapolis, IN, USA, June 2010.

[32] Redis Labs, *memtier_benchmark: A High-Throughput Benchmarking Tool for Redis & Memcached*, Redis Labs, Mountain View, CA, USA, https://redislabs.com/blog/memtier_benchmark-a-high-throughput-benchmarking-tool-for-redis- memcached/, 2013.

[33] Redis Labs, *How Fast is Redis?*, Redis Labs, Mountain View, CA, USA, https://redis.io/topics/benchmarks, 2018.

[34] Twitter, *rpc-perf–RPC Performance Testing*, Twitter, San Francisco, CA, USA, https://github.com/AbdallahCoptan/rpc-perf, 2018.

[35] Postgresql, "pgbench—run a benchmark test on PostgreSQL," https://www.postgresql.org/docs/9.4/static/pgbench.html, 2018.

[36] A. Labs, "HTTP_LOAD: multiprocessing http test client," https://github.com/AbdallahCoptan/HTTP_LOAD, 2018.

[37] Apache, "ab—Apache HTTP server benchmarking tool," https://httpd.apache.org/docs/2.4/programs/ab.html, 2018.

[38] The Regents of the University of California, "iPerf—the ultimate speed test tool for TCP, UDP and SCTP," https://iperf.fr/, 2018.

[39] S. Varrette, P. Bouvry, H. Cartiaux, and F. Georgatos, "Management of an academic HPC cluster: the UL experience," in *Proceedings of the 2014 International Conference on High Performance Computing & Simulation (HPCS 2014)*, pp. 959–967, Bologna, Italy, July 2014.

A Case Study Analysis of Clothing Shopping Mall for Customer Design Participation Service and Development of Customer Editing User Interface

Ying Yuan[1] and Jun-Ho Huh ⓘ[2]

[1]Department of Clothing & Textiles, Hanyang University, Seoul, Republic of Korea
[2]Department of Software, Catholic University of Pusan, Busan, Republic of Korea

Correspondence should be addressed to Jun-Ho Huh; 72networks@pukyong.ac.kr

Guest Editor: Jaegeol Yim

Following the development of networking and mobile devices, the technology of managing the offline information online is being conducted widely. Also, as the social services have become much more active, users are registering and managing their personal information on online websites and sharing it with other users to acquire the information they need. For modern people living in a smart city, the planning of smarter services is required. The convergence of ET and IT or advanced scientific technologies such as AI or Big Data is often mentioned whenever the smart city is discussed. Nevertheless, smart services that could introduce smart solutions to conventional industries or change existing lifestyles should also be considered. Therefore, this paper discusses a service related to the convergence of the traditional clothing industry with IT and a service wherein CT is converged with systems that allow customers to participate in the design work and share the designs they have created. In other words, this study is a case study of CT and IT services in the clothing industry and is inclusive of an apparel shopping mall service that encourages customer participation in design, a customer-oriented editing user interface, and a copyright management system. The results show that both production method and production capacity largely affect the user interface of apparel platform services, with customer freedom significantly correlated with their functional roles. Moreover, the lead index is shown to be one of the factors restraining customer freedom. With this analysis, an apparel shopping mall wherein customers participate in the design work has been developed especially for clothes with more complex designs. The shopping mall emphasizes functionality from the perspective of customer use. At the same time, an online environment for an apparel service appropriate for the smart city has been implemented.

1. Introduction

Following the development of networking and mobile devices, the technology of managing the offline information online is being conducted widely. Also, as the social services have become much more active, users are registering and managing their personal information on online websites and sharing it with other users to acquire the information they need. Recently, the application whose intelligent agent recognizes a user's habits or a lifestyle and provides proper information is increasingly appearing in the online market. For example, a service that provides the information about the places the user often goes or the available means of transportation to user's destination is provided currently.

This is achieved by searching suitable information from the accumulated data pertaining to the user's history of visiting particular places. People nowadays pay more attention to their appearances especially when it concerns clothes, wearing different clothes depending on with whom they will be meeting or avoiding not to wear the same clothes the next day. Currently, a method which conveniently acquires the data on user's clothing habits and provides useful information to him/her based on the accumulated data is not available. Thus, to discuss a service where the existing clothing industry, IT-converged services, and CT are integrated to allow customers to participate in the design process and share their designs with others, an application which recommends a suitable design to the designer by accumulating the history of clothing choices of these customers to grasp a particular customer's clothing habits or inclination has been proposed. All of these tasks can be performed and managed with a smart device.

Technical development has accelerated the change of social structure and contemporary lifestyles [1]. Technical advances are transforming us from a postmodern to a participating generation [2]. The normalization of technology and the reduction of computer equipment cost focus on an efficient creation method between developers and users, who tend to focus on the recent small-quantity batch production [3]. Such customization is also being attempted in costumes. Current clothing items for customizing do not create diversity.

Historically, designers used to deliver a one-sided message that has changed in contemporary times, however. This is caused by the democratic desire of customers who have led the way in changing traditional business models [4]. Nonetheless, it is doubtful whether there are participation and creativity of a design for the current clothing design items in customer-participating services. Despite the good customer-participating structure, there is a lack of professional leadership for the service. In 2012, Armstrong and Stojmirovic argued [2] that the point of inducing customer participation is that it must be easy, fun, and less burdensome in terms of price, and it must give excitement to produce creative works under the leadership of an expert. Note, however, that the current t-shirt customizing is insufficient to produce passionate creativity as customers exercise their creativity to the full extent. Such is partly due to the lack of Back End technology. Thus, Design U developed a DTP printing image extracting system that restores the image from the model where the customer designed before. Technically, it prepares a presumption of customer design participation for various clothing items. From the customers' point of view, it is necessary to study the development of a user interface that can be edited in a more complicated content format and a contents format that can communicate with customers. If the designer designs a creative clothing item and suggests lead content, source contents need to be designed for customers to edit. This part can be made by uploading a customer-created image or sharing through the purchase of fee-paid copyright. The design source is sharing content wherein customers can easily and joyfully complete designs. Content sharing with a small amount of copyright is a method created by Richard Stallman who supported the copyright opposition movement and created a concept of free distribution of information. Lessig [5] also supports the flexible copyright method that reuses content information since it directly affects the participating culture.

This study first analyzed the customer editing screen user interface cases for customer design participation in the clothing shopping mall, investigated the pattern copyright cases used for customer editing, and finally developed a customer-participating editing user interface application for more complicated clothing items.

2. Related Research

Examining the clothing product development process is an essential part in preparing and supplying quality products to a promising market in a timely manner at a reasonable price. The common steps involved in designing clothing products are research, design development, and production [6]. The design development step starts with design sketching and sample production, and most of those who are participating in this process use computer-aided design (CAD) for efficiency and accuracy (Figure 1). The existing CAD systems are often effective in downstream production wherein grading, and marker planning processes take place continuously. Owing to a series of research works carried out by modern engineers and designers, Virtual Sampling, which allows a 2D pattern to be applied to a 3D human model to evaluate its wearability and appearance, has become a standard procedure when designing clothing products [7–10].

Cloth simulations are usually performed to assess the effect of geometrical variation or physical aspects. In most cases, the former draws faster results without considering the physical properties of the cloth being used. This makes it difficult to reproduce the dynamics of the clothes [12]. The latter allows more realistic simulation in understanding the dynamics and provides better accuracy as the cloth material's structural properties will be considered for the simulation. In other words, both the law of dynamics and the law of mechanics are based on discrete dynamics, fluid dynamics, or elasticity theories, all of which determine the cloth behavior and its interaction with external environments [13, 14]. Various methods often categorized as either a continuous physics-based or a discrete physics-based approach have been studied and proposed till now, emphasizing realism or computational efficiency [15]. The former introduces a rigorous, strict representation of a cloth in accordance with the continuum mechanics often adopting either a finite element (FE) or a finite difference model to produce a solution [16–18].

Meanwhile, by using the continuous Lagrange equations to represent the displacements from equilibrium positions, Terozopoulos et al. [16] modeled a surface deformation of a cloth, whereas Eischen et al. [19] employed the nonlinear shell theory and Li and Volkov [20] depicted the image of a cloth immersed in a quasistationary viscous fluid in terms of fluid dynamics. For this, a nonlinear FE method was applied to derive the system equations. This method aimed to produce various types of physical models for computer animation, which are effective in generating the behaviors instead of modeling a certain deformable cloth with high degree of accuracy. This method allowed the qualitative reproduction of similar behaviors without requiring a large number of computations [21]. The instability and high expense are major problems for the continuous physics-based methods, whereas the discrete physics-based methods represent cloth as a grid of particles interacting with each

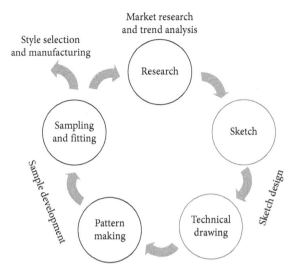

FIGURE 1: Fashion product development cycle [11].

other as well as with the external world following either the force-based Newtonian dynamic laws [22, 23] or the Lagrangian dynamic laws or the energy minimization criteria, all of which are energy-based approaches [21]. Characteristics such as low complexity and simple implementation are the underlying reason for the popularity of the mass-spring system that is also known as the particle system) [15, 24–27]. Nonetheless, an important issue remaining for this system is its accuracy as the physics of a cloth deformation is based on the approximation that is often represented by the mesh topology with a certain discrete physics-based method that influences cloth simulation. A number of meshing and remeshing methods were proposed in the past including Lienhardt [28], Praun and Hoppe [29], and Attene and Falcidieno [30]. Various forms of mesh topologies have been studied by Lu et al. [31], who then validated and proposed an optimal pipeline that is quite adequate for the preparation of the mass-spring model in a scanned garment reconstruction. Meanwhile, some other researchers turned their attention to pattern designing or making. An interesting interactive coevolutionary CAD system for the parametric pattern design was introduced by Chen et al. [32] who produced garment patterns using a neural network along with an immune algorithm.

A fuzzy logic-based optimization of garment pattern design was achieved by Chen et al. [32], whereas Lu et al. [31] proposed an expert knowledge-based approach that is helpful to customized pattern designing. Guo et al. [33] contributed to providing a detailed review of AI applications in the fashion industry. The previous studies suggest that sketch design is still an unexplored field of cloth design and development. Although the use of commercial CAD software such as Adobe Illustrator™ or CorelDraw™ has become a common practice in the sketching process, and their efficiency and effectiveness have been proven, the original idea of a designer starts by creating some sketches, consuming much time and effort. The survey conducted among Hong Kong fashion designers clearly showed that they are continually looking for a user-friendly design support system to

reduce their working hours when designing new clothing [34]. As one of the methods to achieve this, several companies established their own special digital fashion library like SnapFashun™ from which designers can borrow their desired elements to create new designs.

There was a unique development when Mok et al. [35] introduced a customer-oriented design system that allows general customers to create their sketches, adopting some evolutionary computational techniques. Further notable work came from Wan et al. [36] who claimed to have used some shape deformation techniques to create realistic sketches based on standard technical sketches. Fashion illustration and technical sketches are the two main pillars of the design industry (Figure 2). Specifically, fashion illustrations focus on drawing the products that the seller wishes to show and sell to the customers by showing how the products can be arranged and what their uniqueness is. This system translates the technical sketches into fashion sketches without omitting any details. By fitting the same clothing to a different fashion figure that can assume several different positions, the customers could understand the concept and may find the products attractive [11, 37–42].

3. Customized DTP Clothing: Case Analysis

Customized DTP clothing service is drawing much attention, and solutions for copyright issues were analyzed in this study using actual cases. Subjects of the analysis include service methods, user interfaces for the user-edited screens, and pictures used for printing. Among the DTP being serviced, five cases with clear distinctions have been selected for the analysis in order of introduction: My T, Snap T, Design U, Printing Factory, and Adidas. Among these, My T, Snap T, and Printing Factory are mainly selling T-shirts, so they can be regarded as products that originated from IT companies or printing factories instead of being clothing brands. Design U is a service provided by contemporary Korean traditional clothing brand TS, and Adidas is originally one of the clothing brands. In this aspect, the T-shirt business is often run by nonclothing brand firms, and the requirement or level of difficulty of production pertaining to the design of their clothes is quite low. Their interests lie in the files that they need to print. If a DTP service has to be provided for complexly designed clothes like women's clothing, the level of production difficulty will be high, and higher understanding of their design is required.

The DTP mentioned in the case analysis refers to digital textile printing, a method which replaces the conventional dyeing method and saves the time required for cutting patterns so that this method is quite suitable for the modern customized services. The user interface allows convenient communications with the system the user wish to select, aiming to reach the level of communication the customer desires.

Also, from the perspective of recently developing information communication and design technologies, the user interface is an interactive space for the various types of computer-based equipment. Operating the surrounding products in our everyday lives is a normal phenomenon in

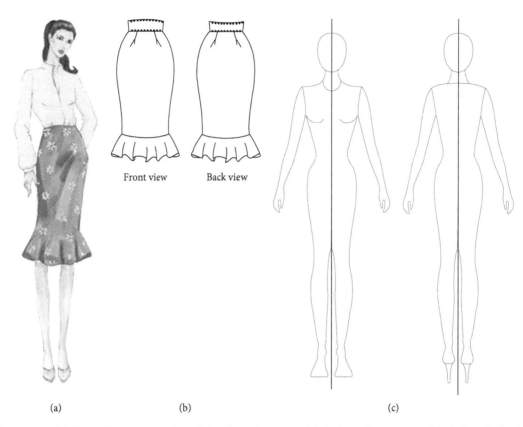

Front view   Back view

(a)   (b)   (c)

Figure 2: Illustration of fashion illustration, technical sketch, and croquis. (a) Fashion illustrations. (b) Technical sketches (flats). (c) Croquis (human figure template).

the environment created between products and user. For instance, the user just needs to use the hardware or the software of the vehicle he/she owns through the embedded user interface even if he/she does not know how it works or what is the principle of it.

A good user interface design makes it easier for the people to operate the products they encounter in their everyday environment. The design includes not only arranging the composition of a computer/similar device's screen or the elements of hardware operation in a convenient way but also includes all the designs of the things the user experiences with the products.

Moreover, since the people who will be providing such service should have a high degree of understanding of the design and form, the work is much different from simply concentrating on the printing files. Although there are many more services utilizing DTP printing, these five services have been selected because they offer customer-participating editing screens (Figures 3–8).

My T (Figure 3) is an app that helps produce personalized T-shirts. It provides an internal service platform that introduces the printable pictures created by artists from various fields and receives a royalty (T-mileage) once they have been sold. The printing house can be arranged, and the optimal printing method will be automatically provided. This service targets consumers who look for T-shirts that are highly individualized but affordable. The customer editing screen is arranged as follows:

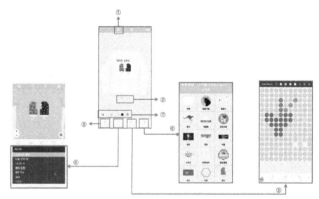

Figure 3: My T [11]. ① Layer control. ② Area selection. ③ Import image. ④ Text editing. ⑤ Dart editing. ⑥ Load Sticker. ⑦ Select background color.

(i) Button ①: *Layer adjustment*. Distinguishes the pictures to be placed in the foreground or background when several pictures overlap.

(ii) Button ②: *Area selection*. This service can be used for only two areas (front or back), selecting the designated rectangle domains. This shows that the clothing will be produced prior to printing.

(iii) Button ③: *Image import*. This function usually imports images (jpg, without transparent area) from the customer's mobile phone.

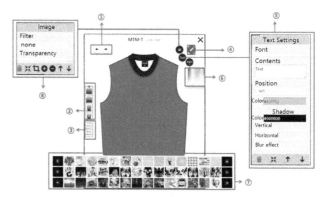

FIGURE 4: Snap T [38]. ① Upload image. ② Select background color. ③ Go down and see various examples.

FIGURE 6: Printing Factory [40]. ① Revert/revive. ② Area selection. ③ Simulation view (2D). ④ Start work. ⑤ Text editing. ⑥ Select background color. ⑦ Image selection. ⑧ Image editing.

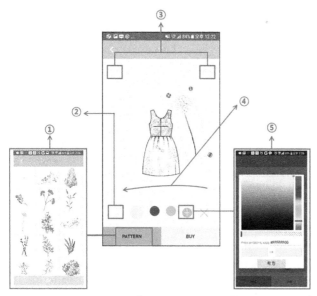

FIGURE 7: My Adidas [41]. ① Color selection. ② Material selection. ③ Area selection.

FIGURE 5: Design U [39]. ① Import image. ② Eyedropper. ③ Revert/revive. ④ Swipe: front, back. ⑤ Color selection.

(vii) Button ⑦: *Background color selection*. Each T-shirt has its own fixed background color. In the case above, however, the color selection option is quite limited as there are only four background colors.

(iv) Button ④: *Text button*. Font selection options will be displayed once clicked.

(v) Button ⑤: *Dart editing button*. Once the button is clicked, the screen on which the dart can be edited will appear.

The size can be adjusted, and certain figure or letter(s) can be imprinted by coloring each dart. In this area, some game factors can be reflected as well.

(vi) Button ⑥: *The sticker is imported*. Sticker refers to any of the artworks provided by various artists, and it is mainly a PNG file with a transparent domain. Images and stickers have apparently been distinguished in this service based on the source of the image (i.e., user's mobile phone, the platform itself, etc.) or file format (i.e., transparent or nontransparent).

Meanwhile, Snap T (Figure 4) is a service initiated by one of the Hong Kong companies, aiming to reflect collective amusement to Instagram or T-shirt design. Its editing screen has been a little more simplified, but the social function has mainly been strengthened. With this service, the customer can disclose or present his/her design; if another customer purchases the design, a 10% sales royalty will be collected. The editing screen is straightforward as shown in Figure 4: one's image is uploaded by clicking Button ①; for the uploaded image, various types of draft designs will be displayed when the customer drags the mobile phone screen downward. This is made possible by marking the image with different geometric shapes and inserting varied transparency levels in advance, giving a dream-like design effect. The background can be selected with Button ②; since it will be printed on the preproduced T-shirt, only the existing colors can be selected.

Design U (Figure 5) is an app developed to load clothing contents more intricate than T-shirts, aiming for better communications between the designers preparing a certain designer brand and customers. At the time of development, designers of contemporary Korean traditional clothing had loaded these dress items reflecting modern trends. This service adopts the "print first, produce later" method. Around 100 arbitrary images matching the items have been

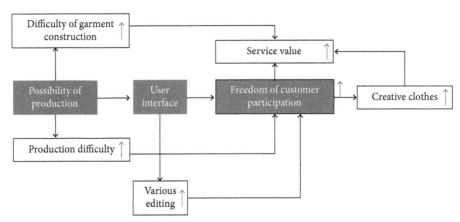

FIGURE 8: Factors affecting the user interface.

uploaded to the app. The customer needs to download the desired design or pattern to import it when editing. The customer editing screen is shown in Figure 5. ① is the image import button with which the customer can apply the downloaded image. ② shows the eyedropper function that elaborately leads the customer when he/she chooses a color. It could be difficult for the customer to find a sensible color in a short period of time among the colors suggested in ⑤. Nevertheless, the customer can have the desired color by importing the image file that includes the same color from the pattern samples and eye-dropping in the desired area. This function can be used to pick out a subtle color shade from a beautiful pattern. There are three circles on the right side of button ⑤ acting as a palette. The customer can start designing by preparing three color combinations in advance with an eyedropper or a color selection function. When the eyedropper in ② has been activated, the palette changes to the color designated by the eyedropper; when this is inactivated, however, it is possible to paint the clothing with the desired color by putting one of the colors in the palette in any area. The function that reverts or restores the working content can be performed with ③. With ④, the front or back of the dress is displayed when the customer moves the screen to the left or right so that the back side can be painted after the front side by shifting the screen to the side.

Although this service provides only two selections (front and back), it is possible to apply the service to all the areas. In other words, the edited arbitrary pattern will be adjusted following the outline of the dress drawing by clicking the check button on the upper right side.

Despite its designing convenience and better looks (prettiness), this method consumes much time since the images have to be put on the clothing sample one by one. In the picture, even though the pattern seems to be pieced together between the different domains, it is little challenging to make the image a perfectly joined and extended one because these domains are actually those of different pieces of fabrics. Therefore, it will be helpful if the work can be performed separately (subdivide) for each domain.

Printing Factory is a Korean T-shirt printing website who undertakes the editing/printing of costumes for fashion shows or K-pop idols by partnering with the relevant companies. On their website, they offer a customer-editable screen as shown in Figure 6. They are currently planning to extend their service to a DIY business wherein customers can make their clothes. Although their basic business strategy is to cooperate with as many clothing companies as possible continuously, their business is still limited to T-shirts and sportswear. ① is the button for undoing or restoring the performed work. ② selects the area to be designed. For T-shirts, four areas (i.e., front, back, right, and left sleeves) are usually available. In addition, each area is not defined by the rectangular box so that all areas of the cloth can be designed freely. ③ is used to check the simulation.

As the present screen, it is replayed in 2D mode; the difference is that the front (back) side of the front (back) sleeve will be shown just next to the front (back) face after the design has been completed so that the customer will be able to see how the T-shirt will look like after the sleeves have been attached. This is because the designed sleeve area is not recognized as just an ordinary arm part of the clothing but as an unfolded design (i.e., coloring the sleeve pattern). The merit of direct designing on a sleeve pattern is that the customer's design details can be reproduced without any data loss when printing them on a cloth. This offers many conveniences in the production process while reducing customer complaints. The customer may modify the design after seeing the simulation, but this process can be a little inconvenient as well. When the operation start button ④ is pressed, the three major functions such as Image, Text, and Color will appear. Button ⑤ is for text editing that supports font selection, contents input, sorting selection, color coating input, shadow code input, horizontal/vertical position adjustment, and blur effect.

The designing freedom of text itself has been enhanced. For the T-shirts, the text function has been improved as texts are often used more widely compared to women's or men's wear. Button ⑥ is for background color selection. The desired color can be selected from the RGB color scheme. Button ⑦ selects an image. The images are provided by theme, but one's images can be uploaded. Button ⑧ is used for editing the image by enlarging, reducing, and rotating it. The image setting window is provided separately for image cutting, image filter selection, or transparency adjustment.

Figure 7 shows My Adidas, a personalized shoe production service by Adidas. According to Adidas, they have commercialized this online service reflecting the trend among consumers who pursue their unique fashion and personality. This pertains to the highest level of Maslow's Hierarchy of Needs, so this personalized service is a prospective service targeting modern people who can feed and house themselves. When designing the Adidas shoes, the colors can be selected from the fixed set of colors and applied within the fixed areas. The materials can be selected from the given set of materials as well. When selecting an area, the Screen View rotates the shoe 360° degrees to show each area, whereas the applied colors can be viewed through 3D shoe design every time they have been applied. Adidas adopts many fixed functions, especially for the color selection. This is to help the customers match colors more professionally. Although customer freedom may be reduced, limiting customer choice by offering special color combinations would assist them better in terms of designing. The elements affecting user interface in the DTP customer-participating type clothing platform are shown in Figure 8. In the first place, production possibility is the base of the user interface. If possible items are different, the items and functions of the user interface vary. The user interface determined by such production possibility determines the degree of freedom of customer participation. If freedom of customer participation is high, more creative work can be produced. Production possibility, user interface, and customer's degree of freedom form the relation below. Firstly, if production possibility is high, complicated clothes have increased service value. If the production process using DTP is difficult, the customer's degree of freedom increases. In terms of user interface, if the diversity of screen function increases, the degree of freedom increases. If the customer's degree of freedom increases, the creativity of output increases together with the service value. In other words, if the customer's degree of freedom increases, production becomes complicated, and user interface needs a complicated function, but it increases service value. In contrast, the lead index restricts customer participation. It is intended to apply some restriction for professional and refined outputs by the customer with the help of professionals. It is expressed in the form of restriction of the customer selection area and fixing of the location. In the case analysis, ID has a high lead index. Customer participation and leading index apply mutual restriction. While the customer's degree of freedom is intended to restrict creative works, the lead index enables customers lacking creativity to express designs professionally by leading the method of restriction. Moreover, even if there is high freedom of customer participation, it does not have a positive effect on customer participation.

Customer participation may vary by generation based on the experience of Internet device. Generally, younger generations with high level of freedom have a positive reaction to customer participation. Therefore, it is ideal to keep a suitable level of service and price by maintaining the appropriate degree of freedom of customer participation concerning the market. There is a high level of freedom of customer participation and service, so it is important to create value for various customer groups. If easy clothing items are focused on, it creates severe competition within the item and decreases the value. Therefore, it is necessary to apply various clothing items to the DTP customer Half Design participation service. The complexity of cloth, diversity of customer participation, degree of freedom, and evaluation of user interface are listed in Figure 9. The diversity of user interface is related to customer participation's degree of freedom. Thus, the lead index is not evaluated in Figure 9. Likewise, 5 cases with clear differences including App and Web are analyzed for the evaluation of the index above to figure out the service situation in this industry. There are three analysis cases: App for My T (Korean), Snap T (China), and Design U. As for the web, Printing Factory and Adidas are analyzed.

4. Analysis Result

4.1. Result of Case Analysis for the DTP Clothing Service User Interface. The analysis table below is expected to be used as a reference to evaluate the platform function of DTP customizing clothing service editing screen and production method. In the evaluation items, factors affecting customer participation's degree of freedom are marked in bold. If the computer and mobile environments support customer editing screen in the development environment, the customer's degree of freedom is high. In the design areas, when it moves down, the degree of freedom is high, and the degree of freedom of Whole is highest. In analyzing the user interface editing and customer's degree of freedom, if there is no color restriction, the degree of freedom is high. The degree of freedom for Free Color is highest. In the selectable color, the degree of freedom is high if more than one color can be selected. If there is a spoilt, it has a high lead index. In the image, seven factors including Upload Customer's image, Provide image, zoom/Rotating image, Image repeat, Image game, and Text are considered. If there are many factors, it has a high degree of freedom. Next is a service on production complexity. It is related to service value, which includes the following. If the difficulty of Item goes down, the production complexity of clothing will increase. If Design Area moves down, the complexity of DTP editing will increase. If the number of pattern pieces for design increases, the production complexity in DTP printing editing increases. If there are many pattern pieces overall, production complexity will increase. Regarding the production method, preprinting and postproduction are closer to a traditional customizing production process than preproduction and postprinting. Compared with preproduction and postprinting, the difficulty of production is three times higher.

4.2. Result of Analyzing Creation-Sharing Contents Copyright Cases. Figure 10 shows a chart of the analysis result of the copyright management system with which the creations belonging to each brand can be shared. The shareable creations are divided mainly into art graphics that can be used for designing purposes and finished clothes. In the chart, the

DTP Clothing Half Design Online Platform			1. My T	2. Snap T	3. Design U	4. Printing Factory	5. Adidas
Environment		App/Web	App	App	App	Web	Web
		Computer				O	O
		Mobile	O	O	O		O
Difficulty of garment Construction	Item	T-Shirt, Eco Bag	O	O		O	
		Sports / Spandex					O
		Fashion Clothes			O		
	Design Area	Limited Area	O	O			O
		Free Area					
		Whole			O	O	
	Designable Pattern Quantity	1 sheet		O			
		Within 1/3 Quantity					
		Within 1/2 Quantity	O				
		All			O	O	O
	Whole Pattern Quantity		4	4	7~12	4~6	4
Editing function	**Back ground Color**	Color Select	No selection					
		Choose from provided colors	O	O			O
		Free Color			O	O	
		Color Spuit			O		
		Selectable degrees of freedom	Choose only one color	O	O			
		Choice of 1, or more colors			O	O	O
	Image	Upload Customer's image	O	O	O	O	X
		Provide Image	O	X	O	O	X
		Zoom, Rotating image	O	X	O	O	X
		Image Effect	O	O	X	X	X
		Image repeat	X	X	X	X	X
		Image Game	O	X	X	X	X
		Text	O	X	O	O	X
Production method		Production –> Printing	O	O			O
		Printing ->Production			O	O	

FIGURE 9: Case analysis of DTP clothing half design platform.

five brands and an additional brand, Real Fabric (Korean fabric printing web service), are included. This service has a robust copyright management system. Even though My T claims to be paying regular returns to the artists when using their designs (pictures), the process is not explicitly shown on their service screen. Meanwhile, Snap T offers a simple editing function, but they have a relatively proper social role for sharing creations and returns. They publicize the clothes design results, and the profits are shared in case of any purchases made. They do not provide art graphics, so the designers should upload their creations from their own devices. In this aspect, it is an art graphic sharing service. Real Fabric has a selection box showing the art graphics presented by the expert artists and which can be applied to

Copyright management		1. My T	2. Snap T	3. Design U	4. Printing Factory	5. Adidas	6. Real Fabric
Revenue share system	Graphic Design	○	○				○
	Clothing Design		○				
Copyright Flows			○				○
Revenue share automation			○				

FIGURE 10: Analysis of copyright management system by brand.

the fabric, sharing the profit with the artist. The profit-sharing follows their reimbursement rules. Since the copyright holders can be checked in the services provided by Snap T and Real Fabric, their services can be said to offer a better environment wherein the graphic artists can openly create artworks and socialize with their coartists and customers.

5. Development Concept Map

5.1. Development of DTP Clothing Service User Interface Web (Excluding Knitwear). Figure 11 shows the designer-participating customer user interface development DB UML for the application of various clothing items. It mainly consists of customer information, image upload information, other information related to image editing, and color information.

5.2. Creative Work Copyright Application Planning for DTP Half Design Platform. Figure 12 is a plan for the method of uploading works to use in the platform. If customer A uploads work, customer B buys the work at a minimum copyright fee and saves it to his/her own web file in the platform. While the internal contents of the web file can be applicable to half design participation, the original copy cannot be downloaded. As the method of paid pattern application by customer B in the design participation, when a customer performs design, there is an open image column on the editing screen. Here, free or fee-paid creative works appear in an image for selection. If the desired works are selected, it can be reused for the customer's work immediately by editing. Such method is quite reasonable as the copyright holder is explicitly specified and payments are made safely when the designs are downloaded to the customer folder.

6. Development Result

6.1. Result of Development of Customer Editing User Interface Applicable to Various Clothes. Currently, the case studies show that customized/personalized DTP clothing services are limited to T-shirts and sportswear. Although Design U once attempted to do women's wear lines, much time was required to reproduce the clothing sample after reflecting the customer design information since the area distinctions were not made clear, but the method used by Printing Factory increases the customer participation level by providing the list of areas for each clothing piece. Note, however, that they use one view for each clothing area. This method does not

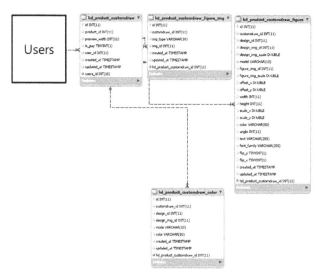

FIGURE 11: User interface web of UML of DTP clothing service.

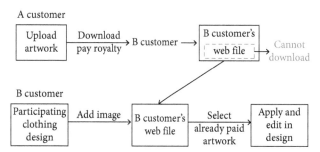

FIGURE 12: Copyrighted artwork in half design platform.

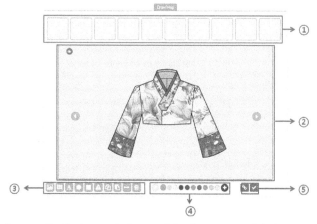

FIGURE 13: Customer design participation screen of design user interface.

allow the customer to see the overall effect, so they added the simulation view function separately. Considering this, for more complicatedly designed clothing, the Design U's method which connects all the areas in a single view would be more suitable. Therefore, for the customer view, this method has been adopted as in ② of Figure 13. ① in Figure 14 is an Area, Selection List. When an arbitrary area has been selected, the area becomes semitransparent

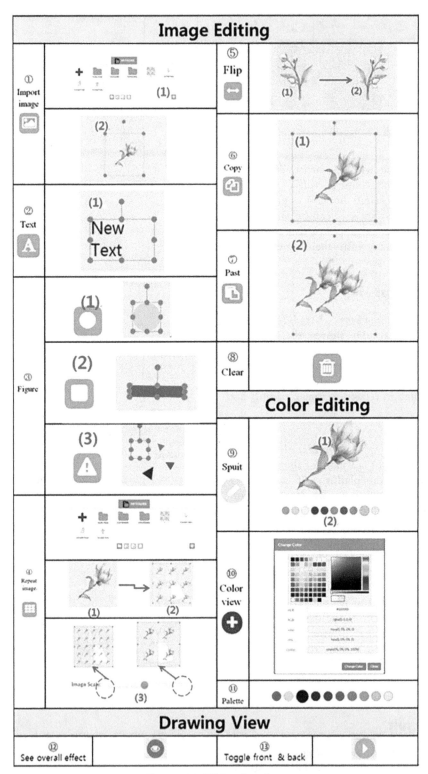

FIGURE 14: Editing function.

(Figure 14) so that the customer can recognize where the selected area is in the picture. When an area has been selected, the entire work output stays in that area. As a result, by pressing the eye-shaped button on the upper left after completing all the areas, the screen will show the result as in Figure 13. In other words, all the images designed in each area will be masked to show them neatly so that the customer can easily image his/her finished work. This method is being applied to complicated wear with many selection options with which the customer can select each area from the complexly set up areas and represent technology designs on a single screen.

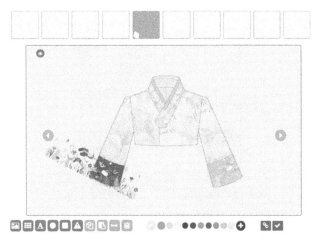

FIGURE 15: Customer design participation screen of design user interface.

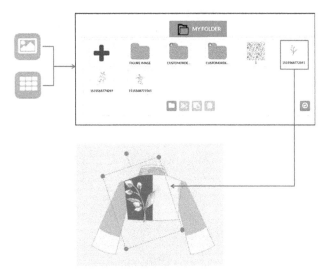

FIGURE 16: Development screen.

The customer design participation screen of the design user interface is divided into five (Figure 13): ① selection area, ② drawing area, ③ image editing, ④ color editing, and ⑤ save. It is divided into selection, coloring, image editing, color editing, and finally saving. Also Figure 15 shows customer design participation screen of design user interface.

As shown in Figure 14, the customer editing screen consists of ① Area Selection List, ② Drawing Area, ③ Image Editing Button, ④ Color Editing Button, and ⑤ Storing and Sharing Buttons. The details of each button are summarized in Figure 14. Button ① is an image import button. Once the button is pressed, the screen shifts to the customer folder where the paid or free pictures have been downloaded. The desired picture can be selected here. (1) of ① allows the user to import the files from his/her mobile phone or PC. (2) of ① shows the picture selected from the folder that has been generated on the screen. (3) is an editing tool generated for each design element, performing five functions such as selection, enlargement, reduction, rotation, and shift. The editing

tool automatically appears when a single item is selected from the elements such as text, geometric pattern, and pattern repetition. Thus, the pattern selected in (1) can be freely edited with editing tool (3) and displayed in (2) by designating the desired location or size.

② is a text button used to input text by clicking the button and positioning it anywhere on the screen. The editing tool is automatically generated when the text has been entered as an element, performing a function similar to image editing. (1) of ② shows the fonts. After the text is selected, one of the available fonts will be selected to decorate the text. ③ shows patterns from which the simplest ones such as (1) circle, (2) rectangle, and (3) triangle were selected. The pattern can be represented freely by drawing it as one big shape or several small ones. By clicking the desired color, the pattern color can be changed. ④ is an image repeating button, and its image import method is the same as ①; after importing an image, however, an arbitrary square should be drawn together with the click-drag function. By doing so, the image selected from (1) repeatedly appears in the square as in

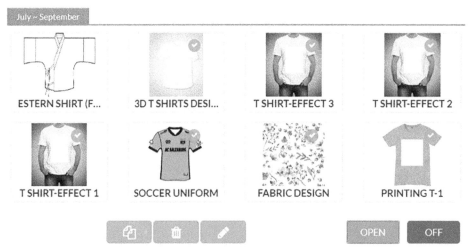

FIGURE 17: The service/product management screens on which the products to be tested are being uploaded.

Example: My T			
Features	1. Service Area: 2 2. Service Area Form: Square 3. Clothing Background Color: Assignable		
Simulation Graphic	Picture of item laid flat		
My T			
Representation on the platform	 Before application	Adaptation process	After application

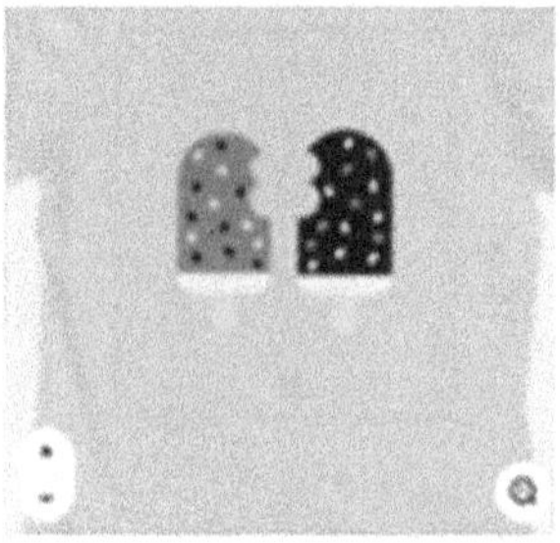

FIGURE 18: The result obtained from restricting an image within the square.

(2). If one wishes to adjust the size, it will become more substantial when the scale button is moved to the right side. The editing tool is also automatically generated for the finished pattern element so that the direction of pattern repetition can be adjusted by the rotation function of the tool. ⑤ is the flip function allowing only horizontal reversal. In other words, when flip has been performed by clicking one of the elements (e.g., image, repeating image, or text), (1) will disappear and (2) will appear.

⑥ and ⑦ are the copy and paste buttons, respectively. By clicking the ⑥ copy button after clicking the image (1) of ⑥, button ⑦ will be activated. The copied image will be pasted when this button is pressed (image (2) of ⑦).

It is possible to develop the copy function without the attach button; in such case, however, there could be some difficulties when one attempts copying in different areas separately. For this reason, both buttons have been developed as separate ones. ⑧ is a Clear button with which the

Example: Snap T		
Features	1. Service Area: 1 2. Service Area Form: Square x graphic 3. Clothing Background Color: Assignable	
Simulation Graphic	Picture of a person wearing the item	
Snap T	 Expressing different effects with the same image	
Result of Application on Platform	Effect 1	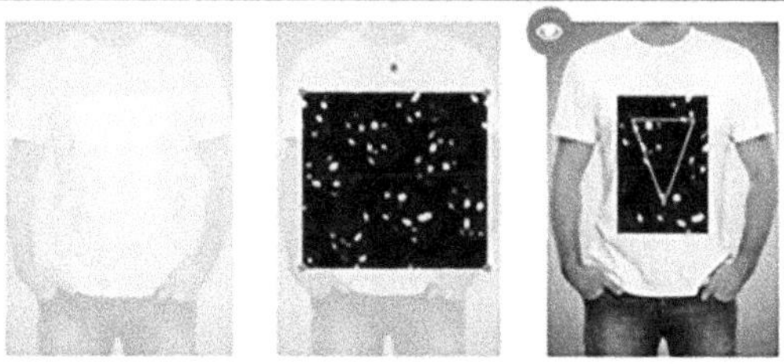 Before Application Adaptation process After application
	Effect 2	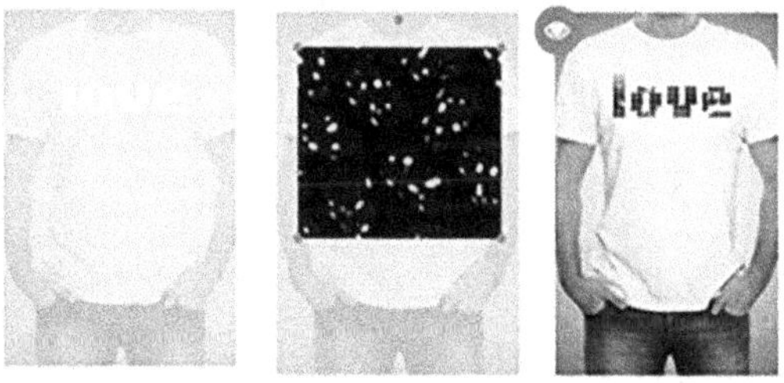 Before Application Adaptation process After application
	Effect 3	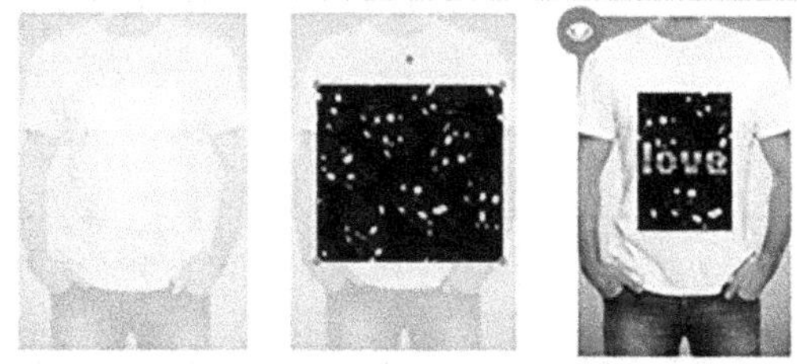 Before Application Adaptation process After application

FIGURE 19: The results obtained when the effects have been brought into the restricted area.

design element selected arbitrarily can be deleted. ⑨ is an eyedropper. With this button, the arbitrary elements and all the colors in the image can be put into the palette ⑪. For instance, when one wishes to use a color combination consisting of Pink, Burgundy, and Green for image (1) of ⑨, one needs to activate the eyedropper; after clicking the palette, click the color to be applied on the pattern. Then, as shown by (2), the colors are classified on the palette so that a natural arrangement of colors will be created. In addition, if one wishes to use the color collected with the eyedropper for the background, one needs to click the eyedropper button again to deactivate the eyedropper. ⑩ is the color view with which one can freely select new colors or input a color code in the HEX box to get the desired color. ⑪ is the palette. There are a total of ten palettes, and the color can be changed in each palette. Customers can have a user-friendly palette by clicking each palette and entering the colors they often use or prefer. ⑫ performs the See Overall Effect. When this button is clicked, the screen in Figure 13 where each area has been selected in advance will be changed into the screen in Figure 15, showing the image applied with all the patterns and colors.

Example: Design U App	
Features	1. Service Area: Entire Area 2. Service Area Form: 2D Form 3. Clothing Background Color: Free Choice
Simulation Graphic	Flat design with a black line (item)
Design U App	
Result of Application on Platform	Before Application Adaptation Process After Application

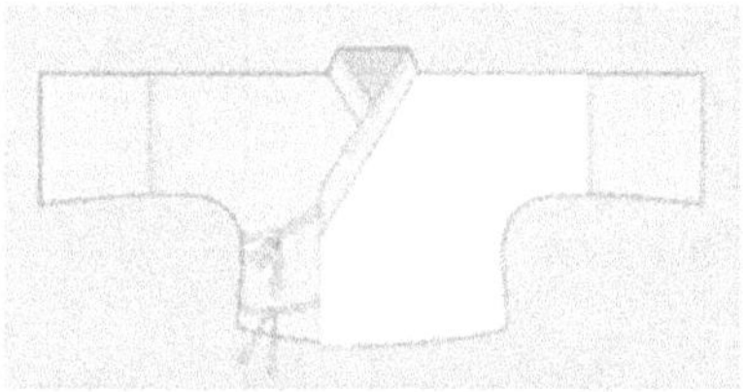

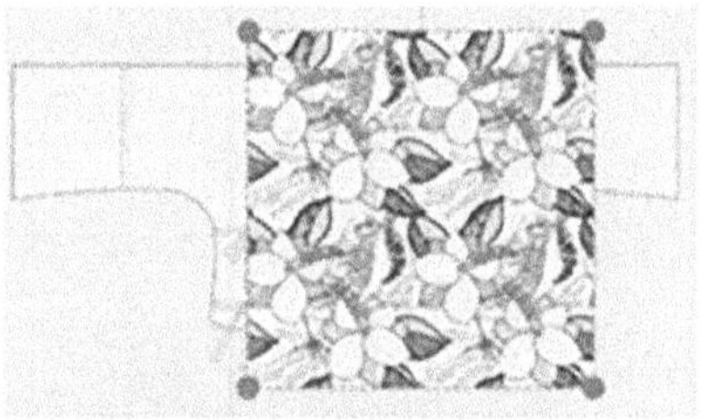

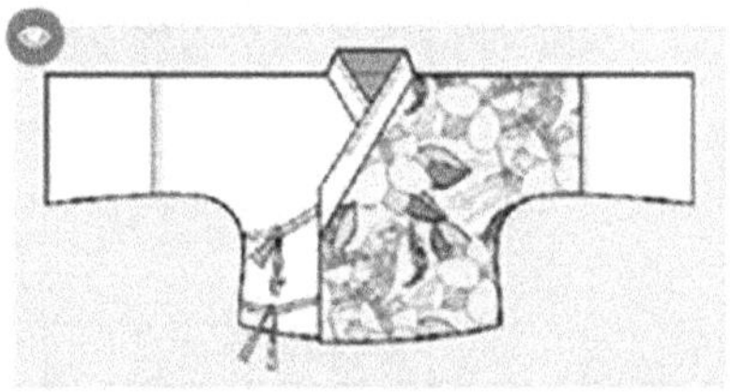

FIGURE 20: The result obtained when it was possible to carry out the design in the entire area freely.

The customer editing screen consists of the Area Selection List, Drawing Area, Image Editing Button, Color Editing Button, and Store/Share Buttons.

6.2. Copyright Image-Applied Screen. In Figure 16, when clicking the image open and pattern buttons, a customer web folder appears. In the folder, images that customers want are gathered, but they are not available for download. The pattern is free or copyrighted. Moreover, in the copyrighted patter, it can apply the design to the customer web folder.

7. Test Evaluation

To test the efficiency of the platform developed, the methods used for the services have been applied to the platform to confirm their validity. As examples, the brands such as My T, Snap T, Design U, Real Fabric, and Ninetyplus have been selected as each of them has a distinctive individuality in their respective service function. All of their service approaches were implemented on the platform. Figure 17 shows the service/product management screens on which the products to be tested are being uploaded.

The main feature of My T is that the image is restricted within the squared area (Figure 18). Although such a service does not allow an elevated level of customer freedom, it offers a much more convenient and cost-saving production method, leading to a better marketability because of its direct effect on the customer preference in low-cost products. Through the test, it was verified that the method of defining

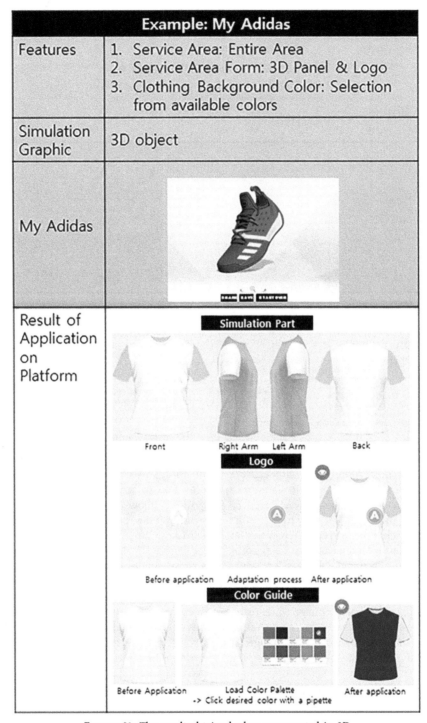

Example: My Adidas	
Features	1. Service Area: Entire Area 2. Service Area Form: 3D Panel & Logo 3. Clothing Background Color: Selection from available colors
Simulation Graphic	3D object
My Adidas	
Result of Application on Platform	

FIGURE 21: The result obtained when represented in 3D.

service area in a single/multiple squared areas is viable on the platform developed in this study.

Snap T shows various types of effects within the squared area (Figure 19). This is to allow customers to create an effect like a professional artist. Such an effect can be achieved with a png file which determines the transparency, color, and the shape of the area. Through above test, it was verified that the service which allows customers to create an artistic effect just by giving various types of effects on a single photo is viable on the platform developed.

The App version of Design U allows customers to design in every corner of the clothing (Figure 20). This is also a service method in the table presented in Figure 19, which embraces a high degree of customer freedom. Through the test, it was possible to verify that such a service approach is also viable on the platform developed.

The My Adidas product in Figure 21 was represented in 3D where customers cannot insert any images. However, it is possible to set the colors automatically for the logo or the geometric patterns in the system such that the customers

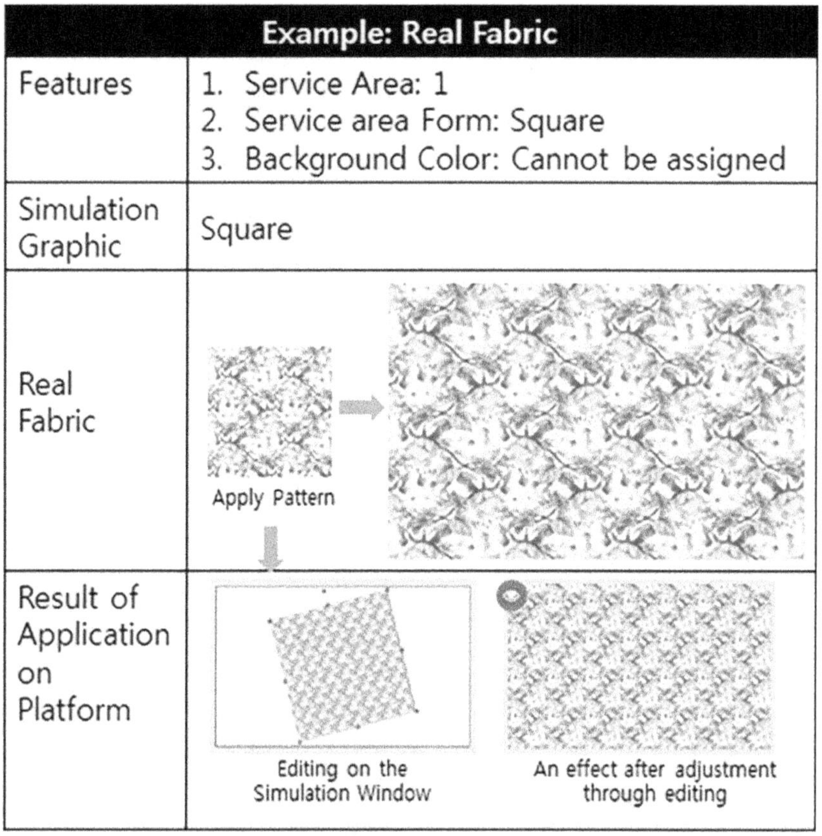

Example: Real Fabric	
Features	1. Service Area: 1 2. Service area Form: Square 3. Background Color: Cannot be assigned
Simulation Graphic	Square
Real Fabric	 Apply Pattern
Result of Application on Platform	 Editing on the Simulation Window An effect after adjustment through editing

FIGURE 22: The result obtained from the application on a fabric.

without expert knowledge in designing can also achieve professional-like color matching by making selections from the colors or color combinations prepared by experts. To perform a similar service, a 3D image was used for a T-shirt while making it possible to select the logo color separately on the platform. Although color selection has not been completely limited, the customers will be able to select some professional color combinations based on the color reference given. That is, they can pick their desired colors with a pipette function from the color palette image while using the editing feature. The test result revealed that even though the 3D-based simulation did not produce detailed images, it was verified that such a service can be achieved on the platform.

Figure 22 shows the result obtained from the application on a fabric. The service by Real Fabric can only be applied to fabrics and its simulation screen has a square shape. The information related to the proportion of a pattern based on its length and width can be checked in advance and it is possible to make changes for a single-pattern element as many times as the customer wants as well as the adjust of sizes. The same fabric used for the Real Fabric's simulation was used on the platform for testing. As a result, it was not possible to change the length or the width of the fabric instantaneously such that this problem was solved by selecting a fabric with the desired length and width from the product list in advance. Also, editing was freely performed on the pattern by using the functions (i.e., Enlarge/Reduce, Shift, Rotate functions) having a "Pattern Element Refit"

function. Also, compared to the Real Fabric service where only one pattern can be applied at a time, the proposed platform allows to add several patterns on top of another, making it a superior platform as a customizing service for fabrics.

Figure 23 shows the result obtained from performing a simulation on the existing customized service offering no simulation function. Finally, Ninetyplus (90+) is a company specializing in customized soccer uniforms. This company does not have any simulation systems and adopts an ordering system which requires a customer to download an Excel style order form from the company website and upload it after filling in the specifics. Then, they show the customer their draft design repeatedly for three times. Such a relatively complicated procedure has to be taken as they lack the simulation system. An uploading test was performed on the platform with product of this company by using the simulation technique and the result showed that all the problems can be solved at the same time. One major feature of the color guide used by Ninetyplus is that only a couple or triple of colors can be used. This is to prevent the customers from creating a rainbow-colored design by coloring every corner as they please. By contrast, the proposed platform allows all the predesignated areas will be painted with the same color simultaneously once a single arbitrary area has been painted. This validates that the service similar to one that is being offered by Ninetyplus (i.e., developing a design with just two or small color distribution) can also be achieved on the platform developed in this study.

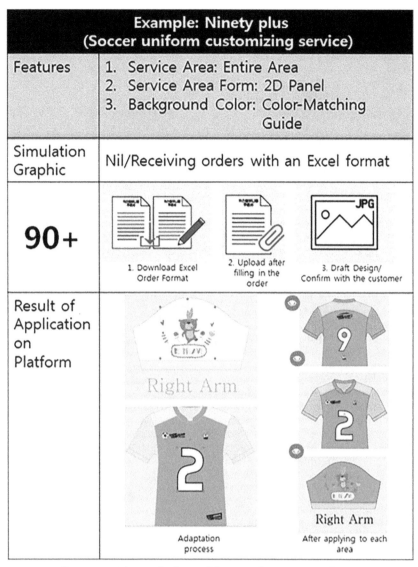

FIGURE 23: The result obtained from performing a simulation.

8. Conclusion

In this study, research was conducted on the instances of IT-integrated services in the clothing industry to develop an upgraded service version. The case analysis involved the user interface for customer editing and the management system including copyrights. Based on the research, a DTP half design service website with an integrated function to which various methods can be applied has been developed.

Thus, in this study, the DTP clothing half design service was developed. The existing DTP clothing half design service mainly focused on t-shirts. It is related to the production method, and the editing screen user interface by production method is associated with the degree of freedom of customer. In this study, with the premise that there is no DTP design participation service on trendy fashion items other than t-shirts, the user interface to express trendy fashion is developed. Based on Figure 7,

trendy fashion includes inner fabric, which has a high level of difficulty in terms of the composition of cloth. The developed user interface also supports the overall design of pattern, free color, and various editing functions. Moreover, the current user interface enables designing t-shirts and fabric. The user interface covers the preprinting and postproduction method. With this development, digital printing is expected to be applied to women's clothing with various complexities as well as t-shirts so that customers can express their creativity and personality and it fills the gap with the market.

Acknowledgments

This work was supported by the research fund of Hanyang University (HY-2014).

References

[1] H. Yoon and J. J. Lee, "A study on ubiquitous fashionable computer design using modules and wear net: with a focus on a detachable modular system," *Journal of Korean Fashion Designers*, vol. 9, no. 1, pp. 1–17, 2009, in Korean.

[2] H. Amstrong and Z. Stojmirovic, *Participatory Design: The Design Created by Users and Designers*, E. A. Choi, Ed., Beads and Beads, USA, 2012.

[3] C. Anderson, "In the next industrial revolution, atoms are the new bits," *Wired Magazine*, vol. 1, no. 25, 2010.

[4] H. Jenkins, *Convergence Culture: Where Old and New Media Collide*, NYU Press, New York, NY, USA, 2006.

[5] L. Lessig, *The Future of Ideas: The Fate of the Commons in a Connected World*, Vintage, New York, NY, USA, 2002.

[6] P. Sinha, "The mechanics of fashion," in *Fashion Marketing: Contemporary Issues*, T. Hines and M. Bruce, Eds., pp. 165–189, Butterworth-Heinemann, Oxford, UK, 2001.

[7] B. K. Hinds and J. McCartney, "Interactive garment design," *The Visual Computer*, vol. 6, no. 2, pp. 53–61, 1990.

[8] H. Okabe, H. Imaoka, T. Tomiha, and H. Niwaya, "Three dimensional apparel CAD system," *ACM SIGGRAPH Computer Graphics*, vol. 26, no. 2, pp. 105–110, 1992.

[9] Z. G. Luo and M. M. F. Yuen, "Reactive 2D/3D garment pattern design modification," *Computer-Aided Design*, vol. 37, no. 6, pp. 623–630, 2005.

[10] F. Durupynar and U. Gudukbay, "A virtual garment design and simulation system," in *Proceeding of 11th International Conference Information Visualization*, pp. 862–870, Zurich, Switzerland, July 2007.

[11] My T, August 2018, https://play.google.com/store/apps/details?id=co.snaptee.android.

[12] P. Volino, F. Cordier, and N. M. Thalmann, "From early virtual garment simulation to interactive fashion design," *Computer-Aided Design*, vol. 37, no. 6, pp. 593–608, 2004.

[13] T. J. Kang and S. M. Kim, "Development of three-dimensional apparel CAD system: Part II: prediction of garment drape shape," *International Journal of Clothing Science and Technology*, vol. 12, no. 1, pp. 26–38, 2000.

[14] N. Metaaphanon and P. Kanongchaiyos, "Real-time cloth simulation for garment CAD," in *Proceedings of the 3rd International Conference on Computer Graphics and Interactive Techniques in Australasia and South East Asia*, pp. 83–89, Dunedin, New Zealand, 2005.

[15] F. Han and G. K. Stylios, "3D modelling, simulation and visualisation techniques for drape textiles and garments," in *Modelling and Predicting Textile Behaviour*, X. Chen, Ed., vol. 94, pp. 388–421, Woodhead Publishing Ltd., Cambridge, UK, 2009.

[16] D. Terozopoulos, J. Platt, A. Barr, and K. Fleischer, "Elastically deformable models," *ACM SIGGRAPH Computer Graphics*, vol. 21, no. 4, pp. 205–214, 1987.

[17] P. Volino and N. Magnenat-Thalmann, "Versatile and efficient techniques for simulating cloth and other deformable objects," in *Proceedings of Annual Conference Series on Computer Graphics, SIGGRAPH*, pp. 137–144, Los Angeles, CA, USA, August 1995.

[18] K. Y. Sze and X. H. Liu, "Fabric drape simulation by solid-shell finite element method," *Finite Elements in Analysis and Design*, vol. 43, no. 11-12, pp. 819–838, 2007.

[19] J. W. Eischen, S. Deng, and T. G. Clapp, "Finite element modeling and control of flexible fabric parts," *IEEE Computer Graphics and Applications*, vol. 16, no. 5, pp. 71–80, 1996.

[20] L. Li and V. Volkov, "Cloth animation with adaptively refined meshes," in *Proceedings of the Twenty-Eighth Australasian Conference on Computer Science*, pp. 107–113, Newcastle, NSW, Australia, January 2005.

[21] D. H. House and D. E. Breen, *Cloth Modeling and Animation*, A K Peters Ltd., Natik, MA, USA, 2000.

[22] S. Petrak, D. Rogale, and V. Mandekic-Botteri, "Systematic representation and application of a 3D computer-aided garment construction method," *International Journal of Clothing Science and Technology*, vol. 18, no. 3, pp. 188–199, 2006.

[23] M. Hauth and O. Etzmuss, "A high performance solver for the animation of deformable objects using advanced numerical methods," *Computer Graphics Forum*, vol. 20, no. 3, pp. 319–328, 2001.

[24] D. E. Breen, D. H. House, and M. J. Wozny, "Predicting the drape of woven cloth using interacting particles," in *Proceedings of the 21st Annual Conference on Computer Graphics and Interactive Techniques (SIGGRAPH'94)*, pp. 365–372, ACM Press/ACM SIGGRAPH, Orlando, FL, USA, July 1994.

[25] X. Provot, "Deformation constraints in a mass-spring model to describe rigid cloth behaviour," in *Proceedings of Graphics Interface'95*, pp. 147–154, Canadian Information Processing Society, Québec, QC, Canada, May 1995.

[26] D. H. House, R. W. DeVaul, and D. E. Breen, "Towards simulating cloth dynamics using interacting particles," *International Journal of Clothing Science and Technology*, vol. 8, no. 3, pp. 75–94, 1996.

[27] M. Fontana, C. Rizzi, and U. Cugini, "3D virtual apparel design for industrial applications," *Computer-Aided Design*, vol. 37, no. 6, pp. 609–622, 2005.

[28] P. Lienhardt, "N-dimensional generalized combinatorial maps and cellular quasimanifolds," *International Journal on Computational Geometry and Applications*, vol. 4, no. 3, pp. 275–324, 1994.

[29] E. Praun and H. Hoppe, "Spherical parametrization and remeshing," *ACM Transactions on Graphics*, vol. 22, no. 3, pp. 340–349, 2003.

[30] M. Attene and B. Falcidieno, "Remesh: an interactive environment to edit and repair triangle meshes," in *Proceedings of the IEEE International Conference on Shape Modelling International (SMI'06)*, pp. 271–276, IEEE Computer Society Press, Silver Spring, MD, USA, 2006.

[31] J. Lu, M. Wang, C. Chen, and J. Wu, "The development of an intelligent system for customized clothing making," *Expert System with Application*, vol. 37, no. 1, pp. 799–803, 2010.

[32] Y. Chen, X. Zeng, M. Happiette, P. Bruniaux, R. Ng, and W. Yu, "Optimisation of garment design using fuzzy logic and sensory evaluation techniques," *Engineering Applications of Artificial Intelligence*, vol. 22, no. 2, pp. 272–282, 2009.

[33] Z. X. Guo, W. K. Wong, S. Y. S. Leung, and M. Li, "Applications of artificial intelligence in the apparel industry: a review," *Textile Research Journal*, vol. 81, no. 18, pp. 1871–1892, 2011.

[34] B. C. A. M. Chow, "An investigation on the needs and expectations of the fashion industry in relation to design support systems—a qualitative approach," BA thesis, The Hong Kong Polytechnic University, Hung Hom, Hong Kong, 2012.

[35] P. Y. Mok, J. Xu, X. X. Wang, J. T. Fan, Y. L. Kwok, and H. Xin, "An IGA-based design support system for realistic and practical fashion designs," *Computer-Aided Design*, vol. 45, no. 11, pp. 1442–1458, 2013.

[36] X. Wan, P. Y. Mok, and X. Jin, "Shape deformation using skeleton correspondences for realistic posed fashion flat

creation," *IEEE Transaction on Automation Science and Engineering*, vol. 11, no. 2, pp. 409–420, 2014.

[37] J. Xu, P. Y. Mok, C. W. M. Yuen, and R. W. Y. Yee, "A web-based design support system for fashion technical sketches," *International Journal of Clothing Science and Technology*, vol. 28, no. 1, pp. 130–160, 2016.

[38] Snap T, August 2018, https://play.google.com/store/apps/details?id=com.avennoms.mytee.

[39] Design U, August 2018, https://play.google.com/store/apps/details?id=com.b05studio.designu.

[40] Printing Factory, August 2018, http://printingdiy.co.kr.

[41] Adidas, August 2018, http://www.adidas.com/us/customize.

[42] Y. Yuan and J.-H. Huh, "Customized CAD Modeling and design of production process for one-person one-clothing mass production system," *Electronics*, vol. 7, no. 11, p. 270, 2018.

3

On the Context-Aware, Dynamic Spectrum Access for Robust Intraplatoon Communications

Michał Sybis ⓘ, Paweł Kryszkiewicz ⓘ, and Paweł Sroka ⓘ

Chair of Wireless Communications, Poznan University of Technology, Poznan, Poland

Correspondence should be addressed to Michał Sybis; michal.sybis@put.poznan.pl

Academic Editor: Ioannis D. Moscholios

Vehicle platooning is a promising technology that allows to improve the traffic efficiency and passengers safety. Platoons that use cooperative adaptive cruise control, however, require a reliable radio link between platoon members to ensure a required distance between the cars within the platoon, thus maintaining platoon safety. Nowadays, the communication can be realized with the use of 802.11p or cellular vehicle-to-vehicle (C-V2V), but none of this technology is able to provide a reliable link especially in the presence of high traffic or urban scenarios. Therefore, in this paper, we propose a dynamic spectrum management mechanism in V2V communications for platooning purposes. A management system architecture is proposed that comprises the use of context-aware databases, sensing nodes, and spectrum allocation entity. The proposed robust system design aims to keep only the minimum necessary information transmitted over the conventional intelligent transportation system (ITS) channel, while moving the remaining data (nonsafety, service-aided, or infotainment) to an alternative channel that is selected from the available pool of spectrum white spaces. The initial analysis indicates that the proposed system may significantly improve the performance of wireless communications for the purpose of vehicle platooning.

1. Introduction

One of the main aims of the future, fifth-generation (5G) wireless systems is to provide ubiquitous communications for many different devices. Among many aspects of the envisioned 5G networks is to provide the so-called ultra-reliable communications (URCs) with minimum latency that can be used to ensure various safety or life-saving services [1]. One of the considered use cases for the URC is to provide wireless links for data exchange between vehicles moving on a road, which is known as vehicle-to-vehicle (V2V) communications. The main purpose of introducing V2V transmission is to improve the safety and efficiency of road traffic using dedicated messages that can be used to warn drivers, enable automated safety systems, or even support autonomous car driving capabilities.

The last mentioned use case, namely, the autonomous driving, is foreseen as an enabler for cost-effective and safe vehicle platooning, which is a coordinated movement of a group of autonomous vehicles forming a convoy led by a platoon leader. According to the study described in [2–4], the use of vehicle platooning may result in an increase of road capacity due to the reduction of intercar spacing and, consequently, reduced fuel consumption and CO_2 emission due to the lower air drag.

An example of such a control system is the cooperative adaptive cruise control (CACC), which makes use of the information received from on-board sensors or exchanged with other cars through wireless links. The key factor limiting the performance of platooning algorithms, and hence enlarging the required intervehicle distance, is the delay of the control system and the accuracy of information describing the surrounding environment, obtained via sensors or wireless communications. The CACC helped to increase the reliability of platooning. However, it requires frequent, highly reliable, and low-latency V2V transmission [5–7]. These requirements cannot be met with the state-of-the-art wireless communication protocols using the ITS frequency band (5.9 GHz band dedicated for vehicle-to-anything (V2X) communications: 5.850–5.925 GHz in USA and

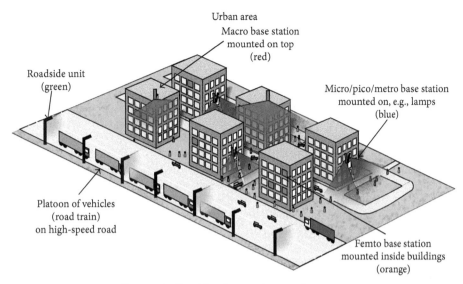

FIGURE 1: Considered scenario: an urban case.

5.875–5.925 GHz in Europe). The main problems are collisions with packets transmitted by other cars utilizing the ITS band or high latency due to medium access control (MAC) layer processing.

A solution to this problem can be to utilize a wider frequency band. However, nearly whole radio frequency (RF) spectrum is assigned to some existing wireless systems. On the contrary, worldwide measurements reveal that real spectrum utilization at a given location in a given time is very limited [8]. Moreover, the potential of use of frequencies above 6 GHz for V2V communication purposes has gained significant interest in recent years [9, 10]. This gives rise to an idea of opportunistic use of these unused spectrum resources (called white spaces) by unlicensed systems, employing the cognitive radio technology [11]. In the considered scenario, a platoon of vehicles could exchange information using a band licensed to another system, acting as a secondary user. This is acceptable from the primary user's perspective only if its transmission is not disrupted. On the contrary, the advantages of in-platoon communications over white spaces are present only if interference from primary users is low or at least foreseen. This requires context awareness of the whole system, being able to sense and adjust to the current situation in a given white space. A necessary step in dynamic band selection is to acquire and process information from different sources, including the context information stored in databases. A role of such a dynamic spectrum access manager can be played by the Geolocation Database (GLDB), REM [12], or even RSM [13]. In what follows, we will refer to the database-oriented system as the Context-Aware Database (CADB).

The generic scenario considered in this paper is visually presented in Figure 1, where one can observe the road train driving autonomously on a high-speed road in various environments (urban and rural). Depending on the location, another frequency band may be occupied; that is, there is a limited set of active services in the rural area compared to the urban case, where the number of present wireless

networks may be high. The communication within the platoon has to be stable and reliable; thus, we claim that the support from surrounding sources of information in the selection of the best frequency band may be highly beneficial.

This position paper is to propose an architecture for 5G systems suitable to solve main problems of V2V communications for platooning purposes using spectrum white spaces. Let us stress that the main aim of this position paper is to propose the design and infrastructure for the platooning support system that realizes the V2X communications with the use of state-of-the-art communication protocols (such as the IEEE 802.11p [14] or 3GPP C-V2X [15]) and spectrum white spaces. The latter ones would be used for intraplatoon communication. For this type of communication, we suggest to use either frequencies lower than 5.9 GHz (reference frequency used by the exemplary 802.11p system) or frequencies higher than 5.9 GHz (e.g., mmWave). Both of these approaches have their advantages and disadvantages. For example, lower frequencies will ensure a broader transmission range (whole platoon will remain within the transmission range), but interference to other systems/platoons may also be higher. On the contrary, the system that works with high frequencies (e.g., mmWave) could be regarded as a remedy for this. The transmission range and interferences are limited, which are positive aspects, but communication within a platoon may suffer from too short transmission range and shadowing; for example, a car-like truck can easily interrupt high-frequency line-of-sight car-to-car transmission.

The new architecture is needed to enable dynamic spectrum management mechanisms based on rich context information collected from different sources. The envisioned platooning support system will be responsible for collecting and storing information describing the spectrum availabilities, and for dynamic spectrum allocation, that will enable ultrareliable exchange of information between platoon vehicles. Hence, definition of the management system entities and the related interfaces is needed.

1.1. Scope and Novelty. This work is a position paper that shows the current status of the V2V communication for platooning purposes with the emphasis on the current system drawbacks, as well as for proposing and discussing a solution based on reusing the spectrum white spaces for intraplatoon communications and the system architecture that would support the dynamic use of the radio spectrum. The main problem formulation is presented in Section 2. This section, apart from presenting the state-of-the-art overview of platooning aided with V2X communications, shows the performance of the platoon in terms of minimal target distances and packet reception rates. Our results clearly present that nowadays, solutions to this kind of communication are inefficient and a new solution should be proposed. In the latter part, we present a possible solution to the problem, that is, possibilities of using the white spaces for V2V communications as well as our spectrum measurements campaign results. In the last part of the paper, we propose new management system architectures that could be used to improve the in-platoon V2V communication.

The novel aspects covered by this paper are the following:

(1) Formulation of the problem of insufficient reliability of V2V communication in particular for in-platoon communication in high traffic/urban areas

(2) Proposition of a system architecture with dynamic spectrum allocation for the dual transceiver communication with the use of 802.11p state-of-the-art V2X communication and white spaces to solve the formulated problem

(3) Considerations on the possible challenges in implementing the proposed system

This paper is organized as follows: Section 2 is devoted to V2V/V2X communication. This section is divided into two parts. The first covers the literature studies in the field of V2V communication, while the second presents the results of simulations of the platooning using the CACC and formulates the problem of reliable in-platoon communications in future V2V systems. In Section 3, the main observations paving the way for dynamic spectrum access are presented. Crucial components of this system are presented including the CADB that can be useful for enhancement of V2V communications within platoons. Existence of white spaces suitable for this purpose in urban areas is shown by the analysis of data coming from spectrum occupancy measurements. Section 4 presents the proposed management system architecture designs that make use of the CADB to dynamically allocate the spectrum for in-platoon communications. In Section 5, the system challenges are presented. This paper ends with conclusions.

2. Contemporary Solutions for V2V and V2X Communications

2.1. State-of-the-Art Overview. A platoon using the CACC has attracted a lot of interest in recent years. A number of empirical studies have been performed to evaluate the performance of the CACC and platooning supported by IEEE 802.11p-based wireless communications. As an example, [16] builds a comprehensive system simulation framework to study the CACC performance in the presence of nonideal communications. The study shows that the CAM broadcast frequency and loss ratio have a strong influence on the performance of the CACC algorithm. A consensus-based approach to the CACC with the use of 802.11p communication is investigated in [17]. In [18], performance comparison of the 802.11p-based (a common control/safety channel and a dedicated service channel) platoon is provided. The performance of the CACC with different CAM beaconing schemes implemented on top of the 802.11p is evaluated in [19]. Further simulation examples presenting the performance of the 802.11p-based CACC are given in [20–22]. The link-level (i.e., block error rate (BLER) versus signal-to-noise ratio (SNR)) and system-level (i.e., packet reception rate (PRR) versus distance) performance of 802.11p and cellular vehicle-to-anything (C-V2X) has been compared in [23–25]. Some examples of research studies that aim to support the requirements of vehicle platooning by improving the existing communication standards can be found in [26, 27]. In [26], the authors study the performance of 802.11p distributed coordination function (DCF) based on a Markov chain model. Moreover, the authors propose a contention window adjustment scheme for 802.11p [26] and resource (i.e., subchannel and power) allocation algorithms for LTE-based multiplatoon communication in [27] to optimize the platoon performance in multiplatoon scenarios.

Investigation of platoon performance is not limited to the simulations only. The SARTRE project [4] conducted an experiment, where a platoon of two trucks and three cars driven autonomously in close formation has been deployed. It turned out that the platoon could drive at speeds of up to 90 km/h with a 5–7 m intervehicle distance. On the contrary, in the framework of the Energy ITS project in Japan, a platoon of three fully automated trucks driving at 80 km/h with a 10 m intervehicle gap was tested on an expressway [28]. In the European Truck Platooning Challenge 2016, automated trucks of six major truck vendors used autonomous driving in platoons on public roads, some of them using IEEE 802.11p for communications [29]. Finally, the PATH Program in the United States demonstrated a platoon of three 802.11p-equipped trucks driving on the busy Interstate 110 freeway in 2017 [30]. The increased interest in platooning is driven by the expected potential revenues. The findings of the SARTRE project show that platooning provides fuel savings from 7 to 15% for trucks travelling behind the platoon leader [4]. Additionally, the fuel savings translate to substantial reduction of CO_2 emission, and according to the study performed in the Energy ITS project [28], when the market penetration of truck platooning increases from 0% to 40% of trucks, CO_2 emission along a highway can be reduced by 2.1% if the gap between trucks is 10 m and by 4.8% if the gap is reduced to 4 m.

The empirical studies described above do not account for one very important factor limiting the use of autonomous driving with the CACC, namely, the reliability of wireless communications in a scenario with a large number of transmitting devices. Therefore, various simulations have been conducted to study the ability of different V2V technologies to meet the requirements of the ITS applications. The analysis performed in [18, 21] shows that the V2V message broadcast frequency and packet loss ratio have a strong influence on the performance of the CACC algorithm. One of the main problems of the CACC when employing the current state-of-the-art wireless communication protocols is the reliability of information exchange between platoon vehicles. A study on the use of the CACC with communications based on the IEEE 802.11p standard revealed that even a moderate increase in road traffic on a motorway can lead to wireless channel congestion and, consequently, prevent the automated controller from reliable and stable operation [31]. A solution to this problem may be the use of the dual-band transceiver that allows to operate simultaneously in two different frequency bands (as an opposite to the single-band transceiver that is able to operate only in one band). However, one should note that even when two subbands of the ITS frequency range are used, due to the high density of communicating vehicles and the large amount of exchanged data (apart from various safety messages, the so-called infotainment data can be transmitted), still channel congestion is possible for both subbands used. In the following section, we present the simulation evaluation of the performance of CACC-driven platooning when IEEE 802.11p is used for V2V communications. We compare the results obtained with a single-band transceiver and a device that simultaneously uses the ITS control channel (CCH) and a dedicated frequency band for in-platoon communications.

2.2. Identification of the Performance Limitations of CACC Schemes: A Single-Radio Case.
To identify the possible drawbacks of the considered system, we used the simulation tool. This allows us to investigate the performance of the platoon itself as well as some selected aspects of V2V platoon communication (e.g., the impact of the message rate or packet collision mitigation mechanism). The obtained results will allow us to answer the question if it is possible to improve the platoon performance. Achieved results are presented in the consecutive subsections.

Let us now consider a scenario where a platoon of ten cars is travelling on the outer lane of a four-lane highway (two lanes in each direction), while the number of non-platooned cars travelling on the remaining three lanes that are not occupied by the platoon is a simulation parameter. All cars on the highway transmit the CAM with the frequency of 10 Hz on the control channel (CCH). As a baseline system, we consider the use of single-radio transceivers for platooned cars that are continuously tuned to the CCH to transmit and receive CAMs.

2.2.1. Simulation Setup.
In the simulations, we assume the 802.11p communication with one transmitting and receiving antenna and the ITU Vehicular-A channel with the path loss according to the Winner + B1 LOS. We consider the densities of nonplatooned cars of 0, 5, 10, 15, and

20 cars/km/lane. Other simulation parameters and assumptions are summarized in Table 1.

2.2.2. Minimum Feasible Target Distance.
The first step in the evaluation aims to determine the minimum feasible target distance between platooned cars that can be safely maintained. The target distance is the input parameter of the CACC controller. In particular, we have assumed that the minimum feasible target distance is the one that provides crash-free platoon operation with 99% probability. Namely, we performed 100 simulation runs for different target distances ranging from 1 m to 15 m and then selected the minimum distance for which no more that one out of 100 simulations ended with a crash to be the minimum feasible target distance. One should note that our simulator does not implement any crash mitigation mechanisms (i.e., emergency braking) that would decrease the probability of hazardous situations in real-world implementations. The CACC alone is not meant to deal with such situations. Therefore, the 99% target is sufficient.

From the results presented in Figure 2 and Table 2, we can see that it is possible to achieve crash-free operation only for the scenario with no cars on neighbouring lanes. With the increase in density of cars on neighbouring lanes, namely, in scenarios with 5 to 20 cars/km/lane with the single-radio transceiver configuration, we selected the target distance that results in approximately the smallest number of crashes (the chosen value ensures that there is no crashes for all higher values). It is caused by the fact that not for all scenarios, we are able to find a value for which the platoon is operating with no (or almost no) collisions (further distance extension gives no improvement in terms of the number of collisions per 100 simulations run). In the rest of this section, we present the results obtained for the target distances indicated in Table 2.

These results reveal that the performance of the platoon with single-radio operation is correlated with the overall traffic intensity. Moreover, only for the smallest traffic, the achievable results are of assumed 99% crash free, while for the higher densities observed, the crash rate is significantly higher.

2.2.3. Reliability.
In the next step, we analyzed the reliability of 802.11p message transmissions. Thus, we run the simulations with the target distances listed in Table 2 and collected the message reception statistics at each car in the platoon. The reception rates of the messages transmitted by the platoon leader are shown in Figure 3. One may notice from Figure 3 that the reception rate of leader messages drops significantly towards the tail of the platoon when a single radio is used. The last car in the platoon was unable to receive more than 5% of messages due to the low reliability of 802.11p communication with the platoon leader.

2.2.4. Fixed BLER Results.
As presented in the previous subsection, the message reception rate drops from 100% for single-line transmission to about 95% for the last vehicle

TABLE 1: Selected simulation parameters.

Parameter	Assumption
Number of simulation runs	100
Simulation time	900 s
Platoon size	10 cars
Actuation lag	20 ms
ACC headway time	0.2 s
CAM message size	300 bytes
CACC message size	16 bytes
Spectrum band	ITS-G5A (5895–5905 MHz)
Channel estimation	Ideal

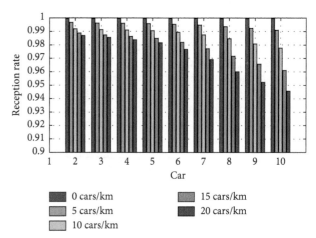

FIGURE 3: Reception rate of platoon leader CAM packets versus the car position in the platoon.

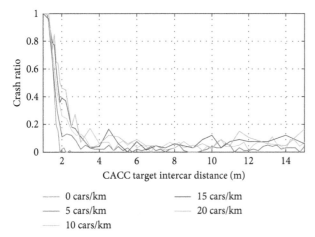

FIGURE 2: Fraction of simulation runs concluding with an in-platoon crash when using only the CCH.

TABLE 2: Selected CACC target distances and the corresponding number of simulations with a crash per 100 runs for a single-radio scenario.

Case (cars/km/lane)	Target distance (m)	Number of crashes
0	2.2	1
5	3.5	2
10	4.0	7
15	5.0	6
20	5.5	7

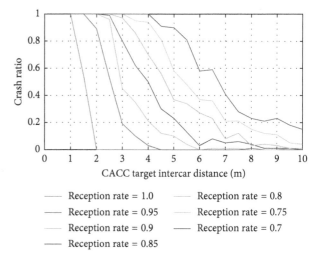

FIGURE 4: Fraction of simulation runs concluding with an in-platoon crash for an abstract system with a fixed reception rate of leader packets.

within the platoon for the highest considered traffic density. To investigate the impact of the decreased BLER on the platoon performance, we simulated the platoon performance using an abstract communication system with a predefined BLER value (in this case, the BLER value is fixed for all cars and is a simulation parameter).

Obtained results are presented in Figure 4. In this figure, curves for the reception rate ranging from 1.0 to 0.7 are shown. For the ideal case, the target distance of 2.0 m ensures the collision-free operation of the platoon. Small decrease in the reception rate (to 95%) degrades the platoon performance significantly (an intercar distance of 4.5 m is required). This increase in the required target distance can be regarded as an additional safety distance in the case of erroneous transmission. Obtained results reveal that the

minimum target distance of the CACC controller increases significantly with the drop in the BLER of packets received from the leader.

Although for almost every case considered here the minimum target distance (the distance that provides the crash ratio no higher than 1%) can be found, one should note that the considered abstract system uses a fixed BLER value for communications; thus, the increase in distance between the transmitter and the receiver is not accounted for. In a real-life system, such an increase in the target CACC distance would result in a further drop of the leader packets' reception ratio and, consequently, might prevent the automated controller from correct operation (which is the situation observed in Figure 2).

2.3. Solutions. To solve the problem of too low message reception rate, the following solutions can be considered:

(i) Decrease the rate of CAM messaging

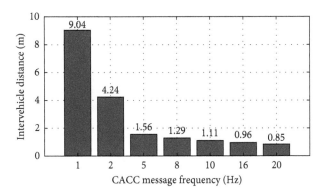

FIGURE 5: Impact of the CAM message rate on platoon performance.

(ii) Apply the decentralized congestion control (DCC) mechanisms

(iii) Use radio transceivers that can operate in two different frequency bands simultaneously

In this section, we will investigate the aforementioned solutions.

2.3.1. Different CAM Message Rates.

The impact of the message broadcast rate on the CACC is shown in Figure 5 with ideal communication assumed. For the purpose of this investigation, message frequencies ranging from 1 to 20 Hz are assumed. For the rates of 1 and 2 Hz, the performance of the platoon is significantly decreased. This is due to the fact that these frequencies correspond to 1 s and 500 ms information update intervals, which at highway speeds translate to a long travelling distance without an information update. For the rates of 8 and 10 Hz, the performance degradation (compared to the 20 Hz performance) is rather small. For the rate of 16 Hz, the platoon performance improvement is insignificant; thus, an increased usage of radio resources has no justification. We therefore conclude that, for the considered scenario, the optimal information update rates are from 8 to 16 Hz (we assume that the increased message rate has a negligible impact on the packet collision probability).

2.3.2. DCC as a Potential Solution.

Another option to improve the communication reliability suggested by the European Telecommunications Standards Institute (ETSI) is to apply the DCC mechanism [32] to reduce the load of the used radio channel. The DCC is a threshold-based mechanism that applies different MAC layer policies based on the estimated occupancy rate of the transmission medium. These include transmit power adaptation, data rate control, or adaptation of the carrier-sensing capabilities of the transmitter. The DCC mechanism restrictions are applied when the estimated channel load is above 20% [32, 33]. In order to estimate the average channel load in considered scenarios, we measured the fraction of time that the channel is considered as busy from the user's perspective, with the averaged results presented in Figure 6. This figure presents an average channel occupancy rate for four considered car

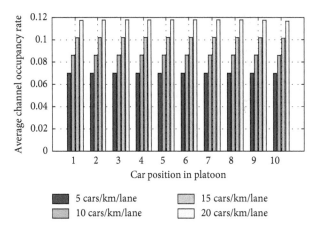

FIGURE 6: Average estimated channel occupancy rate versus the car position in the platoon.

traffic densities, that is, 5, 10, 15, and 20 cars/km/lane, separately for each vehicle within the platoon. One can note that even for the worst-case scenario with 20 cars distributed uniformly on each lane per km, the average channel load is at most 12%, which is below the DCC threshold. Therefore, we can conclude that application of the DCC mechanism will have no impact on the CACC performance, as the reception rate of leader packets will be too low to enable proper automated control before the DCC restrictions are applied.

2.4. Identification of the Performance Limitations of CACC Schemes: A Dual-Radio Case.

The last considered solution is to use the dual-radio transceivers that allow to simultaneously transmit data in the control channel and, additionally, in other frequency channels. In the following part, we consider a dual-radio transceiver, where the first radio is continuously tuned to the CCH to transmit and receive CAMs and the second radio is tuned to another service channel (SCH). We do not consider multichannel operation with a single-radio transceiver (i.e., channel switching), as according to [34], higher collision rates and increased message delays might be experienced in the corresponding alternating mode. Moreover, according to the ETSI specification [35], a vehicle should continuously listen to the CCH for CAMs, thus prohibiting the use of this mode. In order to increase the efficiency of the CACC, we assume that, besides CAMs, every platooned vehicle may broadcast short CACC packets, which contain only speed and acceleration information, on a dedicated SCH. This is possible only with the dual-radio transceiver configuration, since the single-radio transceiver is continuously tuned to the CCH. The transmission frequency of CACC packets is assumed to be 10 Hz (thus, combined with the use of CAMs, we are able to achieve the effective CACC information transmission frequency up to 20 Hz). Moreover, we assume that the additional SCH is reserved for in-platoon communications only, thus preventing other devices from transmitting there.

Similarly, as for the single-radio transceiver, the first step in the evaluation aims to determine the minimum feasible

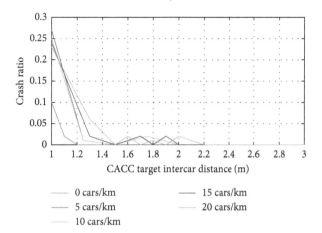

FIGURE 7: Fraction of simulation runs concluding with an in-platoon crash when using the ITS CCH and additional channel reserved only for in-platoon communications.

TABLE 3: Selected CACC target distances and the corresponding number of simulations with a crash per 100 runs for dual-radio transceivers.

Case (cars/km/lane)	Target distance (m)	Number of crashes
Ideal, dual	1.1	0
0	1.2	0
5	1.5	0
10	2.0	0
15	2.0	0
20	2.2	0

target distance between platooned cars for dual-radio transceivers. The target distance is the input parameter of the CACC controller. Obtained results are shown in Figure 7. This figure presents the crash ratio as a function of intercar distance. The best results are obtained if there is no traffic on the neighbouring lanes of the highway. For this case, the distance required between the cars is equal to 1.2 m. If the traffic intensity is increased from 5 to 20 cars/km/lane, then the required distance increases from 1.5 to 2.2 m.

From the results presented in Figure 7 and Table 3, we can see that, in all scenarios with dual-radio transceivers, it is possible to find such a minimum feasible target distance in the 1.1 m to 2.2 m range. Comparing this to the minimum distance for the single-radio transceiver (Table 2), we observe the significant improvement of the platoon performance. In this case, even for the highest considered traffic density, the platoon is able to fulfill the 99% crash-free assumption. In Table 3, we also provided the target distance for the ideal scenario (perfect packet decoding, infinite transmission range) to show the quality of the dual-radio transceiver performance.

In the rest of this section, we present the results obtained for the target distances indicated in Table 3.

One should note that the use of the dual-radio transceiver and special CACC messages transmitted on the dedicated SCH results in a substantial improvement of CACC performance. The minimum feasible target distances

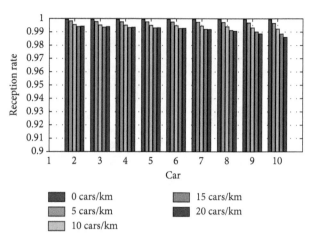

FIGURE 8: Reception rate of platoon leader CAM packets versus the car position in the platoon with a dual radio.

achieved in such a configuration are always better than those in the case of a single radio. This is the result of the increased overall message rate with the dual radio, which is 20 Hz (10 Hz CAM on the CCH plus 10 Hz CACC on the SCH), as well as the improved message reception ratio. While CAM packets might collide frequently, as nonplatooned cars also use the CCH for CAM transmissions, the reliability of CACC messages is much higher because they are transmitted on a dedicated SCH.

To analyze the reliability of 802.11p message transmissions, we collected the message reception statistics at each car in the platoon. The reception rates of the messages transmitted by the platoon leader are shown in Figure 8. For the dual-radio transceiver, the situation is completely different. In this case, the higher transmission reliability of CACC messages may compensate for the loss of CAMs messages to allow crash-free platooning with a shorter target distance, as hardly any CACC packets are lost.

2.5. Conclusions. One can notice that use of an additional frequency channel for in-platoon communications improves significantly the performance of the CACC controller. However, this is valid with an assumption that this additional channel is unoccupied (or at least with extremely low load), and thus, the CACC message transmission is not interrupted. With the current specifications of the IEEE 802.11p standard, only six additional shared channels are available (apart from the control channel) in the ITS frequency band. As low-load requirements stated above cannot be guaranteed with use of these bands, many other services, such as infotainment or nonsafety ones, will use them. Therefore, a very promising solution for IEEE 802.11p-based systems is to search for additional frequency bands that can be used only for platooning purposes.

The reliability problem when using IEEE 802.11p could also be overcome using more advanced communication systems, such as the 3GPP cellular vehicle-to-anything (C-V2X) [15]. It certainly provides higher reliability of the wireless links, especially when working in the assisted mode (mode 3 of C-V2X) with the resource allocation performed by the base station. It is shown in [36, 37] that platooning using the CACC is fully feasible using C-V2X mode 3. However, when working independently (without the aid of infrastructure (mode 4 of C-V2X)), the collisions in medium access are also possible, and thus, the decreased message reception rate can be observed. Moreover, due to the need to allocate resources in advance in C-V2X, the latency of MAC layer processing is much higher than that in case of IEEE 802.11p, which may affect the quality of received information [37]. Furthermore, the centralized allocation of resources performed by the base station can become difficult with very high density of cars requesting for transmission. The worst-case scenario of 20 cars/km/lane (an average intercar distance of 50 m) considered in this paper translates to an average of 80 cars (assuming 4 lanes) requesting for resources in a cell with a radius of approximately 500 m. One can easily imagine a much worse scenario with a higher number of lanes and a traffic jam, where even hundreds of cars may be served by a single base station. This can result in reduced effectiveness of the system as the necessary control information needs to be transmitted. Therefore, the framework proposed in this paper, although presented in the context of IEEE 802.11p, can be relevant also for the C-V2X or future 5G system, as it provides additional spectrum opportunities for intraplatoon communications, where the resource allocation can be performed independently of the base station.

3. Potential of Dynamic Spectrum Access with Coordination via Context-Aware Databases

It is prominent from the discussion made in the previous section that reliable operation of CACC algorithms requires a wireless channel where collision with any other system or device will be negligible. A straightforward solution would be to provide a number of separate licensed channels, each utilized by a single vehicles' platoon. However, foreseeing future platoon densification, the number of available wireless channels may not be enough in one region, while the licensed spectrum may be vacant in other regions. This is a phenomenon common for most of the wireless systems and observed during many spectrum measurement campaigns [8, 38, 39]. Even though most of the frequencies, for example, up to 3 GHz, are licensed by various wireless systems, in practice, many of them are unused in a given time, at a given location. This gave rise to an idea of a cognitive radio [11], where one can access spectrum resources licensed to another system provided there is no interference generated to it. After years of research on reliable detection of unused frequency resources (called white spaces) [40] and efficient utilization of these white spaces [41], a number of practical solutions have been proposed [42]. The most significant conclusion is that the highest spectral efficiency and low probability of intersystem interference can be obtained only if devices are connected to the spectrum management system utilizing the dedicated database. Field trials of cognitive radio systems using databases for the optimization of spectrum utilization were carried out, for example, in London by the British

communications regulator the Office of Communications (Ofcom) [43] and in many other locations. Commonly, GLDB/REM/RSM (covered hereafter under the common name CADB) is proposed for this purpose [12, 13]. The CADB is a database storing data on the radio environment (e.g., propagation conditions) and its users (e.g., parameters of its transmission and sensitivity to interference.). Depending on the scenario, the set of information stored in the CADB can vary. Typically, radio information is stored together with a geographical tag so that it is a kind of geolocation database. The CADB system is typically divided into three parts: the database(s), the CADB manager responsible for analyzing available data and assignment of transmission parameters to wireless devices, and data acquisition and sensing function responsible for obtaining the current status of the radio environment utilizing various information sources. In our case, we call them the CADB, spectrum manager (SM), and measurement capable device (MCD), respectively, as detailed later in the following section. The data stored in the database can vary from relatively static policies on the utilization of a given frequency (based on some regulatory frameworks or in agreement with licensed spectrum users) to rapidly changing status of a given frequency utilization and observed interference. Static data can be obtained from some external database, for example, national regulator, or based on some analytical modeling, for example, received power distribution based on the base station location and power. A more challenging one is acquisition of dynamic, real-time data about a given radio environment. It can be obtained by analyzing control information reported by data-transmitting devices in a given frequency (e.g., user equipments or base stations) or sensing nodes employed purely for this purpose. An example of sensing nodes in the existing dynamic spectrum access standard is a radar-detecting sensor in the Citizens Broadband Radio Service (CBRS).

There are many different reasons why CADB subsystems should be employed. On the one hand, it can allow some existing systems (e.g., LTE) to offload some traffic from the licensed band. On the other hand, it can allow many systems to capitalize advantages of a given frequency band, for example, low path loss at low frequencies. In the case of vehicle platoons, the most important is lack of interference resulting in high reliability of transmission. The CADB can manage frequency assignment in a wide frequency range, providing nearly an unlimited number of separated wireless channels for in-platoon communications; thus, at least one channel should always be detected as unoccupied. Moreover, geolocation-based frequency management can allow many distanced platoons to use the same frequency channel. From the licensed users' perspective, it is important that the in-platoon communications require relatively low power, and even if some marginal interference is observed, it is temporary as the platoon moves.

3.1. Measurement Campaign and Results Analysis. In order to justify the usage of white spaces for V2V communications, let us now analyze the results of the conducted spectrum measurements. Our goal is to verify if the detected white spaces could reliably be used for V2V communication and if their characteristics (amount, stability, etc.) are suitable for this purpose. The spectrum occupancy measurements were carried out on relatively busy roads in order to keep results possibly close to the V2V scenario.

An omnidirectional antenna AOR DA753 was mounted on a rooftop of a car. It was connected to the spectrum analyzer Rohde & Schwarz FSL6 via a low-loss cable (H155). A laptop with MATLAB and Instrumental Control Toolbox software was connected to the spectrum analyzer via an Ethernet cable in order to configure, trigger, and store measurement results. Measurements were carried out in the frequency range from 75 MHz to 1 GHz with a resolution bandwidth (RBW) of 30 kHz. Ten consecutive power samples at a given frequency were averaged before sending to the PC in order to initially decrease the thermal noise influence. Each measurement trace is tagged with a timestamp and a location obtained from an external GPS module. The measurements were carried out in Poznan, a city in Poland, of around 500000 inhabitants, during a typical working day. The car drove a path of 22 km, mainly in the city center, obtaining 600 power spectral density (PSD) traces, each of 30834 frequency points. The resultant PSDs showing the minimum, mean, and maximum of 600 obtained samples are presented in Figure 9. There are just a few frequency ranges of high received signal power, for example, FM broadcasting around 100 MHz, TETRA and CDMA communications at around 450 MHz, and the GSM band at around 950 MHz. At many other frequencies, the mean received power equals about −110 dBm/30 kHz, being a noise floor of the utilized spectrum analyzer. However, the maximum PSD shows that there are timestamps/locations where the instantaneous received power is high above the mean level. Although it can be a result of a high-power thermal noise, the more probable is an existence of a local transmission. Most interesting from V2V communications' point of view is the terrestrial TV band. The TV transmission is relatively stable in time and covers relatively a wide space. However, the other licensed systems utilizing this band are Programme-Making and Special Events (PMSE) devices, for example, wireless microphones. These can transmit low-power, narrowband (200 kHz) signals that should be detected in order not to be disturbed by V2V transmission. A measured power of each 8 MHz channel in the ultrahigh frequency (UHF) band indexed {21, 22, . . ., 69} (center frequencies {474, 482, . . ., 858} MHz) is shown as a function of time in Figure 10. It is visible that at most 4 channels, that is, nos. 23, 27, 36, and 39, have power higher than −80 dBm. In order to estimate the number of UHF channels not utilized by DVB-T receivers, a methodology presented in [44] can be used. The minimum DVB-T receiver input power can be calculated as the thermal noise over the effective channel bandwidth of 7.61 MHz at a temperature of 290 K, that is, −105.2 dBm, can be increased by a noise figure of 7 dB, and required a carrier-to-noise (C/N) ratio. According to [45], the required C/N ratio in the Rayleigh channel (worst-case scenario) for 64-QAM modulation and code rate of 3/4 (mode commonly used in Poland) is 21.7 dB. In

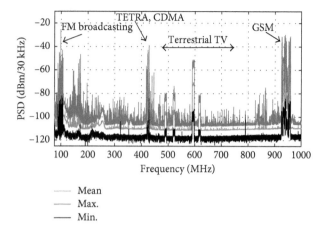

Figure 9: Minimum, mean, and maximal PSDs over the measurement path of 22 km.

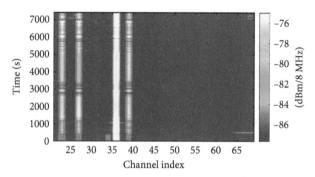

Figure 10: Power of TV channels (indexed from 21 to 69) over measurement time.

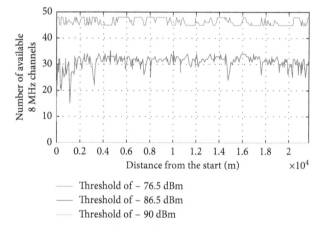

Figure 11: Number of UHF channels available for V2V transmission depending on the minimum required power threshold.

this case, the minimum required DVB-T signal power is −76.5 dBm. All channels of the DVB-T signal power smaller than this value can be treated as unused and potentially be utilized by V2V transmission (although this rough calculation does not consider a hidden node problem or the influence of the chosen V2V transmission power). The results in Figure 11 show that, in any point of the observation path, the number of available channels is greater than or equal to 45 (out of 49 in total). If the V2V communications struggle to support dynamic band switching, there are still 43 channels that are not occupied at any point of the test path. However, DVB-T receivers commonly use directional antennas located significantly higher over the ground than the 0 dBi antenna used in the measurement. In order to mimic this gain, the threshold was decreased to −86.5 dBm and −90 dBm, giving approximately 30 and 0 UHF channels available, respectively. In the case of continuously vacant channels, the number is 2 and 0, respectively. The significant decrease of the available number of channels is caused by the thermal noise floor of the spectrum analyzer exceeding the threshold. As such, the number of available UHF channels is underestimated. Measurements utilizing a more advanced setup, for example, a high-gain antenna, a low-noise preamplifier, and an advanced sensing algorithm, can be carried out in the future to increase the reliability of these results.

The results of spectrum occupancy measurements presented above prove that there is a vast amount of the underutilized spectrum in the urban areas in the frequency range up to 1 GHz. Limitation of the analysis only to DVB-T channels shows that there are 43 (out of the total of 49) channels permanently unoccupied by the licensed systems in the city of Poznan. It constitutes 344 MHz of the spectrum that can be used potentially for V2V communications within platoons. However, such an application requires an architecture that guarantees sufficient QoS both for licensed users and in-platoon communications. A proposal of such a design will be presented in the next section.

4. A Proposed System Design for Context-Aware Platooning

Based on the analysis outlined in Section 2 on the performance of platooning using the CACC and the results of spectrum measurements presented in Section 3, we can formulate a thesis that the use of spectrum white spaces will improve the CACC operation by increasing the reliability of wireless V2V communications. Hence, in this work, we propose the idea of V2V communications for platooning purposes acting as a secondary system in the cognitive radio concept that will dynamically select the used frequency band based on sensing results and context information acquired from the database. However, the use of spectrum white spaces for in-platoon V2V communications requires coordination and management of different functions, including detection of the available spectrum and selection of the used frequency band.

4.1. Spectrum White Spaces Detection. First, sources of information on the occupancy of the potential frequency bands need to be provided. These will include both the sensing nodes (called the MCD), providing short-term information, and databases (called the CADB), comprising the long-term context information. In the investigated scenario, the following sources of information are considered:

(i) Platoon vehicles that perform spectrum sensing in a cooperative way, where the platoon leader acts as

an aggregation node. These can exchange information with other platoon vehicles and the infrastructure (roadside unit (RSU) or base station (BS)) using the V2X communications (e.g., dedicated short-range communication (DSRC) or C-V2X).

(ii) Other vehicles that can be involved in sensing white spaces, with the information reported to the RSU or cellular BS.

(iii) The RSU or cellular BS (in general, any fixed access point) that acts as aggregation nodes for information provided by all vehicles and platoons in their service area. These nodes can also be involved in the spectrum-sensing process.

(iv) The CADB (including REM, RSM, or GLDB) that will provide the long-term information on the spectrum utilization, registered licensed users, interferences, road traffic status, terrain topology, and so on.

We assume that every sensing node is capable of either providing the decision on the availability of the measured band or sending the raw measurements data to an aggregation node, enabling cooperative sensing mechanisms. However, one should note that extensive exchange of the raw data will put further burden on the radio network, so different aggregation levels can be considered. Depending on the current network status, including the data traffic load, the number of available nodes, or the presence of the RSU, different sensing modes will be dynamically selected, with the decision on the spectrum availability being made on the car level, platoon leader level, or RSU/BS level. The decision on the selected sensing mode will be performed at the application level taking into account the global information for the considered sensing region.

4.2. Selection of the Frequency Band. In order to make efficient use of the information stored in the CADB and to select the best frequency band, a functional management architecture of the system is required. This comprises the following elements:

(i) MCD that represents the network nodes capable of providing information on the spectrum availability and reporting it to the CADB.

(ii) Data storage units (CADB) that acquire and store the relevant information provided by different sources, including the MCD.

(iii) An SM unit that is responsible for selecting the optimal frequency band for in-platoon communications; in other words, it makes a decision on the prospective band to be used by the platoon.

(iv) A secondary user, that is, the platoon, that will use the spectrum allocated by the SM for internal V2V communication.

The mapping of listed functionalities to specific elements depends on the architecture of the considered wireless network. Specifically, the MCD will be the nodes that are responsible for providing information on the spectrum

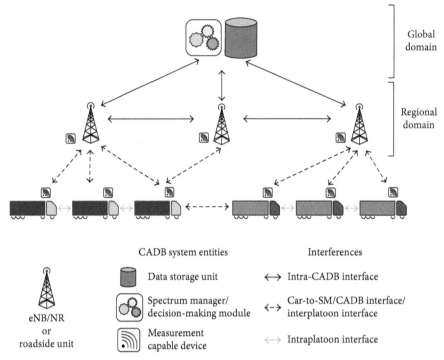

CADB system entities

		Interferences	
Data storage unit		$\longleftrightarrow$	Intra-CADB interface
Spectrum manager/ decision-making module		$\leftarrow \rightarrow$	Car-to-SM/CADB interface/ interplatoon interface
Measurement capable device		$\longleftrightarrow$	Intraplatoon interface

eNB/NR
or
roadside unit

FIGURE 12: Schematic of the centralized management architecture.

availability, that is, the sensing aggregation nodes. The secondary users will be all platoons that request allocation of a dedicated frequency band. However, the location of data storage units and the SM can be associated with different network entities, depending on the selected centralization level. Although the mapping of the data storage units and the SM to specific elements are independent of each other, we assume that these will be treated jointly when formulating the functional architecture. Hence, we can distinguish the following considered management scenarios:

(i) The centralized architecture (presented in Figure 12), where only one global SM is considered. It is connected to a single CADB that is responsible for collecting the context information from all MCDs and other sources. Such an architecture requires deployment of the dedicated network server that will act as the SM and will be connected to the CADB and the RSU/BS using the backbone network. The decision on spectrum allocation for all platoons will be optimized globally in this scenario.

(ii) The distributed architecture (presented in Figure 13), where every platoon is associated with an individual SM that is located in the leader vehicle. In this case, the CADB may be located both centrally (a single database with global data) and locally in the selected RSU/BS (each RSU stores data relevant only to its region). In this scenario, every platoon will decide on the used frequency band independently of the others, based only on the information received from databases and its individual sensing results.

(iii) The hybrid (partially decentralized) architecture (presented in Figure 14), where the regional SM is

located next to the selected RSU/BS. Here, the decision on the frequency band allocation will be made on per-region basis, where every platoon will be provided by information from the corresponding regional SM. Similarly to the distributed case, the CADB may be located here both centrally (a single database with global data) and locally in the selected RSU/BS. One should note here that a communication interface between neighbouring RSUs is needed to coordinate allocation on the boundaries of regions.

Three different communication interface types are defined for the considered management scenarios, namely, the intra-CADB interface, car-to-CADB/SM interface which is also used as an interplatoon interface, and the intraplatoon interface. The intra-CADB communication is assumed to be realized using the fixed wireline or wireless networks (the backbone network) that connect the RSU/BS with the central management entities and with one another (e.g., the S1 and X2 interfaces in LTE). The interplatoon or the car-to-CADB/SM communication can be achieved using, for example, the DSRC, C-V2X [15], or any other V2X systems defined for 5G networks in the future. Finally, the interplatoon interface is intended for further studies due to different frequency bands used here; however, as a baseline, also the use of DSRC or C-V2X can be assumed.

When vehicle platooning is considered, taking into account the results of investigation described in Section 2, the main factors determining the performance of dynamic spectrum management will be the latency of the obtained information and its reliability. Moreover, one should also account for the increase in network load with additional control data and the eventual costs of implementation of

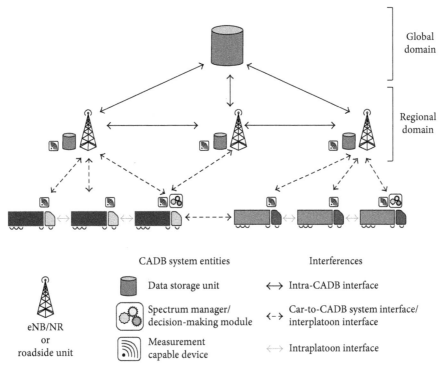

FIGURE 13: Schematic of the distributed management architecture.

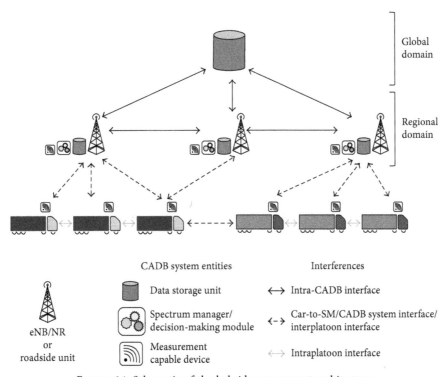

FIGURE 14: Schematic of the hybrid management architecture.

new hardware or software. All of the architecture scenarios described above are characterized with different advantages and disadvantages concerning these aspects.

Obviously, the advantage of centralized management is the global knowledge on the network status and the possibility to jointly optimize the spectrum allocation to all platoons. One should note here that the term "global" refers here to a large geographical area, such as the administrative region, and the management is performed also within its boundaries. Among other advantages, one can distinguish lower load of the RSU/BS with network control duties, as these will be used only to distribute the information. On the

TABLE 4: Summary of selected properties of the proposed management architectures.

Property	Centralized	Distributed	Hybrid
SM location	Central management server	Platoon leader	RSU/BS
CADB location	Central CADB	Central CADB and RSU/BS	Central CADB and RSU/BS
SM-CADB connection	Intra-CADB interface (backbone network)	Car-CADB interface (V2X communications) and intra-CADB interface (backbone network)	Intra-CADB interface (backbone network)
SM-platoon connection	Car-CADB/SM interface (V2X communications) and intra-CADB interface (backbone network)	Direct access (leader)	Car-CADB/SM interface (V2X communications)
Advantages	Global knowledge of the network status, joint optimization of spectrum allocation, and lower management load of the RSU/BS	Lowest latency, decreased load of the network with management data, and deployment of an additional server not required	Lower network load and latency than those in the centralized system, partial (regional) coordination of spectrum allocation, and deployment of an additional server not required
Disadvantages	High load of the management server, high latency, increased network load with management data, and dedicated management server (hardware) required	Incomplete information on spectrum availability, no coordination between platoons (interference), and increased vehicle cost (SM inside the vehicle)	Inter-RSU/BS interface needed for coordination of neighbouring SMs and additional load of the RSU/BS with management tasks

contrary, there are several important disadvantages related to the centralized processing. With the large number of managed platoons, the load of the server will increase dramatically. This may influence the decision timing and, eventually, lead to a situation where outdated decision is received by a platoon. Moreover, the need to provide the sensing information to a central database will further increase the decision latency and the network load.

Contrary to the approach described above, in the distributed scenario, the decision on the spectrum use is made by the platoon leaders. The main advantage of such an architecture is that the SM will use the most recent sensing information provided by the platoon vehicles. Aided with the context information from local and global databases, the decision process will be characterized by a very low latency. Moreover, the load of the network will be lower, as the information on band selection needs to be distributed only locally, and there is no need to deploy a dedicated management server. On the contrary, the main disadvantage of this approach is that the information about the surrounding environment and the expected spectrum white spaces might be incomplete, as it will rely mostly on the short-term information collected locally and the long-term information from databases. Furthermore, there will be no up-to-date knowledge on the frequency bands selected by other platoons, hence resulting in interferences. Finally, the distributed architecture will require implementation of the spectrum management functions inside a vehicle, thus having impact on its cost and energy efficiency.

The third considered architecture, namely, the hybrid one, is aimed at combining the advantages of both the centralized and distributed approaches. First, positioning the SM in the RSU/BS does not necessarily require deployment of dedicated hardware, as existing units may be used for this purpose, with

just a new functionality implemented. Moreover, energy efficiency is not an issue here, as usually such nodes operate with a fixed power supply. Furthermore, the network load and latency will be lower than those in case of the centralized architecture, as only limited sensing information will be fed to the global database, with the most collected in a local (regional) one. Additionally, the spectrum allocation will be optimized within the region, thus reducing the expected interferences between platoons. However, the interference problem may still persist on the region boundaries; hence, a coordination mechanism of neighbouring SMs is necessary. Finally, the additional load of the RSU/BS with management duties may be considered as a disadvantage, compared to the centralized approach.

The main features, advantages, and disadvantages of the considered management system architectures are summarized in Table 4.

5. Challenges

Despite the functional management architecture selected from the options listed in Section 4, the dynamic use of spectrum white spaces for in-platoon communications poses serious challenges that need to be addressed. Among these, one can distinguish the susceptibility to interference, difficulties in estimation of the predicted quality of transmission in the considered frequency band, requirements on the reporting mechanism of sensing results, and, finally, the problem of maintaining the continuity of service when the frequency band switch is performed.

5.1. Interference Problem. The use of the CACC for vehicle platooning requires radio transmission with the highest possible quality that can be regarded as low latency and high reliability. Therefore, the considered system is very susceptible to interference. Among the sources of disturbances, one can consider both the primary (licensed) and secondary users (e.g., other platoons). Therefore, detailed analysis of possible interferences in the investigated system is necessary, including identification of potential sources of interference and exploration of the ways of minimizing their negative impact on signal transmission in the platoon. Such an approach can lead to definition of the possible (acceptable) interference level, as well as the properties of the signal that can be exploited to mitigate its impact on the receiver.

5.2. Estimation of the Channel Quality. The dynamic use of spectrum white spaces has a potential to provide frequency bands characterized by a very low load, thus increasing the reliability of in-platoon communications. However, to fully utilize the possibilities of dynamic spectrum usage, one should consider a wide range of frequencies (e.g., the TV white spaces, frequencies around 3.5 GHz, or the mmWave bands). The main problem with operation using the different frequency bands is the need for reconfiguration of the transceiver and estimation of the predicted channel parameters. When low frequencies are used, the transmission can be characterized by a high range and low impact of the Doppler effect. However, due to small signal attenuation, the extensive use of lower frequencies might result in significant interference. On the contrary, when considering the use of frequency bands above 6 GHz, there are numerous challenges in V2V communications, such as high signal attenuation, high mobility impact on the channel characteristics, significant impact of the Doppler effect, high penetration loss that may result in a blockage of transmission, and the lack of realistic channel models [10]. Some of these will have a minimal impact on transmission in a platoon of vehicles, for example, the Doppler effect (as the relative speed of platoon vehicles is close to zero). Others, such as high signal attenuation, might have either positive or negative impact, as the transmission range will be smaller, but also, the interference will be rapidly diminishing with the increase in distance. On the contrary, penetration loss or frequent channel variations will seriously affect the platoon transmission. Therefore, there is a need to create a channel quality evaluation framework that will allow to compare the advantages and disadvantages of every considered frequency band in the spectrum selection process.

5.3. Reporting of Sensing Results. The use of sensing for detection of spectrum white spaces puts additional burden on the wireless network used for V2V communications as its results need to be reported to the relevant data storage unit and, finally, to the SM. Depending on the selected sensing mode, the exchange of information might be in the form of just decisions on the availability of the spectrum, or, in extreme cases, exchange of raw sensing data. As the data storage units are located in the regional or global domain, this information needs to be distributed using the existing network via the RSU or BS. Therefore, an efficient mechanism of reporting needs to be developed, which, preferably, should be adaptive based on the measured network load. Trade-off studies on the frequency of reporting and expected accuracy of sensing should also be considered.

5.4. Maintaining the Continuity of Service. The idea of dynamic spectrum management introduces a serious challenge for platoon communications which is related to the switch timing and continuity of transmission. The procedure of switching between the spectrum white spaces should take into account the need to maintain reliable links between all platoon vehicles until the switching procedure is finished and all devices are ready to operate in a new band. Similarly, the problem of transmission blockage caused, for instance, by a significant geographical separation of the platoon vehicles (e.g., at traffic lights) and by a significant interference or high penetration loss of obstacles (e.g., other cars) when using above 6 GHz bands, requires measures that will maintain the platoon integrity and ability to operate autonomously. Therefore, dedicated medium access control (MAC) layer protocols are required that will control the procedures of band switching and platoon reorganization.

6. Conclusion

In this article, we have considered the problem of V2V intraplatoon communication for the system based on the IEEE 802.11p/C-V2X communication protocols. We proposed a novel concept of the spectrum management system architecture for the dual-radio transceiver communication utilizing the spectrum white spaces. The aim of the proposed dynamic frequency band allocation and related system design is to significantly improve the platooning performance especially in the high-traffic areas/urban areas. In particular, the proposed system design has been developed to support dual-radio communication for platooning purposes, where critical system information to nonplatoon vehicles or the infrastructure is transmitted with the use of a primary radio (in accordance with the 802.11p or C-V2X communication protocols), while other data (platoon management, noncritical, other services, and infotainment) can be transmitted using an additional frequency band dynamically selected from the spectrum white spaces. Our experimental simulation results of the IEEE 802.11p-based system reveal that introduction of the second link allows to improve the performance of the considered platooning using the CACC mechanism significantly. One should also note that the proposed framework can be applied to any current or future V2X standard as a supplementary dynamic spectrum management system is proposed.

As a future work, we plan to perform deeper analysis of the proposed architectures. These include careful scenario definition, performance estimation, and detailed estimation/calculation of data flow within the proposed structures.

Acknowledgments

This work of Michał Sybis and Paweł Kryszkiewicz has been funded by the Polish Ministry of Science and Higher Education funds for the status activity for the young scientists' project DSMK/81/8125. The authors would like to thank their colleague Adrian Kliks from the Poznan University of Technology for his support in development of this paper.

References

[1] A. Osseiran, F. Boccardi, V. Braun et al., "Scenarios for 5G mobile and wireless communications: the vision of the METIS project," *IEEE Communications Magazine*, vol. 52, no. 5, pp. 26–35, 2014.

[2] J. Lioris, R. Pedarsani, F. Y. Tascikaraoglu, and P. Varaiya, "Doubling throughput in urban roads by platooning," in *Proceedings of the IFAC Symposium on Control in Transportation Systems (CTS 2018)*, Genoa, Italy, June 2016.

[3] Scania. *Annual Report 2001*, 2001, http://www.scania.com/group/en/wp-content/uploads/sites/2/2015/09/sca01EN_tcm10-8527_tcm40-543841.pdf.

[4] *SARTRE Project, D.4.3 Report on Fuel Consumption*, 2014, http://www.sartre-project.eu/en/publications/Documents/SARTRE_4_003_PU.pdf.

[5] J. Ploeg, B. Scheepers, E. van Nunen, N. van de Wouw, and H. Nijmeijer, "Design and experimental evaluation of cooperative adaptive cruise control," in *Proceedings of the 14th International IEEE Conference on Intelligent Transportation Systems (ITSC 2011)*, Washington, DC, USA, October 2011.

[6] S. Santini, A. Salvi, A. S. Valente, A. Pescape, M. Segata, and R. L. Cigno, "A consensus-based approach for platooning with inter-vehicular communications," in *Proceedings of the IEEE Conference on Computer Communications*, Kowloon, Hong Kong, April-May 2015.

[7] D. Jia, K. Lu, J. Wang, X. Zhang, and X. Shen, "A survey on platoon-based vehicular cyber-physical systems," *IEEE Communications Surveys and Tutorials*, vol. 18, no. 1, pp. 263–284, 2015.

[8] A. Kliks, P. Kryszkiewicz, J. Perez-Romero, A. Umbert, and F. Casadevall, "Spectrum occupancy in big cities-comparative study: measurement campaigns in Barcelona and Poznan," in *Proceedings of the International Symposium on Wireless Communication Systems in 2013*, pp. 1–5, Ilmenau, Germany, August 2013.

[9] J. Choi, V. Va, N. Gonzalez-Prelcic, R. Daniels, C. R. Bhat, and R. W. Heath, "Millimeter-wave vehicular communication to support massive automotive sensing," *IEEE Communications Magazine*, vol. 54, no. 12, pp. 160–167, 2016.

[10] M. Giordani, A. Zanella, and M. Zorzi, "Millimeter wave communication in vehicular networks: challenges and opportunities," in *Proceedings of the 6th International Conference on Modern Circuits and Systems Technologies (MOCAST)*, pp. 1–6, Thessaloniki, Greece, May 2017.

[11] J. Mitola and G. Q. Maguire, "Cognitive radio: making software radios more personal," *IEEE Personal Communications*, vol. 6, no. 4, pp. 13–18, 1999.

[12] H. B. Yilmaz, T. Tugcu, F. Alagz, and S. Bayhan, "Radio environment map as enabler for practical cognitive radio networks," *IEEE Communications Magazine*, vol. 51, no. 12, pp. 162–169, 2013.

[13] J. Perez-Romero, A. Zalonis, L. Boukhatem et al., "On the use of radio environment maps for interference management in heterogeneous networks," *IEEE Communications Magazine*, vol. 53, no. 8, pp. 184–191, 2015.

[14] *IEEE Standard for Information Technology–Telecommunications and Information Exchange between Systems Local and Metropolitan Area Networks–Specific Requirements Part 11: Wireless LAN Medium Access Control (MAC) and Physical Layer (PHY) Specifications*, pp. 1–2793, March 2012.

[15] *3GPP TS 36.300: Evolved Universal Terrestrial Radio Access (E-UTRA) and Evolved Universal Terrestrial Radio Access Network (E-UTRAN); Overall Description; Stage 2*, version 15.0.0, pp. 1–338, December 2017.

[16] C. Lei, E. M. van Eenennaam, W. K. Wolterink, G. Karagiannis, G. Heijenk, and J. Ploeg, "Impact of packet loss on CACC string stability performance," in *Proceedings of the 11th International Conference on ITS Telecommunications (ITST 2011)*, St. Petersburg, Russia, August 2011.

[17] S. Santini, A. Salvi, A. S. Valente, A. Pescap, M. Segata, and R. L. Cigno, "A consensus-based approach for platooning with inter-vehicular communications and its validation in realistic scenarios," *IEEE Transactions on Vehicular Technology*, vol. 66, no. 3, pp. 1985–1999, 2017.

[18] A. Bohm, M. Jonsson, and E. Uhlemann, "Performance comparison of a platooning application using the IEEE 802.11p MAC on the control channel and a centralized MAC on a service channel," in *Proceedings of the 2013 IEEE 9th International Conference on Wireless and Mobile Computing, Networking and Communications (WiMob)*, pp. 545–552, Lyon, France, October 2013.

[19] M. Segata, B. Bloessl, S. Joerer et al., "Towards communication strategies for platooning: simulative and experimental evaluation," *IEEE Transactions on Vehicular Technology*, vol. 64, no. 12, pp. 5411–5423, 2015.

[20] P. Fernandes and U. Nunes, "Platooning with IVC-enabled autonomous vehicles: strategies to mitigate communication delays, improve safety and traffic flow," *IEEE Transactions on Intelligent Transportation Systems*, vol. 13, no. 1, pp. 91–106, 2012.

[21] A. Vinel, L. Lan, and N. Lyamin, "Vehicle-to-vehicle communication in C-ACC/platooning scenarios," *IEEE Communications Magazine*, vol. 53, no. 8, pp. 192–197, 2015.

[22] O. Shagdar, F. Nashashibi, and S. Tohme, "Performance study of CAM over IEEE 802.11p for cooperative adaptive cruise control," in *Proceedings of the 2017 Wireless Days*, Porto, Portugal, March 2017.

[23] J. Hu, S. Chen, L. Zhao et al., "Link level performance comparison between LTE V2X and DSRC," *Journal of Communications and Information Networks*, vol. 2, no. 2, pp. 101–112, 2017.

[24] R. Blasco, H. Do, S. Serveh, S. Stefano, and Y. Zang, "3GPP LTE enhancements for V2V and comparison to IEEE 802.11p," in *Proceedings of the 11th ITS European Congress*, Glasgow, UK, June 2016.

[25] A. Papathanassiou and A. Khoryaev, "Cellular V2X as the essential enabler of superior global connected transportation services," *IEEE 5G Tech Focus*, vol. 1, no. 2, 2016.

[26] H. Peng, D. Li, K. Abboud et al., "Performance analysis of IEEE 802.11p DCF for inter-platoon communications with autonomous vehicles," in *Proceedings of the IEEE Globecom'15*, San Diego, CA, USA, December 2015.

[27] H. Peng, D. Li, Q. Ye et al., "Resource allocation for D2D-enabled inter-vehicle communications in multiplatoons," in *Proceedings of the 2017 IEEE International Conference on Communications (ICC)*, Paris, France, May 2017.

[28] S. Tsugawa, S. Kato, and K. Aoki, "An automated truck platoon for energy saving," in *Proceedings of the 2011 IEEE/RSJ International Conference on Intelligent Robots and Systems*, pp. 4109–4114, San Francisco, CA, USA, September 2011.

[29] *European Truck Platooning Challenge*, 2016, https://www.eutruckplatooning.com.

[30] *Volvo Trucks Successfully Demonstrates On-Highway Truck Platooning in California*, 2017, http://www.volvotrucks.us/about-volvo/news-and-events/.

[31] P. Sroka, M. Rodziewicz, M. Sybis, A. Langowski, K. Lenarska, and K. Wesoowski, "Improvement of cooperative adaptive cruise control operation by scheduling of BSM messages," *Przegląd Telekomunikacyjny i Wiadomości Telekomunikacyjne*, vol. 6, pp. 350–355, 2017, in Polish.

[32] *ETSI TS 102 687 Intelligent Transport Systems (ITS); Decentralized Congestion Control Mechanisms for Intelligent Transport Systems Operating in the 5 GHz Range; Access Layer Part*, pp. 1–45, July 2011.

[33] S. Subramanian, M. Werner, S. Liu, J. Jose, R. Lupoaie, and X. Wu, "Congestion control for vehicular safety: synchronous and asynchronous MAC algorithms," in *Proceedings of the Ninth ACM International Workshop on Vehicular Inter-Networking, Systems, and Applications*, pp. 63–72, Lake District, UK, June 2012.

[34] *IEEE Standard for Wireless Access in Vehicular Environments (WAVE)-Multi-Channel Operation*, Revision of IEEE Standard 1609.4-2010, pp. 1–92, January 2016.

[35] *ETSI TS 102 724 Intelligent Transport Systems (ITS); Harmonized Channel Specifications for Intelligent Transport Systems Operating in the 5 GHz Frequency Band*, pp. 1–31, October 2012.

[36] C. Campolo, A. Molinaro, G. Araniti, and A. O. Berthet, "Better platooning control toward autonomous driving: an LTE device-to-device communications strategy that meets ultralow latency requirements," *IEEE Vehicular Technology Magazine*, vol. 12, no. 1, pp. 30–38, 2017.

[37] V. Vukadinovic, K. Bakowski, P. Marsch et al., "3GPP C-V2X and IEEE 802.11p for vehicle-to-vehicle communications in highway platooning scenarios," *Ad Hoc Networks*, vol. 74, pp. 17–29, 2018.

[38] M. A. McHenry, P. A. Tenhula, D. McCloskey, D. A. Roberson, and C. S. Hood, "Chicago spectrum occupancy measurements & analysis and a long-term studies proposal," in *Proceedings of the First International Workshop on Technology and Policy for Accessing Spectrum*, New York, NY, USA, August 2006.

[39] A. Kliks, P. Kryszkiewicz, K. Cichon, A. Umbert, J. Perez-Romero, and F. Casadevall, "DVB-T channels measurements for the deployment of outdoor REM databases," *Journal of Telecommunications and Information Technology*, vol. 2014, no. 3, pp. 42–52, 2014.

[40] K. Cicho, A. Kliks, and H. Bogucka, "Energy-efficient cooperative spectrum sensing: a survey," *IEEE Communications Surveys Tutorials*, vol. 18, no. 3, pp. 1861–1886, 2016.

[41] P. Kryszkiewicz, A. Kliks, and H. Bogucka, "Small-scale spectrum aggregation and sharing," *IEEE Journal on Selected Areas in Communications*, vol. 34, no. 10, pp. 2630–2641, 2016.

[42] H. Harada, T. Baykas, C.-S. Sum et al., "Research, development, and standards related activities on dynamic spectrum access and cognitive radio," in *Proceedings of the IEEE Symposium on New Frontiers in Dynamic Spectrum*, pp. 1–12, Singapore, April 2010.

[43] *TV White Spaces: DTT Coexistence Tests*, Ofcom, pp. 1–100, December 2014.

[44] *BT.1368-10 Planning Criteria, Including Protection Ratios, for Digital Terrestrial Television Services in the VHF/UHF Bands*, ITU-R Standard, 2013.

[45] *Digital Video Broadcasting (DVB); Frame Structure Channel Coding and Modulation for Digital Terrestrial Television*, ETSI 300 744 v1.6.1, January 2009.

4

HealthNode: Software Framework for Efficiently Designing and Developing Cloud-Based Healthcare Applications

Ho-Kyeong Ra [ID],[1] Hee Jung Yoon [ID],[1] Sang Hyuk Son,[1] John A. Stankovic,[2] and JeongGil Ko [ID][3]

[1]*Information and Communication Engineering, Daegu Gyeongbuk Institute of Science and Technology (DGIST), Dalseong-Gun, Daegu, Republic of Korea*

[2]*Computer Science, University of Virginia, Charlottesville, VA, USA*

[3]*Software and Computer Engineering, Ajou University, Yeongtong-gu, Suwon, Republic of Korea*

Correspondence should be addressed to JeongGil Ko; jgko@ajou.ac.kr

Academic Editor: Andrea Gaglione

With the exponential improvement of software technology during the past decade, many efforts have been made to design remote and personalized healthcare applications. Many of these applications are built on mobile devices connected to the cloud. Although appealing, however, prototyping and validating the feasibility of an application-level idea is yet challenging without a solid understanding of the cloud, mobile, and the interconnectivity infrastructure. In this paper, we provide a solution to this by proposing a framework called HealthNode, which is a general-purpose framework for developing healthcare applications on cloud platforms using Node.js. To fully exploit the potential of Node.js when developing cloud applications, we focus on the fact that the implementation process should be eased. HealthNode presents an explicit guideline while supporting necessary features to achieve quick and expandable cloud-based healthcare applications. A case study applying HealthNode to various real-world health applications suggests that HealthNode can express architectural structure effectively within an implementation and that the proposed platform can support system understanding and software evolution.

1. Introduction

The advancement of Internet of Things (IoT) applications has allowed various data points collected from a large number of heterogeneous devices to be gathered in a single repository for application development. Naturally, the distributed nature of these systems concentrates on the cloud infrastructure for achieving novel application designs with the support of integrated data processing and effective resource management.

In particular, healthcare-related IoT applications have developed to the point where it not only preserves the safety and health of individuals but also improves how physicians deliver care. Using smart and mobile devices, healthcare IoT allows the delivery of valuable data to users and lessens the need for direct patient-physician interaction.

Despite their attractiveness, however, implementing such cloud-based applications is not in any way trivial. This is especially true for a large number of nontechnical researchers in the healthcare domain. While health and IoT domains are being revolutionized in convergence, designing an effective cloud application asks the researchers to know various technical domains from the server operation, local and web-application languages, to data communication protocols. For example, for efficient web development, developers typically require knowledge of up to five different programming languages, such as JavaScript, HTML, CSS, a server-side language such as PHP, and SQL.

For overcoming these challenges, a recently introduced software platform called Node.js [1] has gained a significant amount of traction in the developer communities. Specifically, Node.js is a scalable single-threaded server-side JavaScript environment implemented in C and C++ [2]. Developers can build scalable servers without using threading but rather by using a simplified model of event-driven programming that uses callbacks to signal the completion of a task [3]. Owing to its simplicity, industry leaders such as Microsoft [4], IBM [5], Netflix [6], PayPal [7], and Walmart [8] have integrated the support of Node.js into their cloud platforms. Notwithstanding the recent developments, Node.js's approach in developing web applications has made it an attractive alternative to more traditional platforms such as Apache + PHP and Nginx servers.

One of the many benefits of using Node.js is its architecture that makes it easy to use as an expressive, functional language for server-side programming [9]. Although it may be trivial to perform development in Node.js once a developer fully comprehends the language, there are many obstacles prior to being able to implement real cloud-based applications for beginners. For example, Node.js requires multiple initialization steps, such as the configuration of the HTTP package and route.js, which guides where the request is routed. Due to the framework being relatively young and the software yet reaching maturity, developers still face the lack of documentation support and can face troubles in receiving support from the development community.

To the best of our knowledge, there has not yet been a Node.js software design or implementations guideline documentation, where developers are influenced to create well-structured healthcare applications. Nevertheless, for software developers, the fact that organizing the program design with solid boundaries is crucial in a successful real-world deployment. Designing an ad hoc model will stimulate logical complexity and cause difficulties in maintaining and updating the application over time. Moreover, without a sturdy guideline documentation, it takes a significant amount of time and needless effort even to configure the primary application development environment.

In this paper, we present a practical solution for implementing cloud-based healthcare applications by providing a framework called HealthNode, which consists of (1) a software design and (2) essential APIs and implementation guidelines for prototyping. We specifically emphasize on healthcare applications rather than other types of applications because Node.js is capable of supporting the complex requirements that healthcare applications ask, which are the needs to support multiple patient-doctor connections, exchange medical data with a unified language and data format, and allow reusability of the developed medical components. Nevertheless, we put as one of our future work to expand HealthNode to be applicable for other applications.

To simplify the use of Node.js, our software design helps developers to easily observe data flow, create modules, and add in functions even without detailed descriptions, which Node.js lacks. We use a top-down approach that structures our software design with a hierarchy of modules and a divide-and-concur approach that organizes the tasks for each module. The top-down approach helps developers to observe application flow from the main module, which acts as the main class in Java, to other submodules in a top-down manner. It is essentially the breaking down of the application development to gain insight into its compositional sub-modules. The divide-and-concur approach breaks down each module into submodules that specify target tasks for each module. This tactic also enables multiple developers to work on different portions of the application simultaneously.

To guide the development of cloud-based healthcare applications using Node.js, we provide APIs and a guideline that constructs the skeleton of essential components within the implementation. We focus primarily on the back end of the server, which contains the core operational functions. Our envision is that HealthNode would overall influence in developing cloud-based healthcare applications.

The contributions of this work can be summarized in three-fold:

(i) A software design that tackles the challenges of maintaining and updating cloud applications developed using Node.js. This design layout will support the resulting application to comprise well-defined, independent components which lead to better maintainability. In addition, new capabilities can be added to the application without major changes to the underlying architecture.

(ii) APIs and implementation guideline that provides explicit, but straightforward instructions for developing general-purpose Node.js prototype. This guideline instructs how to (1) create prototypes by using a limited set of APIs, (2) divide modules and organize function, (3) allow a client communicate to a server, (4) utilize cloud application database, and (5) handle uploading and downloading files. Through this guideline, developers can focus on implementing application logic.

(iii) Examples of healthcare application that open the possibilities for a new structure of healthcare to be developed. Our software design and guideline can be the basis for developing a variety of cloud-based healthcare applications. We provide examples of systems that can be enabled by our work.

In Section 2, we discuss some of the technical challenges of implementing cloud applications using Node.js. We introduce the overall structure of Node.js architecture, libraries, modules, and functions that make up Node.js applications, as well as HealthNode's software design in Section 3. In Section 4, we present our guideline that instructs the implementation procedures for developing cloud-based healthcare applications and introduce couple potential applications where our work can be applied in Section 5. Finally, we position our work among others in Section 6 and conclude the paper in Section 7.

2. Challenges

Technically, for developing a cloud-based healthcare application using Node.js, there are many materials both online and offline for users to easily get a start on building a software environment. However, based on our experiences, the level of these tutorials mostly remains in the beginner level, and due to its relatively new life cycle, it is fairly difficult to identify the information required to perform more advanced tasks. Furthermore, practical troubles that typically arise within this development phase include the lack of (1) a formal software structure, (2) fundamental guidelines for advanced functionality implementation, and (3) real-application examples. We use the remainder of this section to discuss the importance of such support and the challenges that application developers can face due to such limitations.

2.1. Usability. There is an active community that supports Node.js. More developers watch the repository of Node.js at GitHub than other recently trending software environments. Nevertheless, Node.js is relatively new compared to traditional web-application frameworks such as ASP.NET. Therefore, naturally, in contrast to these older frameworks, Node.js lacks documentation and examples on how to structure the overall implementation. Due to a small number of easy-to-follow guidelines, Node.js is not commonly used to its full capability where developers can create data exchange applications for connected infrastructure such as IoT applications. An ideal application should allow clients to submit data to a server and the server to respond back to the clients to fully utilize the cloud computation power.

2.2. Feasibility. For beginners to use Node.js for implementing a fully functioning prototype, it takes a significant amount of time and effort to feasibly set up the basic application environment, properly route incoming and outgoing information, and format the overall application structure. Ideally, to catalyze the development of various cloud applications, this process should be simple to both understand and implement.

2.3. Maintainability and Extensibility. In designing the back end of a cloud application using Node.js, existing tutorials often promote examples using only a single module containing all the possible functions, regardless of system design. Using a single module limits the tasks to be distributed to other modules and therefore diminishes advantages of designing an organized implementation. Although Node.js supports building a hierarchy of multiple modules, ad hoc plans of software structuring at the hierarchy level can cause increased logical complexity without solid boundaries between heterogeneous modules. Moreover, to maintain and update applications efficiently, proper documentation is essential. However, documenting within a single module or multiple ad hocly planned modules can add additional burden to the code review process. To support these issues, a software design with an organized structure of modules will not only help developers in designing their software but also benefit them in maintaining their applications. Furthermore, such a repository of modules can ultimately benefit the application to be more conveniently extensible over time.

3. Architecture

This section first describes the overall structure of the Node.js architecture, libraries, modules, and functions that make up a typical Node.js application. We then present HealthNode design that uses a top-down approach and the divide-and-conquer strategy, which are both the essence of any software development.

3.1. Background

3.1.1. Architectural Description of Node.js. Well known for its event-based execution model, the Node.js platform architecture uses a single thread for executing application code which simplifies application development. However, heavy calculations and blocking I/O calls that are executed in the event thread prevent the applications from handling other requests. Node.js tackles this issue by using event loop and asynchronous I/O to remain lightweight in the face of data-intensive, real-time applications [10]. The Node.js execution model is different to the thread-based execution model where each client connection is processed by a separate thread. Overall, the platform must coordinate data input and output adaptably and reliably over a different range of distributed systems.

3.1.2. Libraries and Modules of Node.js. Node.js comes with an API covering low-level networking, basic HTTP server functionality, file system operations, compression, and many other common tasks. Moreover, the available external libraries of Node.js can add more capability in a module form. The modules are delivered by public or private package registries. The packages are structured according to the CommonJS package format and can be installed with the Node Package Manager (NPM) [11].

Our software design works off the Express library, which is one of the Node.js packages that support the rapid development of Node.js cloud application [12]. It helps set up middlewares to respond to HTTP requests, specifies a routing table which is used to achieve various action based on the HTTP method and URL and allows to present HTML pages based on passing arguments to templates dynamically. Other than public libraries, local modules also can be referenced either by file path or by name. Unless the module is the main module, a module that is referenced by a name will map into a file path.

3.1.3. Functions of Node.js. In programming, a function is defined as the portion of code that performs a specific task with series of statements. It has a capability of accepting data through parameters for a certain task and returning a result. In Node.js, a function requires extra implementation for routing requests. To support the ease of routing, the Express package enables the capability to create middleware functions. Middleware functions allow setting up a routing path in one line. In addition, middleware functions are only accessible by clients and are not accessible by back-end computation functions. For example, when a client submits and requests data to the server, one of the middleware functions is triggered, and output is returned to the client. For accessing data, middleware functions only access data through shared objects or a library such as MongoDB.

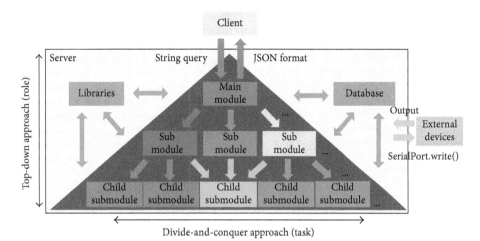

FIGURE 1: HealthNode software design.

3.2. HealthNode. We now detail two software design processes that make up HealthNode. The top-down approach helps developers to couple and decouple modules while the divide-and-conquer approach guides developers to divide the task into simpler modules while enabling multiple module developments concurrently. We use the top-down approach first and then the divide-and-conquer approach in the latter step so that the roles between the modules are first defined before tasks are assigned to each of the roles. Although both of these techniques are commonly used in other types of web and application developments, to the best of our knowledge, there has not been a foundational software design that supports Node.js implementation structure.

3.2.1. Top-Down Approach. The top-down and bottom-up approaches play a key role in software development. The top-down approach is a standard inheritance pattern which decomposes a system to gain insight into subsystems. Each subsystem is then refined in yet greater detail in many additional subsystem levels until the entire specification is reduced to base elements. This process continues until the lowest level of the system in the top-down hierarchy is achieved. In a bottom-up approach, the base elements of the system are first specified. These elements are then linked together to form larger subsystems until a complete top-level system is constructed. Using the bottom-up approach may be beneficial for implementing first-level systems for early testing. However, the bottom-up approach is not suited particularly for our software design due to its requirement of permitting space to grow. Since our study focuses on prototyping, which requires a continuous update, it must be easy to add and couple modules. However, the bottom-up strategy does not allow this, and over time, organization and maintenance issues may exist.

In the software design of HealthNode, a hierarchy structure creates a connection between modules that supports data flow. Figure 1 visualizes how HealthNode maps the top-down perspective by starting from the main module and initializing submodules. Each submodule can also have multiple child submodules with external packages. In the main module, a shared component such as a database is initialized, and necessary submodules are coupled.

The top-down approach is effective when the application idea is clear, and the system design is ready prior to implementation. Looking at Figure 1, a top-down design concentrates on designing vertical hierarchy levels and uses couples to connect data flow. The coupling process happens during the period when modules are added. This process finishes when the submodule is ready to be used after testing and connecting to a higher module. In addition, the coupling process can be used for sharing child submodules. In cases when already implemented child submodule is required by other submodules, each of the submodules can couple to preimplemented child submodule to avoid redundancy. During implementation, coupling is used to create a weak connection between modules. In other words, a submodule is allowed to use a child submodule only once during implementation. The decoupling process can be easily done due to the weak connection between all the modules.

The decoupling process supports maintaining and extending the application over time. When a particular submodule expands, in such case as having two different tasks, the submodule needs to be decomposed. To decompose a module, the decoupling process closes the connection between the higher submodules to the current module. This will lead the decoupling process to stop allowing higher modules to use the current submodule. Overall, the coupling and the decoupling process is completed by initializing and removing the connection between modules and submodules.

3.2.2. Divide and Conquer. Divide and conquer is a concept of recursively breaking down a system into two or more subsystems until these become simple enough to be labeled directly. The outputs to the subsystems are then combined to provide an output to the highest system.

The divide-and-conquer approach is different than the top-down approach in a sense that the top-down approach defines hierarchy levels while the divide-and-conquer method focuses on horizontally dividing a task in each

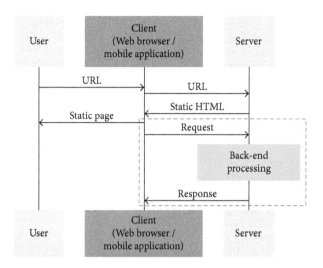

FIGURE 2: Location of HealthNode in a general cloud application.

level to specify a task. The higher the module level is, the application design uses task characteristics to divide modules. The lower the module level is, the design concentrates on the functionality of higher modules' basic requirements. Moreover, in HealthNode's software design, the divide-and-conquer process also reduces redundancy by dividing a module into groups of reusable and unique task-oriented functions. Since one of the functions or middleware functions within the higher module uses submodules, the divide-and-conquer technique can be easily applied using one or two lines of code.

4. Implementation

In this section, we present the implementation components of HealthNode and a step-by-step guideline that instructs the implementation procedures for developing Node.js cloud applications using the proposed framework. Figure 2 shows a general application execution flow from users to clients and then from clients to a server. We focus on the core components located at the back end of the server and the communication components that are used between the clients and the server. Note that we do not look into the details of the server's front end, since there are many existing examples of front-end frameworks available [13–15].

In HealthNode, the proposed implementation guideline follows five essential steps:

(1) A hierarchy structure is set up between modules.

(2) A function is placed on each of the modules to be used for computation.

(3) A middleware function is added into the function, which directly communicates to a client.

(4) During the initialization process of the main module, database information is configured to be used throughout the implementation process.

(5) The client prepares the communication procedures for testing.

Specifically, we describe the implementation details of structuring the hierarchy of the back end, operations that take place in the modules, and the communication process between the client and the server in detail using the following subsections. We will use code snippets to present usage implementations for the concepts in HealthNode.

4.1. Hierarchy. In the main module, all of the necessary libraries required for application development are imported during the initialization stage. As the sample implementation on lines 1 to 10 in Listing 1 shows, the Express Library, along with supporting libraries, is first imported to construct a hierarchical structure of the application along with enabling most of the HTTP utilities. The MongoDB library is then imported to be used for creating a connection between the back-end server and the database.

The main module and the submodules are connected through a coupling process as exemplified on lines 20 and 21 in Listing 1. Commonly required libraries are shared using parameters during coupling process. The coupling process allows a client to access the submodule method. Following the top-down approach, although conventional Node.js application allows methods in the main module, Health-Node recommends methods to be exclusively in the submodules to avoid design complexity. Having functions in the main module can cause documentation and maintenance issues due to its limited role of initializing the core tasks. In contrast, submodules can couple to and call functions in the child submodule as shown on lines 25 to 30 in Listing 1. When calling functions in the child submodule, parameters with information can be passed to acquire necessary information. All modules are coupled with only one or two lines of code, which allows for a simple decoupling process for extending the application when needed. This procedure is the first principle in the divide-and-conquer process.

4.2. Module. As Figure 3 shows, each of the submodules contains initialization blocks and functions that include libraries to be used within the current submodule. Note that each function consists of multiple middleware functions. For the function to be part of a module, it requires each function to be exported as illustrated through Listing 2 (lines 1 to 5). A middleware function is required when a client needs to communicate to a server. The module containing a function without middleware functions usually sits as the lowest child submodule unless it requires assistance from lower level child submodules. In other words, if the function is only used for the back-end computation and called by a higher submodule, the function is accessible by calling the function name such as "ChildSubModuleFunction" as Listing 2 shows on lines 17 to 22. The function can further process computation tasks with the data received through a function parameter. Each submodule requires at least one function that may only have a computation task or have both computation task and middleware functions.

4.3. Middleware Functions. As Listing 3 shows, a middleware function can be added to a function. Each middleware function contains REQUEST and RESPONSE parameters.

```
(1) //Necessary libraries and variables
(2) var express = require( express );
    ...
(9) var mongoClient = require( mongodb ).MongoClient;
(10) var mongoDBuri =  mongodb://localhost:27017/database
    ...
(20) //Coupling sub modules
(21) require( ./SubModuleONe.js )(app, mongoClient,mongoDBuri);
    ...
(25) //Coupling to child sub-module and its function
(26) var ChildSubModule = require( ./ChildSubModuleOne ).FunctionName;
    ...
(29) //Inside of fuction or method call child sub-module function
(30) ChildSubModule(data);
```

LISTING 1: Initializing libraries, submodules, and child.

REQUEST and RESPONSE is a basic operation and key feature for enabling the communication between the clients and the server. When clients need information, a mobile application or a web browser sends a REQUEST message to the server. A request message is sent in the form of a query string. During the REQUEST message processing, the body of the REQUEST is primarily used. The message is then unpacked by using the "stringfy" API function from the "querystring" library. After unpacking the query, the message is translated into values of an object. Values of an object are used for executing tasks or computation. To execute tasks, middleware functions can also call functions offered through other modules to process back-end computation tasks. In addition to the back-end computation, external executables, such as compiled machine learning algorithms, can also be executed using the "child_process" library as shown on lines 1 to 8 in Listing 4. Additionally, lines 10 to 20 in Listing 4 show an example of using the "SerialPort" library from the NPM, so that the development boards enabled with serial communication can be controlled by the WRITE command and output data. Furthermore, REQUEST can be used to receive file-level data. By using "fs" library, an incoming file can be processed and saved to a particular location (cf. lines 26 to 35 in Listing 4).

Each REQUEST from the clients is answered at the server using a RESPONSE. Specifically, RESPONSE has the role of returning messages and is in the form of JavaScript Object Notation (JSON) message. In JSON messages, heterogeneous information can be stored as an object. Within the object, there is a collection of <field name,data> pairs and there can be a single object or an array of objects. These objects can be placed on the JSON message for exchange. Once the JSON message prepares necessary information for the client, information is sent back to the client by using the WRITE function as we present in Listing 5. The uploaded file can be accessed from the client by using the "fs" library shown on lines 41 to 49 in Listing 4.

4.4. Database. To provide or store information, having a database is also essential. We specifically chose to support MongoDB in HealthNode. In contrast to MySQL, MongoDB

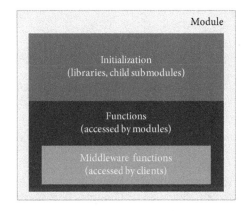

FIGURE 3: General structure of a module.

uses dynamic schemas, which means that records can be created without first defining the structure [16]. In MongoDB, three basic (yet important) operations are INSERT, FIND, and DELETE. To access the database, the "MongoDB" library is initialized in the main module and shared with the submodules. Prior to any database operation, a connection is established by using the CONNECT function along with the URL and port number of the database server. With the established connection, query string information can be inserted, deleted, or used for finding data as we show an example in Listing 6. For updating documents in the database, the documents are replaced by using DELETE and INSERT commands.

4.5. Client. Communicating with the server from a client is also an essential operation. For mobile applications, typically, the communication between the client and the server is accomplished by using a web browser or HTTP library. For a web browser, the client retrieves or sends information by using POST operation with string query. Specifically, the web browser uses JavaScript to embed information in string query and requests POST to the server. After sending REQUEST, a RESPONSE from the server is returned to the client in JSON message format. As presented in Listing 7, the JSON

```
(1) module.exports = function(app, mongoClient,mongoDBuri){
(2)   app.post ( /method , function(req, res){
(3)     //Method contents
(4)   });
(5) };

      ...
(17) function ChildSubModuleFunction(data){
(18)   //Use data
(19)   info = Information ;
(20)   return info
(21) }
(22) exports.ChildSubModuleFunction = ChildSubModuleFunction;
```

LISTING 2: Structure of a middleware function and child submodule function.

```
(1) var querystring = require( querystring );
(2)   app.post ( /method , function(req, res){
(3)
(4)     var postObjectToString = querystring.stringify(req.body);
(5)     var postObject = querystring.parse(postObjectToString);
(6)     //Data on  Info  is accessed from post object(query string)
(7)     var Info = postObject[ Info ];
(8) ...
```

LISTING 3: Example of middleware function.

message is parsed into readable objects and accessed by a client web browser. In a mobile application, specifically for androids, the Apache HTTP library is used to simulate the web browser. The POST operation in mobile applications works similarly as that of the web browser.

4.6. HealthNode API. For designing the API set, we gathered general requirements from a number of previous works on health and home monitoring projects [17, 18] and implemented necessary methods for the HealthNode API. The API uses and extends the Express library which allows developers to add necessary methods by following the conventional Express implementation rules [12]. We include a library for Android mobile and web browser applications to communicate with the HealthNode server applications. Both mobile and web applications API enables sending JSON messages and files by accessing middleware functions on the server. The HealthNode APIs can be installed using the NPM install commands. Starting the server and importing basic libraries such as "MongoDB" and "fs" are already managed by simply importing the API. Other than the previously mentioned libraries, APIs also use the existing "SerialPort" library to communicate with external development boards such as an externally connected Arduino or Raspberry Pi.

4.7. Security. The security of data exchange between a server and a client can be protected using Transport Layer Security (TLS), which is a well-known protocol that provides privacy and data integrity. By following the Express API on TLS,

HealthNode can enhance the security during data exchange. Moreover, the privacy of patient health information is password protected. When a developer implements the data exchange process, login function needs to be used prior to requesting data from the server. For example, an Android application or a web browser client can request data from the server after obtaining login approval.

5. Case Study-Based Evaluation

There are various types of cloud-based healthcare applications that can take advantage of HealthNode's software design and implementation guidelines. Such examples include applications that simply log information to the cloud through the web, exchange medical or healthcare information between mobile devices, or execute physical component actuation through a local network. In general, HealthNode supports the fundamental requirements for developing of cloud applications. These requirements include sending/receiving data, constructing a database to store information, calling external executables for machine learning algorithms, and ensuring space for the application to be expanded.

In this section, we evaluate HealthNode by providing possible application scenarios of how our software design and implementation guideline can be used for different mobile-cloud application development. Note that the case studies benefit from HealthNode due to the framework (1) following intuitive design strategies which help external field members to understand the system design and enhance the medical features of the system, (2) containing a practical

```
(1)  //Execute external program or classifier
(2)  var querystring = require( child_process ).executable;
     ...
(4)  app.post( /method , function(req, res){
(5)    executable( ./execProgram + testFile , function(err, stdout, stderr){
(6)      //output is on stdout
(7)  }
(8)  });
     ...
(10) //For serialport
(11) var SerialPort = require( serialport );
(12) var Readline = SerialPort.parsers.Readline;
(13) var port = new SerialPort( /dev/ttyUSB0 ,{baudrate: 9600 });
(14) var parser = port.pipe(Readline({delimiter: \r\n }));
(15) //Getting data from external board
(16) parser.on( data , function(data){
(17)      console.log(data);
(18) });
(19) //Sending data to external board
(20) port.write( some data );
     ...
(26) //File upload/download
(27) var fs = require( fs );
     ...
(30) //File upload:
(31) app.post( /uploads , function(req, res) {
(32)    fs.readFile(req.files.fileU.path, function(err,data){
(33)      var dirname= /file/dir/location ;
(34)      var newPath = dirname+ /uploads/ +req.files.fileU.originalname;
(35)      fs.writeFile(newPath, data, function (err){
           ...
(41) //Access file on server:
(42) app.get( /uploads/:file , function(req, res){
(43)      file = req.params.file;
(44)      console.log( File requested:  +file);
(45)      var dirname= /file/dir/location ;
(46)      var img = fs.readFileSync(dirname+ /uploads/ +file);
(47)      res.writeHead(200, { Content-Type :  image/jpg });
(48)      res.end(img,  binary );
(49) });
```

LISTING 4: Calling child submodule function, controlling external development board using serial communication, and managing file exchange.

set of medical application-related methods which allows the developer to utilize or alter the given functions to complete cloud-based health application implementation, and (3) allowing these applications to communicate with external sensing systems.

5.1. Case Study: AsthmaGuide. To evaluate HealthNode's design and framework, we use one of our previous works, AsthmaGuide [17], as a case study. AsthmaGuide is a monitoring system for asthma patients in which a smartphone is used as a hub for collecting indoor and outdoor environmental information and physiological data. Specifically for indoor environments, we use Sensordrone [19] to measure information of the patient's surroundings such as the temperature, humidity, and air quality. For outdoor environmental data, we use a national database to gather

information of air quality, pollen count, and asthma index. Furthermore, we collect physiological data from the patients by collecting their lung sounds using an electronic stethoscope, and present questionnaires that patients fill out manually on an Android application. The data collected over time is then displayed through a cloud web application for both patients and healthcare providers to view.

By utilizing HealthNode, AsthmaGuide first gathers requirements, and consequently, each of the requirements is placed into a designated role level with the top-down approach and assigned a job with the divide-and-conquer approach as shown in Figure 4. The requirements are directly linked with middleware functions in which each requirement is responsible for exchanging data between the client and the server.

Figure 4 also shows the list of 26 middleware functions that are required to implement AsthmaGuide. These middleware

```
(1) module.exports = function(app, mongoClient,mongoDBuri){
(2)   app.post( /method , function(req, res){
(3)     res.write( Response back );
(4)     res.end();
(5)   });
(6) };
```

LISTING 5: Example of responding to REQUEST.

```
(1) module.exports = function(app, mongoClient,mongoDBuri){
(2)   app.post( /method , function(req, res){
(3)     mongoClient.connect(mongoDBuri, function(err, database){
          ...
(11)      collection = database.collection( testingCollection );
(12)      //Instead of insert other functions(delete, find) works also
(13)    collection.insert(postObject, function(err, records){});
```

LISTING 6: Example of MongoDB INSERT operation.

```
(1) <script>
(2) var data_server = http://localhost/methodName ;
(3) var requestInfo = {requestInfo: data };
(4) var receivedData = {receivedInfo:  };
(5) $(document).ready(function(){
(6)   $.post(data_server, requestInfo).done(function(data){
(7)     objArr = JSON.parse(data);
(8)     if(objArr != ){
(9)       $.each(objArr, function(key1, obj){
(10)        receivedData.receivedInfo = obj[ receivedInfo ];
(11) });
```

LISTING 7: Example of requesting to the server from a web browser.

functions provide necessary results back to a web browser or a mobile device while accepting incoming data from clients. Each of the middleware functions is mapped to the HealthNode design pattern accordingly to the categorized alphabetic letter. All middleware functions are prebuilt into HealthNode and are accessible by importing the HealthNode library. The library implementation follows the HealthNode design pattern, and a developer can reference the implementation as well as add the middleware functions to enhance their application.

For example, when a patient needs to upload his or her collected indoor and outdoor environmental data as well as physiological data to the server, middleware functions such as "Patientlogin," "PatientRetrieveZipCode," "PatientRetrieveCountryCode," "PatientInsertData," "FileUploadImageFile," "FileUploadWaveFile," "ClassifyLungSound," and "GeneratePatientAdvice" are used. Note that "ClassifyLungSound" and "GeneratePatientAdvice" are AsthmaGuide application-specific functions. During mobile or web-application implementation, the AsthmaGuide

developer calls the needed middleware functions to request and insert data to the server. Besides these simple data upload operations, AsthmaGuide requires far more methods when the system starts interchanging information between a patient and a healthcare provider. Therefore, HealthNode can help reduce the burden of developers by easing the development complexity in the process of implementing and maintaining cloud applications.

5.2. Case Study: Smart Home Automation Framework. Another case study that we apply HealthNode on is a system called the Smart Home Automation Framework (SHAF) [18]. IoT covers a various network of physical objects with actuation and sensing embedded units. Under the hood, the communication between devices is connected through multiple network protocols. One of the domains that take advantage of IoT is home automation. Home automation uses different types of network protocols such as Wi-Fi, Bluetooth, and ZigBee. However, existing home equipment often requires

A. StartServer
B. DoctorRegister
C. DoctorRegisterCheck
D. DoctorLogin
E. DoctorCommentPatient
F. DoctorRetriveComments
G. DoctorRetriveCommentsOfAPatient
H. PatientRegister
I. PatientRegisterCheck
J. PatientLogin
K. PatientRetriveDataByDate
L. PatientRetriveAllData
M. PatientRetriveMostRecentData
N. PatientRetriveAdvice
O. PatientRetriveZipCode
P. PatientRetriveCountryCode
Q. PatientInsertData
R. PatientRetriveCommentsFromDoctor
S. EnviromentRetriveInfo
T. FileUploadImageFile
U. FileUploadWaveFile
V. FileUploadMp4File
W. FileUploadRetriveFile
X. ClassifyLungSound
Y. GeneratePatientAdvice
Z. SendAlert

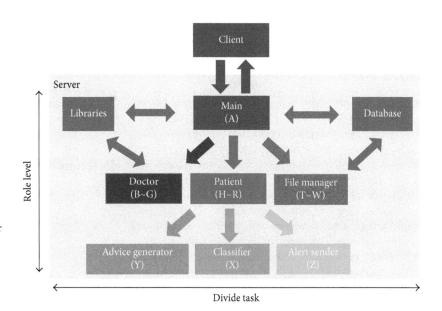

Figure 4: Designing AsthmaGuide according to HealthNode design pattern.

network communication-enabled power plugs or devices that hold a unique communication protocol specified by a manufacturer. While these types of equipment typically follow a standard communication capability, each device is limited to communicate only within the same network protocol. The goal of SHAF is to resolve issues that can be raised due to such limitations.

Specifically, SHAF targets to provide and maintain a comfortable and healthy living environment for patients. For example, the surrounding temperature is a critical metric for those who are sensitive to cold or hot temperatures such as people with chronic health conditions, given that extreme (than normal) temperatures can aggravate various symptoms. Figure 5 illustrates an example of SHAF monitoring and actuating the smart home by using the HealthNode framework. We use Raspberry Pi with ZigBee as a smart central server and an Arduino with ZigBee as sensor nodes. For SHAF ZigBee communication, multihop communication is enabled for larger homes. The server accepts incoming JSON queries where a client can request sensor readings or operate an actuation unit. While our current client application is implemented on Android and Windows smartphones, any programming languages that can support JSON message requests can communicate with the smart home's central server.

SHAF's server should have the capability of handling requirements or multiple middleware functions such as "AddSensor," "RemoveSensor," "RefersehSensorReadings," "ActuateSensor," "LearnHomeUsage," "AutonomusMode," and "HealthyLivingEnvironmentMode." Although the quantity of the middleware functions is not large, the use of HealthNode allows for easier software maintenance since it instructs task-specific submodules rather than one large

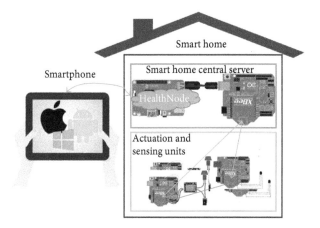

Figure 5: Communication map of SHAF.

module. During maintenance, the developer couples and decouples the submodule by changing one or two lines of code. Specifically, using HealthNode's software design, we structure the architecture of SHAF so that it contains three submodules (e.g., patient, caregiver, and sensor manager). Followed by the architecture, multiple middleware functions are implemented. For example, the patient and the caregiver submodules are accessed from the mobile or web browsers by using JSON message requests. When there is a request, one of the middleware function returns the result to the client followed by the database operation. The server also logs up-to-date sensing and actuation values by using the "SerialPort" library and middleware function in the sensor manager submodule. Furthermore, stored data can be used by a child submodule with a machine-learning algorithm for automatic environment configuration based on a user's preference. Note that all of the middleware functions are prebuilt into HealthNode and are accessible by importing the HealthNode library.

6. Related Work

Existing Node.js applications are spread across various fields of study including IoT or web [20–25], medical [26, 27], transportation [28], and environmental [29] domains. To develop a well-structured application using Node.js, Frees [30] proposes a way to overcome many challenges in teaching web development by placing Node.js in the computer science curriculum. He presents a semester-long, 14-week course outline to allow students to fully understand the use of Node.js and be able to apply it for web development. Although Node.js is easily utilized compared to standard techniques, Node.js still requires a strong background before it can be used to its full capacity. Understanding Node.js development merely at a surface level will result the application to stay simple and be more prone to mistakes and difficulties in maintaining the application over time. A study by Ojamaa and Düüna [31] stated that mistakes are more common with Node.js applications because programmers lack the extensive experience of writing JavaScript application.

Many attempts have been made to solve this issue. There are existing books [9, 32–38] and online tutorials [39, 40] that go into the depths of using Node.js. Although these sources provide extensive guidelines, there has not been a straightforward software design that allows developers to get a complete picture of the Node.js programming structure for general-purpose application prototype.

7. Conclusion

Healthcare applications are emerging at an exponential rate, and application systems in the remote healthcare application domain are becoming critical sectors in IoT research. As Node.js becomes a well-used essential tool for developing cloud-based applications, we propose HealthNode, which is a general-purpose framework for developing healthcare applications on cloud platforms using Node.js. The principal goals of HealthNode are to provide explicit software design, API, and guidelines to achieve quick and expandable cloud-based healthcare application. We specifically tailor Health-Node for healthcare applications due to a significant potential for addressing many of the challenges of providing accessible, cost-effective, and convenient healthcare. With development support systems such as HealthNode, we envision that the development of mobile-connected application systems will inevitably increase in the healthcare domain.

Acknowledgments

This work was supported in part by the DGIST R&D Program of the Ministry of Science, ICT and Future Planning (18-EE-01), the Global Research Laboratory Program through the National Research Foundation of Korea (NRF) funded by the Ministry of Science, ICT and Future Planning (NRF-2013K1A1A2A02078326), the DGIST Research and Development Program (CPS Global Center) for the project "Identifying Unmet Requirements for Future Wearable Devices in Designing Autonomous Clinical Event Detection Applications", and the Ministry of Trade, Industry and Energy and the KIAT through the International Cooperative R&D Program (no. N0002099; Eurostars-2 Project SecureIoT).

References

[1] Joyent Inc., Node.js, http://www.nodejs.org/, 2016.

[2] R. R. McCune, *Node.js paradigms and benchmarks*, Striegel, Grad OS, 2011.

[3] S. Tilkov and S. Vinoski, "Node.js: using javascript to build high-performance network programs," *IEEE Internet Computing*, vol. 14, no. 6, pp. 80–83, 2010.

[4] Microsoft, Microsoft azure, https://www.azure.microsoft.com/en-us/develop/nodejs/, 2016.

[5] IBM, Node.js @ ibm, https://www.developer.ibm.com/node/, 2016.

[6] Netflix, Node.js in flames, http://www.techblog.netflix.com/2014/11/nodejs-in-flames.html, 2014.

[7] PayPal, Paypal node sdk, http://www.paypal.github.io/paypal-node-sdk/, 2016.

[8] Joyent Inc., Node.js at walmart: introduction, https://www.joyent.com/developers/videos/node-js-at-walmart-introduction, 2016.

[9] P. Teixeira, *Professional Node.js: Building Javascript Based Scalable Software*, John Wiley & Sons, Hoboken, NJ, USA, 2012.

[10] T. Capan, Why the hell would i use node. js? a case-by-case tutorial, 2015.

[11] Node Package Manager, Npm, https://www.npmjs.com/, 2016.

[12] Express, Express api, http://www.expressjs.com/en/4x/api.html, 2017.

[13] M. A. Jadhav, B. R. Sawant, and A. Deshmukh, "Single page application using angularjs," *International Journal of Computer Science and Information Technologies*, vol. 6, no. 3, pp. 2876–2879, 2015.

[14] N. Jain, P. Mangal, and D. Mehta, "Angularjs: a modern mvc framework in javascript," *Journal of Global Research in Computer Science*, vol. 5, no. 12, pp. 17–23, 2015.

[15] V. Balasubramanee, C. Wimalasena, R. Singh, and M. Pierce, "Twitter bootstrap and angularjs: frontend frameworks to expedite science gateway development," in *Proceedings of the 2013 IEEE International Conference on Cluster Computing (CLUSTER)*, p. 1, Indianapolis, IN, USA, September 2013.

[16] MongoDB, Mongo db and mysql compared, https://www.mongodb.com/compare/mongodb-mysql, 2016.

[17] H.-K. Ra, A. Salekin, H.-J. Yoon et al., "AsthmaGuide: an asthma monitoring and advice ecosystem," in *Proceedings of the 2016 IEEE Wireless Health*, pp. 128–135, Charlottesville, VA, USA, October 2016.

[18] H.-K. Ra, S. Jeong, H. J. Yoon, and S. H. Son, "SHAF: framework for smart home sensing and actuation," in *Proceedings of the 2016 IEEE 22nd International Conference on Embedded and Real-Time Computing Systems and Applications (RTCSA)*, p. 258, Daegu, Republic of Korea, August 2016.

[19] Sensorcon, Your smartphone can do much more with sensordrone, http://www.sensorcon.com/sensordrone, 2014.

[20] Y. Jiang, X. Liu, and S. Lian, "Design and implementation of smart-home monitoring system with the internet of things technology," in *Lecture Notes in Electrical Engineering*, pp. 473–484, Springer, Berlin, Germany, 2016.

[21] H. Lee, H. Ahn, S. Choi, and W. Choi, "The sams: smartphone addiction management system and verification," *Journal of Medical Systems*, vol. 38, no. 1, pp. 1-10, 2014.

[22] T. Steiner, S. Van Hooland, and E. Summers, "Mj no more: using concurrent wikipedia edit spikes with social network plausibility checks for breaking news detection," in *Proceedings of the 22nd International Conference on World Wide Web companion*, pp. 791–794, Rio de Janeiro, Brazil, May 2013.

[23] I. K. Chaniotis, K.-I. D. Kyriakou, and N. D. Tselikas, "Proximity: a real-time, location aware social web application built with node.js and angularjs," in *Proceedings of the Mobile Web Information Systems: 10th International Conference, MobiWIS 2013*, Paphos, Cyprus, August 2013.

[24] S. K. Badam and N. Elmqvist, "Polychrome: a cross-device framework for collaborative web visualization," in *Proceedings of the Ninth ACM International Conference on Interactive Tabletops and Surfaces, ITS'14*, pp. 109–118, New York, NY, USA, September 2014.

[25] T.-M. Grønli, G. Ghinea, and M. Younas, "A lightweight architecture for the web-of-things," in *Mobile Web Information Systems*, pp. 248–259, Springer, Berlin, Germany, 2013.

[26] J. Kim, E. Levy, A. Ferbrache et al., "MAGI: a Node. js web service for fast microRNA-Seq analysis in a GPU infrastructure," *Bioinformatics*, vol. 30, no. 19, pp. 2826-2827, 2014.

[27] T. Di Domenico, E. Potenza, I. Walsh et al., "Repeatsdb: a database of tandem repeat protein structures," *Nucleic Acids Research*, vol. 42, no. D1, pp. D352–D357, 2013.

[28] A. Nurminen, J. Järvi, and M. Lehtonen, *A Mixed Reality Interface for Real Time Tracked Public Transportation*, Helsinki Institute for Information Technology (HIIT), of Aalto University and University of Helsinki, Helsinki, Finland, 2014.

[29] K.-L. Wang, Y.-M. Hsieh, C.-N. Liu et al., "Using motion sensor for landslide monitoring and hazard mitigation," in *Intelligent Environmental Sensing*, pp. 111–127, Springer, Berlin, Germany, 2015.

[30] S. Frees, "A place for Node.js in the computer science curriculum," *Journal of Computing Sciences in Colleges*, vol. 30, no. 3, pp. 84–91, 2015.

[31] A. Ojamaa and K. Düüna, "Security assessment of Node.js platform," in *Proceedings of the Information Systems Security: 8th International Conference, ICISS 2012*, Guwahati, India, December 2012.

[32] A. Mardan, "Publishing Node.js modules and contributing to open source," in *Practical Node.js*, pp. 261–267, Springer, Berlin, Germany, 2014.

[33] G. Rauch, *Smashing Node.js: JavaScript Everywhere*, John Wiley & Sons, Hoboken, NJ, USA, 2012.

[34] J. R. Wilson, *Node.js the Right Way*, Pragmatic Programmers, Dallas, TX, USA, 2014.

[35] C. Gackenheimer, *Node. js Recipes: A Problem-Solution Approach*, Apress, New York, NY, USA, 2013.

[36] C. J. Ihrig, *Pro Node.js for Developers*, Apress, New York, NY, USA, 2013.

[37] M. Thompson, *Getting Started with GEO, CouchDB, and Node.js*, O'Reilly Media Inc., Newton, MA, USA, 2011.

[38] S. Pasquali, *Mastering Node.js*, Packt Publishing Ltd., Birmingham, UK, 2013.

[39] TutorialsPoint, Node.js-express framework, http://www.tutorialspoint.com/nodejs/nodejs_express_framework.htm, 2016.

[40] M. Kiessling, Node beinner book, http://www.nodebeginner.org/, 2016.

5

Hybrid Aspect of Context-Aware Middleware for Pervasive Smart Environment

J. Madhusudanan,[1] S. Geetha,[2] V. Prasanna Venkatesan,[2] U. Vignesh (ID),[3] and P. Iyappan[4]

[1]Department of Computer Science and Engineering, Sri Manakula Vinayagar Engineering College, Puducherry, India
[2]Department of Banking Technology, Pondicherry University, Puducherry, India
[3]Department of Computer Science and Engineering, K L University, Guntur, Andhra Pradesh, India
[4]Department of Computer Science and Engineering, Manonmaniam Sundaranar University, Thirunelveli, Tamil Nadu, India

Correspondence should be addressed to U. Vignesh; vigneshbun@gmail.com

Academic Editor: Habib M. Fardoun

Pervasive computing has made almost every device we see today to be communicated and function in collaboration with one another. Since the portable devices have become a part of our everyday life, people are more involved in a pervasive computing environment. They engage with many computational devices simultaneously without knowing the availability of their existence. The current world is being filled with more and more smart environments. These smart environments make them to be attracted towards the new technological emergence in the field of pervasive computing. Various researches are being carried out to improve the smart environment and their applications. Middleware plays a vital role in building the pervasive applications. The pervasive devices act based on the context of the situation, that is, they do their actions according to the environment of the application. They react to the situations smartly as they can take their own decisions based on the context developed for that specific application. Most of the pervasive applications were using its own middleware that is specific towards their need. As today, most of the applications are using their own middleware with their specific requirement, which leads us to unearth out their common features and their scope of using it. In this paper, a survey on the various hybrid aspects of the different context-aware middleware has been done. This middleware is classified based on service, context, and device aspects. Merits and demerits are identified from the existing smart environments, and future perspective of their development such as generic context-aware middleware need has been discussed.

1. Introduction

Pervasive computing is drastically shifting the day-to-day activities by bringing "anytime, anywhere, and anything" computing potential into the living environment. Computers spread across different environments and tend to merge and disappear into day-to-day objects [1]. They are capable of sensing the context of the environment, communicating with different objects and providing the right information to the user. Due to these rich computing powers, more pervasive computing applications are developed, and the need of the pervasive application is sensed today. As the demand of pervasive application grows enormously, more and more research are undertaken in the industry and in academia.

Most of the researches in pervasive computing focus on addressing the specific domain application development, for example, smart home-like aware home [2], UbiHome [3], and ThinkHome [4]. The objectives of these studies are to pattern the user behavior and to provide different services for matching the pattern that supports the "comfort of the user." From the existing applications, researchers identified the major challenge like "how to handle different devices and different contexts" and also proposed middleware as a solution to it.

2. Evolutions of Pervasive Computing

According to Gabriel et al. [5], before three decades, "computer" is totally referred as mainframe systems which simply

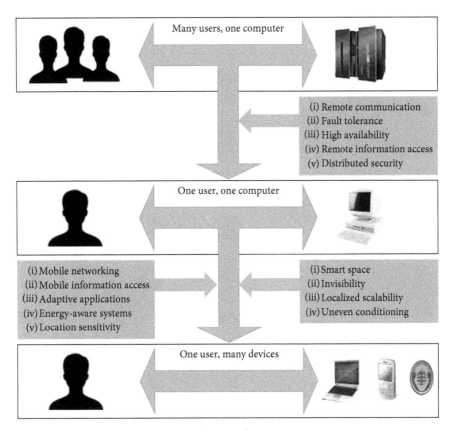

FIGURE 1: Evolution of computing era.

processes input and produces output. This period is referred as the first wave of computing where the user and computer will have less interaction. The first era of computing is termed as "many users-one computer"; that is, one computer is shared by many users. By the rapid growth in the computing vicinity, personal computer came into existence. This period is referred as the second wave of computing, and it is termed as "one user-one computer"; that is, one user uses one computer. Personal computer is very powerful as a standalone system, and it takes part as a node when the network is evolved.

Enormous growth in smart devices/gadgets gave rise to the new computing dreams called ubiquitous computing; this was initiated by Mark Weiser who is called as the father of ubiquitous computing. This period is referred as the third wave of computing, and it is also termed as "one user-many devices"; that is, one user is supported by many devices such as computer, smart phone, and tab as shown in Figure 1.

Ubiquitous computing is a combination of pervasive computing and mobile computing [6]. Making the computing service available in any location context is the primary objective of mobile computing, whereas the primary objective of pervasive computing is to acquire the context information from the environment.

Ultimate focus of mobile computing and pervasive computing is on context-aware computing [7, 8]. So, the term "ubiquitous computing" is also called as pervasive computing, and it can be used either way [9, 10]. Ubiquitous computing is also referred as disappearing computing or everywhere computing or ambient intelligence or invisible computing or internet of things [11].

Pervasive computing has slowly and steadily taken the computing world into a new dimension with its features such as adaptive system, location sensitivity, energy-aware system, and context-aware system.

3. Pervasive Applications

There is a rapid growth in the device world where most of the devices support smartness and the environments are constructed with the intention of supporting smartness. By the integration of smart devices (e.g., microdevices and complex devices) and smart environments (e.g., embedded devices in the physical environment such that a wall can sense camera is recording), a new environment is created that supports any environment and any device, which will work together to produce the pervasive smart environment as shown in Figure 2.

Many applications are developed with the pervasive smart environment in different domains such as

(i) buildings (smart home-like aware home [2], UbiHome [3], ThinkHome [4], SM4ALL [12], smart office-GreenerBuilding [13], and smart classroom and smart meeting hall [14]),

(ii) industries [15],

(iii) healthcare (smart monitoring system, patient tracking system, staff management system, and asset tracking system [16]),

(iv) transportation [17],

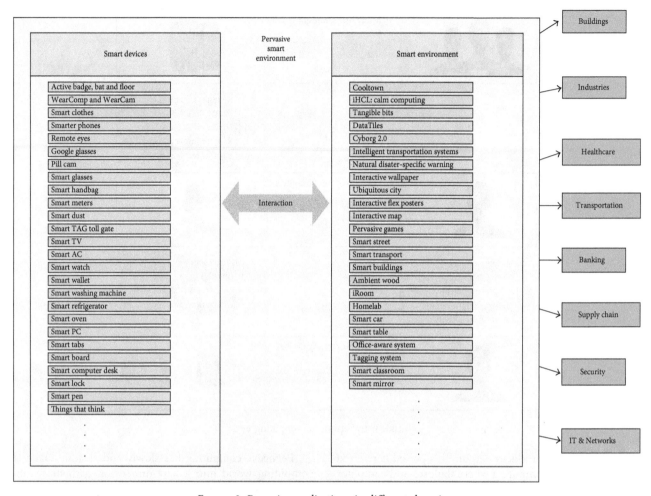

FIGURE 2: Pervasive applications in different domains.

(v) banking (smart ATM, smart counter, and smart wall [18]),

(vi) supply chain (smart shelf, smart billing, and tagging of goods [19]),

(vii) security and IT [20],

(viii) networking [11].

4. Classifications of Pervasive Architecture

Architecture is the first step in the development of any system; pervasive architecture is classified into three types based on its function, data, and its working as shown in Figure 3 [8].

4.1. Direct Sensor Access Architecture.

This architecture was widely used in the earlier stage of pervasive computing when devices were available for direct access. The client software gathers the required information from the available devices without any additional layer to acquire the information from the device. Separate drivers are hardcoded into the specific application. Hence, this can be used in different applications as there will be different drivers for different devices. Therefore, this architecture is well suited for standalone

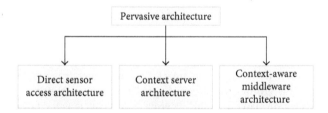

FIGURE 3: Classification of pervasive architecture.

systems which will directly access the devices as it is capable of handling homogeneous devices.

4.2. Context Server Architecture.

Context server architecture supports many client systems to access the remote data server concurrently. Data acquired are stored in the context server, and they are used according to the context of the application and contexts are monitored using different devices. Most of the devices used in context-aware systems are portable gadgets which are limited to power energy, computation power, memory space, and so on. These are vital features for a centralized context server. It will consider different issues like fault tolerance, quality of services, and scalability of dynamic devices.

4.3. Context-Aware Middleware Architecture. Most of the software designs use decomposition as a technique to split the layers, for example, business logic, data logic, and presentation logic. Hence, the middleware is one which can individually take a different logic to work. Context middleware layer is introduced with the intention of hiding the lower level complexities. It has an advantage of reusing the lower level coding (hardware-dependent device coding), and it is flexible to adapt to any change in the application logic.

More studies on pervasive architecture are being conducted in different dimensions such as "interoperability of service," "interoperability of the device," "context modeling," and "building smart environment". The primary objective of pervasive computing is "context-aware computing" which deals with the different contexts of the application like location context, user context, and time context. These contexts are helpful to provide appropriate service to the user. Context can be defined as "any information used to predict the state of an object, where the object can be user, location, or any physical object." Today, middleware for context-aware computing is one of the further steps for pervasive computing to move towards veracity.

5. Related Works on Context-Aware Middleware

According to Bandyopadhyay et al. [21], context-aware middleware is essential for three reasons:

(i) To develop a common method to interact between heterogeneous devices belonging to a different domain.

(ii) Middleware acts as a relationship to merge different heterogeneous components together.

(iii) Hiding of complexity can be achieved in different applications with different domains.

Last two decades have been the peak periods in the development of context-aware middleware [22], where most of the middleware are classified into seven categories [23]: agent-based middleware, reflective middleware, metadata-based middleware, tuple space middleware, adaptive middleware, objective middleware, and OSGI-based middleware.

There are several surveys carried out in relation to the context-aware middleware field. The study and analysis of the survey are given in Table 1. Table 1 specifies the functional and nonfunctional characteristics of the existing context-aware middleware based on literature study [7, 24].

As per the result of the survey mentioned in Table 1, the working aspects of middleware can be classified into three levels of aspects, namely, service aspect, context aspect, and device aspect, from the context perspective as shown in Figure 4. Service aspects act as an interface between the application layer and the context aspects. It converts the requirement into service. Context aspects play a vital role in bridging the gap between the service aspects and the device aspects, where context information is derived from the device aspects. Device aspects act as a logical device to share information between context aspects and physical devices.

From the literature survey, it is obvious that the context-aware middleware is growing as an indispensable architecture component in building any pervasive application. From the study mentioned in Table 1, important parameters used for evaluating any context-aware middleware are identified as device management, interoperability, context awareness, security and privacy, autonomy, intelligence, and adaptability. These parameters are used to analyze the context-aware middleware.

A study on the various context-aware middleware is carried out, in order to understand the working of different aspects of the context-aware middleware and to make sure the number of domains that the context-aware middleware supports. Metrics for evaluating the existing context-aware middleware are achieved through the number of domains that the same middleware can support. The "multi" effects (multidomain, multiapplication, and multiculture) are to be addressed at the architecture level [32, 34]. In order to categorize the existing context-aware middleware, the following criteria are proposed:

(i) If it supports three or more domains, then it is rated as high (H).

(ii) If it supports two domains, then it is rated as medium (M).

(iii) If it supports one domain, then it is called as low (L).

(iv) If it supports no domain, then it is said as not applicable (N).

These metrics are verified on different aspects such as service aspects, context aspects, and device aspects.

5.1. Service Aspects. Service-oriented device middleware and service-oriented context middleware are two different middleware available to provide service aspects in the perspective of device aspects and context aspects, respectively. This leads the researcher to move towards a middleware to manage both device and context functionalities in a particular middleware. Existing middleware are developed to provide service for the specific domain application, and there are no provisions to alter the available service to suit or accommodate to a different domain application.

Considering an example to illustrate how a same service works differently in diverse domain applications, consider the tracking as a service and it is applied in different domain applications, such as in the following:

(i) In Scenario 1, an application of the transport system, location is tracked using GPS or some other techniques. Here, location is the context, and GPS is a device used to trace the location [33].

(ii) In Scenario 2, an application of hospital, location of the doctor, patient, and object can be traced using RFID tag. Here, location, users, and objects are context, and RFID acts as a device [35].

(iii) In Scenario 3, an application of home, location of the user/object can be identified using RFID.

TABLE 1: Contribution towards context-aware middleware.

Authors	Parameters or attributes	Finding/report
Messer et al. [25]	Mobility, adaptability, power management, security, and QOS	How new ideas are converted to everyday project. Its major contribution is towards device aspect.
Kjær [24]	Environment, Storage, reflection, quality, adaption, migration, and composition	No single middleware is appropriate to handle different contexts. Its major contribution is towards service and context aspects.
Romero [26]	Adaptation, communication, context management, service discovery, and persistency	Most of the context-aware middleware are focusing on specific aspect of pervasive computing. Its major contribution is towards context and device aspects.
Lee et al. [27]	Decentralized, autonomy, context awareness, lightness, external sensory capture, intelligence, and web enable	Context is the key for smart environment application development. Its major contribution is towards context and device aspects.
Davidyuk et al. [28]	Specification, composer, interoperability, and end user involvement	Constant monitoring of user activity and representing in the standard format to support the development of middleware, which can act as a common interface. Its contribution is towards context aspect.
Saeed and Waheed [29]	Architecture style, location transparency, aspect oriented, decomposition, fault tolerance, interoperability, service discovery, and adaptability	Most of the research takes place in the fixed domain, and scope for the development of generic middleware is addressed. Its major contribution is towards service aspect.
Bandyopadhyay et al. [21]	Device management, interoperability, platform portability, context awareness, and security and privacy	Most of the middleware focus on specific aspect and need for generic middleware. Its major contribution is towards device and context aspects.
Vasanthi and Wahidabanu [30]	Interoperability, discoverability, location transparency, adaptability, context awareness, scalability, security, and autonomous management	Need to interface the heterogeneity protocol used in heterogeneous networks. Its major contribution is towards device aspect.
Nadia Gámez and Fuentes [31]	Services, implementation, heterogeneity, network technology, network prototype, and new services	Need to support heterogeneous devices to work together and to capture the context change and to adapt to it. Its major contribution is towards service aspect.
Bellavista et al. [7]	Context data model, processing, dissemination, routing overlay, and run-time adaption support	Context-aware application needs to work in different domains. Its major contribution is towards context aspect.

Here, location, users, and objects are considered as context, and RFID belongs to device aspects [36].

(iv) In Scenario 4, an application of supply chain management, transporting of goods from one location to another can be monitored using RFID and GPS techniques. Here, goods, location, and so on act as a context, and RFID and GPS act as some devices [37].

(v) In Scenario 5, an application of the security system, movement of users and abnormal activities will be monitored using camera and identification sensors. Here, movement of users and abnormal activities are considered as context, and camera and sensors act as devices [38].

From the above scenarios, it is evident that different tracking services are used for each pervasive application. Even though tracking is a common service, it varies only with the domain-specific features, and also, there is no architecture or framework or middleware to make the service common to all domain applications as a generic service.

In service aspects, there exist five middleware. They are OASIS [39], SAMI [40], SOCAM [41], NEXUS [42], and KASOM [43] as shown in Table 2. Each middleware works

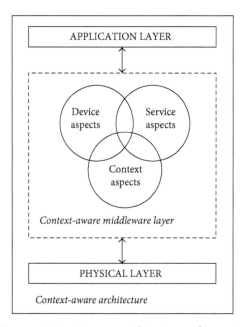

FIGURE 4: Context-aware architecture with aspects.

TABLE 2: List of context-aware middleware on service aspects.

Serial number	Middleware	Device management	Interoperability	Platform Portability	Context awareness	Security and privacy	Autonomy	Intelligence	Adaptability	Domain specific
1	NEXUS	N	M	L	N	N	N	M	N	SOA service discovery
2	SOCAM	M	N	M	M	L	L	L	M	Event-based application
3	OASIS	M	N	M	N	N	M	N	N	SOA middleware for sensor networks
4	SAMI	M	M	N	N	N	M	N	N	Share information among devices
5	KASOM	M	M	L	M	M	M	N	M	SOA-based application

for a specific application like sharing of information among devices in smart home, interface for network service in smart office, and so on. Thus, all the service aspects of context-aware middleware support for a specific domain application.

5.1.1. NEXUS. It finds a new way to merge service-oriented architecture and pervasive computing to create a smart, healthy, and flexible network to access different active service resources [42].

5.1.2. SOCAM. Pervasive application can be developed using service-oriented context-aware middleware platform. It mainly works with two layers. The first layer is used to control and to manage context information, whereas the second layer helps to frame different policies to support various client applications [41].

5.1.3. OASIS. It is object-centric service-oriented middleware to support different sensor devices, and it works according to the need of the service. It also supports automatic discovery of the service and the registry of service [39].

5.1.4. SAMI. It is a self-adaptive information sharing middleware to locate different devices, and it is used for sharing the device information despite its dynamic location. Device information files are grouped as data [40].

5.1.5. KASOM. The knowledge-aware and service-oriented middleware (KASOM) was illustrated as a new idea to combine pervasive computing and service-oriented cloud computing to create a novel era in the computing generation. The major purpose of knowledge-aware and service-oriented middleware is to provide sophisticated and enhanced pervasive services to everyone connected to the smart environment.

From the above discussion, to support the diverse domain applications, the different domain services are to be maintained, and they should be provisioned to discover the suitable service and also to maintain the hierarchy of service [43].

5.2. Context Aspects. Context-aware applications acclimatize according to the context of the user, context of the location, nearby people in the environment, objects, and the reachable devices in the environment, and changes take place to those objects over a period of time. A context-aware application with these potentials will analyze the environment and adapt according to the context of the environment [44].

Context of the application is monitored and processed with the support of smart devices, which acquires context information of the environment or its surrounding. In order to make context acquisition in reality is a demanding task. In order to achieve this, context aspect of the middleware is used, and it undergoes different challenges. They are as follows [45]:

(i) What are the different contexts that are available in general?

There are different types of context available like the location, user, device, time, and activity [24]. Context cannot be fixed in general, but it can be fixed for a specific application.

(ii) From where context is discovered?

Context information is taken from the available environment of the application like the location, user, and time and is monitored with the help of smart devices.

(iii) How to process the context information?

Context is processed generally in four different ways [7]. They are

(a) context data history,
(b) context data aggregation through logical reasoning and probabilistic reasoning,
(c) context data filtering using both time-based filtering and change-based filtering,
(d) context data security.

(iv) How context is represented?

Context information is represented in any one of the five different models [7]. They are

(a) key-value model,
(b) markup schema model,
(c) object-oriented model,
(d) logic-based model,
(e) ontology-based model.

(v) How the conflict in the context is handled?

Addressing two contexts of the application at the same time is a crucial decision to be made. It can be addressed in different ways [46]:

(a) Conflict resolution Policy
(b) QOC parameter
(c) Priority to the context

(vi) How same contexts differ from one domain to another domain?

Same contexts will change from one domain application to another. For example, consider a location context which changes according to the domain. In smart hospital application, location contexts will be the doctor's room, ICU, general ward, and so on. In smart bank application, location contexts will refer to the manager's room, locker room, counter, and so on. In smart home application, location contexts will include bedroom, kitchen, hall, and so on.

In the context aspects, there exist twenty-four middleware, to mention few such as AURA [47], CARMEN [48], and CARISMA [49] as shown in Table 3. Here, different context-aware middleware work in different domains like smart home, smart hospital, and smart office. Each context-aware middleware works for a specific application.

TABLE 3: List of context-aware middleware on context aspects.

Serial number	Middleware	Device management	Interoperability	Platform Portability	Context awareness	Security and privacy	Autonomy	Intelligence	Adaptability	Domain specific
1	AURA	M	N	N	M	M	M	N	M	Support context-aware application
2	GAIA	M	M	M	M	M	L	L	M	Metaoperating system
3	CARISMA	N	H	N	H	N	H	N	H	Mobile application
4	COOLTOWN	M	H	N	H	N	H	N	N	Interfacing devices with web-enabled environment
5	COBRA	N	M	M	M	N	L	M	N	General middleware
6	TOTA	N	L	L	H	N	N	N	N	Representation of context information
7	JADABS	M	M	M	M	L	N	N	M	Mobile application
8	COMPACT	M	N	N	M	N	N	N	N	Managing context-aware data
9	MARKS	M	N	N	M	L	N	N	L	Mobile application
10	CARMEN	M	N	N	M	N	M	M	M	Mobile agent for context application
11	CORTEX	M	M	N	M	M	M	L	M	Context-aware home application
12	MIDDLEWHERE	N	N	N	H	H	N	N	N	Home-based application
13	MOBILPADS	L	L	L	L	N	L	N	L	Mobile application
14	MIPEG	M	M	N	M	N	L	N	L	Grid application
15	CAMPS	M	M	M	M	N	N	L	N	Smart classroom application
16	AMiCA	M	N	N	M	L	M	N	N	Hospital application
17	IBICOOP	L	L	L	L	L	N	N	N	Home server application
18	CDTOM	M	M	N	M	N	L	N	N	Elder home automation
19	UBIROAD	M	M	M	M	M	L	N	L	Context-aware application
20	UMOVE	M	M	N	M	N	M	N	M	Mobile application
21	SMEPP	M	N	M	M	M	N	N	M	Context management for context-aware application
22	CAMEO	N	N	N	H	N	H	H	H	Mobile-based application
23	AWARE	N	N	N	H	N	H	H	H	Mobile-based application

5.2.1. AURA. AURA is a context-aware middleware to design and develop a system to work as "personal information atmosphere" regardless of their location, and it acquires support of wearable computing, handheld device, sensor, desktop computers, users, and infrastructure computers [47].

5.2.2. GAIA. GAIA is a distributed middleware system and acts as a metaoperating system to coordinate the different types of devices in the pervasive environment. As an operating system, it allocates resources, manages different file systems, and establishes communication with different resources [50].

5.2.3. CARISMA. CARISMA uses policies to work under dynamic context of the environment, and it adapts to the need of the application. Conflicts in the policies are solved during run time [49].

5.2.4. COOLTOWN. COOLTOWN middleware is used to connect the web resources by means of physical objects, and also, it starts interaction with different gadgets to share information according to the context of the environment [51].

5.2.5. COBRA. COBRA is an agent-based context-aware middleware to support context-aware application in the pervasive environment. It is the responsibility of the agent to collect the context information from the devices, environments, and other agents [52].

5.2.6. TOTA. TOTA ("Tuples on the Air") is a new middleware for supporting the programming model of a application, which adapts to the context-aware information received from the pervasive computing environment. It works only for the specific application domain [53].

5.2.7. JADABS. JADABS middleware works in the dynamic mobile environment and to adapt the user behaviors. It is built to cope up with changing devices in different environments. It also stores the context information of the environment [54].

5.2.8. COMPACT. COMPACT middleware is mainly for context representation which is an important aspect of pervasive computing systems. In this middleware, data are retrieved regularly from the environment and processed. The processed data are stored as dynamic context information for future use (Maria Strimpakou et al., 2006).

5.2.9. MARKS. MARKS middleware provides information on how to use the knowledge inferred from the users in the given system, how to discover the resources available in an environment, and how to handle a resource failure and provide solution for it (Moushumi Sharmin et al., 2006).

5.2.10. CARMEN. CARMEN is a middleware for handling context-aware resources in a wireless network, and it supports

to create resource metadata. Metadata help to find the dynamic context of the application and work according to it [48].

5.2.11. CORTEX. CORTEX middleware is based on sentient objects, which sense and observe the behavior of its surrounding environment and objects. From the observation, it generates the context information and works according to it [26].

5.2.12. MIDDLEWHERE. MIDDLEWHERE is a context-aware middleware which gives the location information well in advance to applications, and it integrates a different range of location-sensing methods. A separate reasoning engine determines the context of the environment from the location information [26].

5.2.13. MOBILPADS. MOBILPADS is a context-aware middleware for mobile environment. It is a service entity, which can migrate between different mobile entities. It works in the principle of the client-server system, and it observes the changes in the entities to keep the context information updated [26].

5.2.14. MIPEG. MIPEG is a middleware for pervasive grid applications, and it supports for integration of various mobile devices at anytime. It extracts the context information from the devices and executes the task according to the user task [55].

5.2.15. CAMPS. CAMPS is an agent-based middleware to support context-aware service in the smart environment. Its primary objective is creating a standard to represent the context information and the method to process it [56].

5.2.16. AMiCA. Ambient middleware for context awareness is used to continuously and implicitly adapt to the environment to meet evolving user expectations. Up-to-date valid context information is the key requirement for successful transparent interaction between the smart environment and the middleware [27].

5.2.17. IBICOOP. IBICOOP middleware is developed to support many challenges like abstraction of device configuration, connectivity of devices, and heterogeneity of devices [57].

5.2.18. CDTOM. Context-driven task-oriented middleware focuses on the user's work or task in different situations, rather than focusing on the various devices and services in the environment. It also makes the system to work intelligently as per the user request [58].

5.2.19. UBIROAD. UBIROAD is a middleware developed to handle the smart road environment. Here, context information is acquired from the vehicle and road in which the vehicle is travelling (road condition, traffic, another vehicle detail, etc.) to create a smart environment for the user [21].

5.2.20. UMOVE.
UMOVE middleware will consider every device as an entity, and also, it works based on the entity-layered architecture. It collects the information from the entity as entity data and processes the context of the entity data according to the application [59].

5.2.21. SMEPP.
SMEPP is a middleware to support security in the pervasive environment. It creates various security policies to access the context information which is acquired from the smart environment. According to the policy, middleware ensures the system is secured [21].

5.2.22. Cameo.
CAMEO middleware works for mobile devices and its context. It collects the context from the environment and analyze the context information. It provides a common programming interface to connect the smart environment. CAMEO middleware architecture also supports nonfunctional qualities like integrability, reusability, and ease of creation [60].

5.2.23. AWARE.
AWARE middleware is developed to collect the context from the environment. Accumulated context is managed and shared. It supports to present the context information to the stakeholders of the application. AWARE middleware architecture also supports nonfunctional qualities like scalability, integrability, reusability, and ease of creation.

Most of the context-aware middleware work for the context representation such as the key-value model, logic-based model and ontology-based model, context acquisition, and context processing such as logical reasoning and probabilistic reasoning. There is no middleware to support different domain applications in terms of all the aspects.

The main objectives of the existing context-aware middleware are to investigate how contexts are acquired, processed, and modeled for a specific domain application. From the existing context aspect middleware, it is evident that there is no generic context-aware middleware to support context aspects. Therefore, the need for the generic context-aware middleware is reported [61].

5.3. Device Aspects.
Currently, middleware is available to support specifically the device aspects of the domain which help in device management of a specific application, and in few middleware, generic device aspects are also supported. Hydra [62] and SenseWrap [63] are the middleware to support diverse domain applications, and both these middleware concentrate on

(i) how to interface the heterogeneous physical device with the application software;

(ii) how to handle the heterogeneous network protocol;

(iii) device management.

It provides a device platform for context-aware middleware; despite this, it does not support context aspects. It uses virtual entity to map the physical devices. The virtual entity interfaces the physical devices to maintain the device properties. Any changes observed in the physical devices will be reflected in the virtual entity. These mappings of the virtual entity help to achieve abstraction in the device management aspect.

Scope for interfacing the context aspect with the device aspect is shown in the study of Lowe. R et al. (2013). Both device aspects and context aspects combine together to provide a domain service to the user.

In device aspects, there exist twenty middleware such as HYDRA [21], MADAM [64], and PERLA [65], as shown in Table 4. Each middleware works for a specific application like interfacing the different devices in the specific environment of a domain.

Different middleware work in different domains like smart home, smart hospital, and smart office. Device aspect acts as a bridge between the physical device and context aspect of the system. It manages the devices in the smart environment to coordinate and act as a group to solve the different tasks of the application.

5.3.1. PICO.
PICO is developed to hide the complexity of integrating the heterogeneous devices of the smart environment. Its objective is to provide device-integrated service to any place at anytime. It is developed to support hospital-based application [66].

5.3.2. HOMEROS.
HOMEROS middleware coordinates different devices with different properties in the pervasive smart environment. It provides the right service at right time to facilitate the context-aware application [67].

5.3.3. M-ECHO.
M-ECHO is used to support performance of the system even when available devices are connected through a weak network. It also adapts the morphing step to tune the peer-to-peer performance of the system and focuses on the data that are acquired [68].

5.3.4. MADAM.
MADAM mainly works to address the two issues such as to support mobile clients and heterogeneous devices. It also provides support to heterogeneity for both the client and the server. In the client, different context structures are framed to support different users, and in the server, specific services are created to support different requests [64].

5.3.5. GAS-OS.
GAS-OS is a middleware to separate the application logic from the complex pervasive environment by applying abstraction on the working of the heterogeneous device and environment [69].

5.3.6. GSN.
GSN is a middleware to provide virtual devices as a simple and high-level abstraction. It uses an XML

TABLE 4: List of context-aware middleware on device aspects.

Serial number	Middleware	Device management	Interoperability	Platform Portability	Context awareness	Security and privacy	Autonomy	Intelligence	Adaptability	Domain specific
1	PICO	M	M	M	L	L	L	L	M	Hospital-based service application
2	HOMEROS	H	H	H	L	L	N	N	H	Mobile application
3	M-ECHO	M	M	N	M	N	M	M	N	ROBOT application
4	GAS-OS	M	M	M	N	N	M	N	L	Interface the device
5	MADAM	M	M	N	N	N	N	N	L	Mobile application
6	GSN	H	N	H	N	M	L	N	L	Device management
7	SATIN	M	M	M	L	N	N	N	N	Mobile application
8	MUNDOCORE	L	N	L	L	N	N	N	N	Communication middleware
9	SAMI	H	H	N	N	N	H	N	N	Share information among devices
10	ATLAS	M	M	M	N	N	L	N	L	Device management
11	SANDMAN	H	N	N	N	N	N	N	N	Device management
12	PANOPLY	H	H	N	N	M	N	N	N	Communication interface
13	MISSA	M	N	N	N	N	M	N	L	Stream of data from devices
14	UBISOAP	M	M	M	N	N	M	N	N	Communication middleware
15	ISMB	M	N	M	N	N	M	N	M	RFID-based application
16	HYDRA	H	H	H	N	N	N	N	H	Hospital application in the device level
17	SIRENA	M	M	M	N	L	L	N	L	Device management
18	ASPIRE	L	N	L	N	N	L	N	L	RFID-based application
19	PERLA	H	H	H	N	N	M	N	M	Communication interface
20	PECES	M	M	M	L	L	L	L	L	Smart home application

representation to represent the various devices and is also used to connect the devices in the smart environment [70].

5.3.7. SATIN.
SATIN is a component-based middleware developed to support various mobile computing devices to handle the dynamic scenario in the smart environment [71].

5.3.8. MUNDOCORE.
MUNDOCORE is designed to support the communication between the networks and devices. It provides the platform-independent environment to support heterogeneous devices [72].

5.3.9. SAMI.
SAMI is a self-adaptive information sharing middleware which works in a dynamic pervasive computing environment. It is used to share the information among the devices regardless of their network and environment [40].

5.3.10. ATLAS.
ATLAS is a sensor and actuator middleware platform to support integration of devices in the pervasive smart environment. Its main challenge is to handle heterogeneous devices and make them to work together to support different pervasive applications [73].

5.3.11. SANDMAN.
SANDMAN middleware is developed to reduce the energy used in the smart environment by means of using the efficient protocol stack for communication. It also deactivates the unused devices to sleep mode which helps to reduce the power consumption [74].

5.3.12. PANOPLY.
The main objective of PANOPLY is to group the devices in the smart environment to solve the particular task according to the context of the pervasive smart environments [75].

5.3.13. MISSA.
The primary objective of MISSA is to separate the service logic from the data stream. Data stream is given more importance for the collection of context information from various devices and the proper standard for processing it [76].

5.3.14. UBISOAP.
UBISOAP middleware supports ubiquitous (anytime services) device services in the pervasive smart environment. It is achieved only by making the devices as a service provider and a service receiver [77].

5.3.15. ISMB.
ISMB is a middleware to support communication between devices and networks that can maintain one-to-one, one-to-many, and many-to-many relationships. It automatically discovers the device in the environment and also information maintenance of the mobile device [21].

5.3.16. HYDRA.
HYDRA middleware supports its application by allowing heterogeneous physical devices to flexibly adapt into their environments. It manages the devices in the environment and considers each device as a separate entity [21].

5.3.17. SIRENA.
SIRENA is mainly developed for the integration of smart devices in different domain applications. Devices are seamlessly connected in the smart environment. It is demonstrated with two domains, namely, industries and smart homes [21].

5.3.18. ASPIRE.
ASPIRE is a middleware to support tracking application along with many RFID devices and tools. It helps to track the number of goods in the environment [21].

5.3.19. PERLA.
PERLA is developed to provide support for managing and controlling the heterogeneous smart devices in the pervasive environment. It has its own language to describe the specification of the smart devices used in the application [65].

5.3.20. PECES.
Pervasive Computing in Embedded Systems (PECES) project is developed to create a framework to combine different devices of the smart environment to cooperate and work together to solve a particular task. It is demonstrated with the help of smart home application.

Most of the context-aware middleware work for the device management, and establishes communication between the various devices and middleware, and its focus is on device aspects. There is no middleware to support different domain applications in terms of all the aspects [78].

Device aspects provide support to the physical layer and start with how to connect different devices without any problem and to maintain an active observer/listener to monitor the device and its changing information/data. Device data/information will be represented in a common way and stored in a common format for generic sharing of information. Devices are categorized as a group in order to interact with each group and solve a task effectively and smartly.

In order to support device aspects, a separate list of live devices is to be maintained, to retrieve and control the device information. The live list will give the exact number of devices working in the environment along with its properties. Different devices are grouped to achieve a common goal, and also, it should be ready to handle many requirements at the same time.

6. Existing Tool-Based Development Efforts

In order to test the working of the application model/architecture, a tool-based method is needed for the developer. The development tool acts as a test bed for most of the application developers. In recent times, pervasive application developers have started using various pervasive developer tools for their applications. The Context Toolkit [79] model is made up of different context widgets, and it is placed in a distributed environment. These context widgets are used to access the context information and also hide the context-sensing procedure. Context widgets are software components that are used for developing pervasive application. This model deals only with the context-sensing aspect.

The tool Olympus [80] is to specify the devices, locations, and user in a high-level description form which in turn is converted into actual active space components that are used to build programs. Since this model specifies the context in a high-level description, it is difficult for the developer to understand. It is domain specific, so there are more difficulties in defining the specification of the application.

Archface is a tool [81] that acts as an interface between the architecture and methods and to implement the application. It has its own architecture description language to describe the specification of application, and also, it has programming level interface to convert the specifications into the implementation. The major drawback of the tool is that it deals only with architecture perspective and not with the various components that are used to develop the application.

PervMl [82] is a domain-specific language tool used to define the requirement specification in the metamodel design. This metamodel is in turn mapped to convert it into program that hides the high-level complexities involved in the development of the pervasive application. This model defines the specifications of the tool in a descriptive form, which makes the developer difficult for defining the specifications without predefined knowledge about the domain.

Diasuite [83] is a tool that allows the designer to define the taxonomy of specific application environment by using the existing application model. It has its own simulation tool to simulate the working of the application with the pre-written specification of the application. These tools are taken as reference for the tool-based development.

7. Limitations of the Existing Work

From the observation of the literature survey, context-aware middleware is working on three different aspects as shown in Figure 5. The first aspect of the context-aware middleware concerns on domain on which it is operating, such as smart home, smart office, smart class, and smart hospital. The second aspect of the context-aware middleware focuses on the various devices operating on the pervasive smart environment and devices involved are dynamic or fixed to what extend is the great challenge.

The third aspect of the context-aware middleware deals with the context of the diverse domain environment. Context predicts the situation of the pervasive smart environment and adapts according to it. The different contexts are the location, time, device, operation, and so on, and the main objective is to assess how many contexts can the application support.

From the study, it may be observed that there are mainly three reasons to state middleware is specific to the application:

(1) Context-aware middleware architecture is specific to the pervasive applications. There is no generic context-aware middleware architecture to support diverse domain applications.

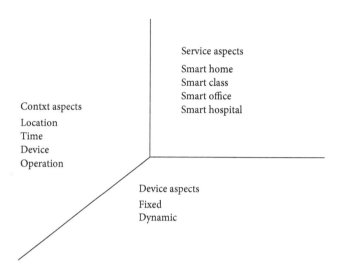

FIGURE 5: Different aspects of context-aware middleware.

(2) Existing context-aware middleware implicitly states the limited nonfunctional qualities. From the literature survey, existing context-aware middleware architecture exhibits the limited number of nonfunctional qualities, and also, they are implicitly stated.

(3) Context-aware middleware is biased to the specific aspects. Most of the existing context-aware middleware are biased of specific aspects like service aspect alone or only to context aspect or particularly to the device aspect.

8. Conclusions

The evolution of pervasive computing is presented. Various pervasive applications are obtained from the literature and classified according to their domain. The different architectures for pervasive computing are analyzed, and it is found that context-aware middleware architecture provides a better solution to build pervasive application. In order to better understand the demands of stakeholders of context-aware middleware, a set of different aspects of context-aware middleware are derived from the literature and presented.

Context-aware middleware is classified based on three aspects such as context aspect, device aspect, and service aspect. Classification of the context-aware middleware has been given to obtain a clear understanding about the requirements of the different aspects. On analyzing the requirements of the context-aware middleware, the limitations are obtained, and the scope for the development of generic context-aware middleware is reported.

References

[1] M. Weiser, "The computer for the 21st century," *Scientific American*, vol. 265, no. 3, pp. 94–104, 1991.

[2] C. D. Kidd, R. J. Orr, G. D. Abowd et al., "The aware home: a living laboratory for ubiquitous computing research," in *CoBuild'99, Proceedings of the Second International Workshop on Cooperative Buildings, Integrating Information, Organization, and Architecture, LNCS*, vol. 1670, pp. 191–198, Springer-Verlag, ISBN: 978-3-540-665960, October 1999.

[3] Y. Oh, S. Jang, and W. Woo, "User centered context-aware smart home applications," *Journal on Korea Information Science Society (KISS): Software and Applications*, vol. 31, no. 2, pp. 111–125, 2004.

[4] C. Reinisch, M. J. Kofler, and W. Kastner, "ThinkHome: a smart home as digital ecosystem," in *Proceedings of the 4th IEEE International Conference on Digital Ecosystems and Technologies (DEST)*, pp. 256–261, Dubai, UAE, April 2010.

[5] P. Gabriel, M. Bovenschulte, E. Hartmann et al., *Pervasive Computing: Trends and Impacts*, BSI Godesberger Allee, Bonn, Germany, 2006, https://www.bsi.bund.de/SharedDocs/Downloads/EN/BSI/Publications/Studies/Percenta/Percenta_elay_pdf.pdf?__blob=publicationFile&v=4.

[6] K. Lyytinen and Y. Yoo, "The next wave of nomadic computing: a research agenda for information systems research," *Information Systems Research*, vol. 13, no. 4, pp. 377–388, 2002.

[7] P. Bellavista, A. Corradi, M. Fanelli, and L. Foschini, "A survey of context data distribution for mobile ubiquitous systems," *ACM Computing Surveys*, vol. 44, no. 4, pp. 1–45, 2013.

[8] M. Baldauf, S. Dustdar, and F. Rosenberg, "A survey on context-aware systems," *International Journal of Ad Hoc and Ubiquitous Computing*, vol. 2, no. 4, pp. 263–277, 2007.

[9] M. Satyanarayanana, "Pervasive computing: vision and challenges," *IEEE Personal Communications*, vol. 8, no. 4, pp. 10–17, 2001.

[10] R. Zhao and J. Wang, "Visualizing the research on pervasive and ubiquitous computing," *Scientometrics*, vol. 86, no. 3, pp. 593–612, 2011.

[11] D. Cook, J. Augusto, and V. Jakkula, "Ambient intelligence: technologies, applications and opportunities," *Journal of Pervasive and Mobile Computing*, vol. 5, no. 4, pp. 277–298, 2009.

[12] I. Georgievski, V. Degeler, G. Andrea Pagani, T. Anh Nguyen, A. Lazovik, and M. Aiello, *Optimizing Offices for the Smart Grid*, Distributed Systems Group Institute for Mathematics and Computer Science, University of Groningen, Groningen, Netherlands, 2011.

[13] V. Degeler and A. Lazovik, "Architecture pattern for context-aware smart environments: creating personal, social, and urban awareness through pervasive computing," *Advances in Human and Social Aspects of Technology*, pp. 108–130, ISBN: 978-1-4666-4695-7, 2014.

[14] S. S. Yau, S. K. S. Gupta, F. Karim, S. I. Ahamed, Y. Wang, and B. Wang, "Smart classroom: enhancing collaborative learning using pervasive computing technology," in *Proceedings of the ASEE 2003 Annual Conference and Exposition*, Nashville, TN, USA, June 2003.

[15] L. Atzori, A. Iera, and G. Morabito, "The internet of things: a survey," *Journal of Computer Networks*, vol. 54, no. 15, pp. 2787–2805, 2010.

[16] Y.-W. Wang, H.-L. Yu, and Y. Li, "Internet of things technology applied in medical information," in *Proceedings of the International Conference on Consumer Electronics, Communications and Networks (CECNet)*, pp. 430–433, Xianning, China, April 2011.

[17] Y. Chen, J. Guo, and X. Hu, "The research of internet of things' supporting technologies which face the logistics industry," in *Proceedings of the International Conference on Computational Intelligence and Security (CIS)*, pp. 659–663, ISBN: 978-1-4244-9114-8, Shanghai, China, December 2010.

[18] G. Chong, L. Zhihaoand, and Y. Yifeng, "The research and implement of smart home system based on internet of things," in *Proceedings of the International Conference on Electronics, Communications and Control (ICECC)*, pp. 2944–2947, ISBN: 978-1-4577-0320-1, Ningbo, China, September 2011.

[19] L. W. F. Chaves and C. Decker, "A survey on organic smart labels for the internet-of-things," in *Proceedings of the Seventh International Conference on Networked Sensing Systems (INSS)*, pp. 161–164, E-ISBN: 978-1-4244-7910-8, Kassel, Germany, June 2010.

[20] K. Markantonakis and K. Mayes, *Secure Smart Embedded Devices, Platforms and Applications*, Springer Science and Business Media, Berlin, Germany, 2014.

[21] S. Bandyopadhyay, M. Sengupta, S. Maiti, and S. Dutta, "A survey of middleware for internet of things," in *Communications in Computer and Information Science*, vol. 162, pp. 288–296, Springer, Berlin, Germany, 2011.

[22] L. Chen, J. Hoey, C. Nugent, D. Cook, and Z. Yu, "Sensor-based activity recognition: a survey," *IEEE Transactions on Systems, Man, and Cybernetics, Part C*, vol. 42, no. 6, pp. 790–808, 2012.

[23] H.-L. Truong and S. Dustdar, "A survey on context-aware web service systems," *International Journal of Web Information Systems*, vol. 5, no. 1, pp. 5–31, 2009.

[24] K. E. Kjær, "A survey of context-aware middleware," in *Proceedings of the 25th conference on IASTED International Multi-Conference: Software Engineering*, pp. 148–155, Innsbruck, Austria, February 2007.

[25] A. Messer, H. Song, P. Kumar, P. Nguyen, A. Kunjithapatham, and M. Sheshagiri, "InterPlay: A Middleware for Integration of Devices, Services and Contents in the Home Networking Environment," in *Proceedings of the 3rd IEEE Consumer Communications and Networking Conference (CCNC 06)*; Las Vegas, NV, USA, 7–10 January 2006, pp. 1083–1087.

[26] D. Romero, "Context-aware middleware: an overview," *Revista Electrónica Paradigma en Construcción de Software*, vol. 3, pp. 1–11, 2008.

[27] K. Lee, T. Lunney, K. Curran, and J. Santos, "Ambient middleware for context-awareness (AMiCA)," *International Journal of Ambient Computing and Intelligence (IJACI)*, vol. 1, no. 3, pp. 66–78, 2008.

[28] O. Davidyuk, N. Georgantas, V. Issarny, and J. Riekki, "MEDUSA: middleware for end-user composition of ubiquitous applications," in *Handbook of Research on Ambient Intelligence and Smart Environments: Trends and Perspectives*, pp. 197–219, IGI Global, Hershey, PA, USA, 2009.

[29] A. Saeed and T. Waheed, "An Extensive Survey of Context-Aware Middleware Architectures," *American Journal of Computer Architecture*, vol. 1, no. 3, pp. 51–56, 2012.

[30] R. Vasanthi and R. S. D. Wahidabanu, "A middleware platform for pervasive environment," *International Journal of Computer Science and Information Security*, vol. 9, no. 4, 2011.

[31] N. Gámez and L. Fuentes, "FamiWare: a family of event-based middleware for ambient intelligence," *Personal and Ubiquitous Computing*, vol. 15, no. 4, pp. 329–339, 2011.

[32] R. Cloutier, D. Verma, M. Bone, and K. Sommer, "System architecture entropy," in *Proceedings of the INCOSE International Symposium*, Singapore, July 2009.

[33] L. Kovács, P. Mátételki, B. Pataki, and M. Sztaki, "Service oriented context-aware framework," *Young Researchers Workshop on Service-Oriented Computing*, pp. 15–26, 2009.

[34] J. M. Neighbors, *Software Construction Using Components*, Ph.D. thesis, ICS-TR-160, University of California at Irvine, Irvine, CA, USA, 1980.

[35] L. Catarinucci, R. Colella, A. Esposito, L. Tarricone, and M. Z. appatore, "RFID sensor-tags feeding a context-aware rule-based healthcare monitoring system," *Journal of Medical Systems*, vol. 36, no. 6, pp. 3435–3449, 2011.

[36] E. U. Warriach, "State of the art: embedded middleware platform for a smart home," *International Journal of Smart Home*, vol. 7, no. 6, pp. 275–294, 2013.

[37] A. Kalmar and R. Vida, "Extracting high level context information using hierarchical temporal memory," in *Proceedings of Conference of Advances in Wireless Sensor Networks*, pp. 27–34, ISBN: 978-963-318-356-4, Qingdao, China, October 2013.

[38] A. Ema and Y. Fujigaki, "How far can child surveillance go? Assessing the parental perceptions of an RFID child monitoring system in Japan," *Surveillance & Society*, ISSN: 1477-7487, pp. 132–148, 2011.

[39] I. Amundson, M. Kushwaha, X. Koutsoukos, S. Neema, and J. Sztipanovits, "OASiS: a service-oriented middleware for pervasive ambient-aware sensor networks," *Pervasive and Mobile Computing Journal on Middleware for Pervasive Computing*, 2006.

[40] A. Negash, V.-M. Scuturici, and L. Brunie, "SAMi: a self-adaptive information sharing middleware for a dynamic pervasive computing environment," in *Proceedings of the IADIS International Conference Wireless Applications and Computing*, Lisbon, Portugal, July 2007.

[41] T. Gu, H. K. Pung, and D. Q. Zhang, "A service-oriented middleware for building context-aware services," *Journal of Network and Computer Applications*, vol. 28, no. 1, pp. 1–18, 2005.

[42] N. Kaveh and R. G. Hercock, "NEXUS-resilient intelligent middleware," *BT Technology Journal*, vol. 22, no. 3, pp. 209–215, 2004.

[43] I. Corredor, J. F. Martínez, M. S. Familiar, and L. López, "Knowledge-aware and service-oriented middleware for deploying pervasive services," *Journal of Network and Computer Applications*, vol. 35, no. 2, pp. 562–576, 2011.

[44] B. Schilit, N. Adams, and R. Want, "Context-aware computing applications," in *Proceedings of the Workshop on Mobile Computing Systems and Applications*, Santa Cruz, CA, USA, December 1994.

[45] S. Presecan, *Pervasive Computing: Collaborative Architecture Applied in Business Environment*, Ph.D. thesis, Babeş-Bolyai University, Cluj-Napoca, Romania, 2011.

[46] A. Manzoor, H. L. Truong, and S. Dustdar, "Using quality of context to resolve conflicts in context-aware systems," in *Proceedings of the 1st International Conference on Quality of Context (QuaCon'09)*, vol. 5786, pp. 144–155, Stuttgart, Germany, June 2009.

[47] D. Garlan and J. P. Sousa, "Aura: an architectural frame work for user mobility in ubiquitous computing environments," in *Proceeedings of the IFIP 17th world computer congress-TC2Stream/3rd IEEE/IFIP conference on software architecture: system Design, Development and Maintenance*, pp. 29–43, Montréal, QC, Canada, August 2002.

[48] A. Toninelli, *Semantic-Based Middleware Solutions to Support Context-Aware Service Provisioning in Pervasive Environments*, Dissertation thesis, 2008.

[49] L. Capra, W. Emmerich, and C. Mascolo, "CARISMA: context-aware reflective middleware system for mobile applications," *IEEE Transactions on Software Engineering*, vol. 29, no. 10, pp. 929–945, ISSN: 0098-5589, 2003.

[50] M. Román, C. Hess, R. Cerqueira, A. Ranganat, R. H. Campbell, and K. Nahrstedt, *Gaia: A Middleware Infrastructure to Enable Active Spaces*, National Science Foundation, grant NSF 98-70736, NSF 9970139, and NSF Infrastructure grant NSF EIA 99-72884, 2002.

[51] D. Saha and A. Mukherjee, "Pervasive computing: a paradigm for the 21st century," *Computer*, vol. 36, no. 3, pp. 25–31, 2003.

[52] H. Chen, T. Finin, and A. Joshi, "An intelligent broker for context-aware system," in *Proceedings of the Adjunct Proceeding of Ubicomp'03*, pp. 183-184, Seattle, WA, USA, 2003.

[53] M. Mamei and F. Zambonelli, "Programming pervasive and mobile computing applications with the TOTA middleware," in *Proceedings of the Second IEEE Annual Conference on Pervasive Computing and Communications (PERCOM'04)*, pp. 263–273, Orlando, FL, USA, March 2004.

[54] A. R. Frei, *Jadabs–An Adaptive Pervasive Middleware Architecture*, Doctoral dissertation, Swiss Federal Institute of Technology, Zurich, Switzerland, 2005.

[55] A. Coronato and G. De Pietro, "MiPeG: a middleware infrastructure for pervasive grids," *Journal of Future Generation Computer Systems*, vol. 24, no. 1, pp. 17–29, 2008.

[56] N. Sahli, "Survey: agent-based middlewares for context awareness," *Electronic Communications of the EASST*, vol. 11, 2008.

[57] A. Bennaceur, P. Singh, P.-G. Raverdy, and V. Issarny, "The iBICOOP middleware: enablers and services for emerging pervasive computing environments," in *Proceedings of the IEEE Middleware Support for Pervasive Computing Workshop, PerWare*, pp. 1–6, , October 2009.

[58] H. Ni, B. Abdulrazak, D. Zhang, and S. Wu, "CDTOM: a context-driven task-oriented middleware for pervasive homecare environment," *International Journal of UbiComp*, vol. 2, no. 1, 2011.

[59] P. Bruegger, *uMove: A Wholistic Framework to Design and Implement Ubiquitous Computing Systems Supporting User's Activity and Situation*, Ph.D. thesis, University of Fribourg, Fribourg, Switzerland, 2011.

[60] V. Arnaboldi, M. Conti, and F. Delmastro, "CAMEO: a novel context-aware middleware for opportunistic mobile social networks," *Journal of Pervasive and Mobile Computing*, vol. 11, pp. 148–167, 2013.

[61] D. Ferreira, *AWARE: A Mobile Context Instrumentation Middleware to Collaboratively Understand Human Behavior*, Doctoral dissertation, University of Oulu Graduate School, University of Oulu, Faculty of Technology, Department of Computer Science and Engineering, ISBN: 978-952-62-0189-4, 2013.

[62] M. Eisenhauer, P. Rosengren, and P. Antolin, "A development platform for integrating wireless devices and sensors into ambient intelligence systems," in *Proceedings of the 6th Annual IEEE Communications Society Conference on Sensor, Mesh and Ad Hoc Communications and Networks Workshops*, pp. 1–3, Rome, Italy, June 2009.

[63] P. Evensen, *Event Processing Applied to Streams of TV Channel Zaps and Sensor Middleware with Virtualization*, Ph.D. dissertation, University of Stavanger, Stavanger, Norway, 2013.

[64] X. Zhi, L. Rong, and W. Tong, "Adaptive middleware for uniform access to legacy systems for mobile clients," in *Proceedings of the International Symposium on Pervasive Computing and Applications*, pp. 46–50, Xinjiang, China, August 2006.

[65] F. A. Schreiber, R. Camplani, M. Fortunato, M. Marelli, and G. Rota, "PerLa: a language and middleware architecture for

data management and integration in pervasive information systems," *IEEE Transactions on Software Engineering*, vol. 38, no. 2, pp. 478–496, ISSN:0098-5589, 2012.

[66] M. Kumar, B. A. Shirazi, S. K. Das, B. Y. Sung, and D. Levine, "PICO: a middleware framework for pervasive computing," *IEEE Transactions on Pervasive Computing*, vol. 2, no. 3, pp. 72–79, 2003.

[67] S. W. Han, Y. B. Yoon, H. Y. Youn, and W.-D. Cho, "A new middleware architecture for ubiquitous computing environment," in *Proceedings of the Second IEEE Workshop on Software Technologies for Future Embedded and Ubiquitous Systems (WSTFEUS'04)*, pp. 117–121, Vienna, Austria, May 2004.

[68] H. Raj, K. Schwan, and R. Nathuji, "M-Echo: a middleware for morphable data-streaming in pervasive systems," in *EESR '05: Workshop on End-to-End, Sense-and-Respond Systems, Applications, and Services*, pp. 13–18, ISBN: 1-931971-32-3, Berkeley, CA, USA, 2005.

[69] N. I. Drossos, C. A. Goumopoulos, and A. D. Kameas, "A conceptual model and the supporting middleware for composing ubiquitous computing applications," *Journal of Ubiquitous Computing and Intelligence*, vol. 1, no. 2, pp. 1–13, 2006.

[70] K. Aberer, M. Hauswirth, and A. Salehi, *Middleware Support for the Internet of Things*, Universitt Stuttgart, Stuttgart, Germany, GI/ITG KuVS Fachgesprch "Drahtlose Sensornetze", 2006.

[71] S. Zachariadis, C. Mascolo, and W. Emmerich, "The SATIN component system-a metamodel for engineering adaptable mobile systems," *IEEE Transactions on Software Engineering*, vol. 32, no. 10, pp. 910–927, 2006.

[72] E. Aitenbichler, J. Kangasharju, and M. Muhlhauser, "MundoCore: a light-weight infrastructure for pervasive computing," *Journal of Pervasive and Mobile Computing*, vol. 3, no. 4, 2007.

[73] A. Helal, H. Yang, J. King, and R. Bose, *Atlas–Architecture for Sensor Network Based Intelligent Environments*, ACM, New York City, USA, 1073-0516/01/03 00-0034, 2007.

[74] G. Schiele, M. Handte, and C. Becker, "SANDMAN: an energy-efficient middleware for pervasive computing," in *PERCOM '08-Proceedings of the 2008 Sixth Annual IEEE International Conference on Pervasive Computing and Communications*, pp. 504–508, Washington, DC, USA, 2008.

[75] K. Eustice, V. Ramakrishna, N. Nguyen, and P. Reiher, "The smart party: a personalized location-aware multimedia experience," in *Proceedings of the 5th IEEE Consumer Communications and Networking Conference (CCNC)*, pp. 873–877, Las Vegas, Nevada, USA, January 2008, E-ISBN: 978-1-4244-1457-4, 2008.

[76] S. Kang, Y. Lee, S. Ihm, S. Park, S.-M. Kim, and J. Song, "Design and implementation of a middleware for development and provision of stream-based services," in *Proceedings of the IEEE 34th Annual Conference on Computer Software and Applications*, pp. 92–100, Seoul, Republic of Korea, July 2010.

[77] M. Caporuscio, P.-G. Raverdy, and V. Issarny, "ubiSOAP: a service oriented middleware for ubiquitous networking," *IEEE Transactions on Services Computing*, vol. 5, no. 1, pp. 86–98, 2010.

[78] R. Zhao and N. Speirs, "Development tools for context aware and secure Pervasive Computing in Embedded Systems (PECES) middleware," *Journal of Networks*, vol. 8, no. 1, 2013.

[79] A. K. Dey, G. D. Abowd, and D. Salber, "A conceptual framework and a toolkit for supporting the rapid prototyping of context-aware applications," *Journal of Human-Computer Interaction*, vol. 16, no. 2, pp. 97–166, 2001.

[80] A. Ranganathan, S. Chetan, J. Al-Muhtadi, R. H. Campbell, and M. D. Mickunas, "Olympus: a high-level programming model for pervasive computing environments," in *Proceedings of the IEEE Third International Conference on Pervasive Computing and Communications*, pp. 7–16, ISBN: 0-7695-2299-8, Kauai Island, HI, USA, March 2005.

[81] N. Ubayashi, J. Nomura, and T. Tamai, "Archface: a contract place where architectural design and code meet together," in *Proceedings of the 32nd ACM/IEEE International Conference on Software Engineering*, vol. 1, pp. 75–84, Cape Town, South Africa, May 2010.

[82] J. Muñoz, V. Pelechano, and C. Cetina, "Software engineering for pervasive systems: applying models, frameworks and transformations," in *Proceedings of the IEEE International Conference on Pervasive Services (ICPS)*, pp. 290–294, E-ISBN: 1-4244-1326-5, Istanbul, Turkey, July 2007.

[83] D. Cassou, J. Bruneau, C. Consel, and E. Balland, "Toward a tool-based development methodology for pervasive computing applications," *IEEE Transactions on Software Engineering*, vol. 38, no. 6, pp. 1445–1463, 2012.

[84] L. Bass, P. Clements, and R. Kazman, *Software Architecture in Practice*, Pearson Education, Third Indian Reprint, London, UK, 2002.

[85] M. Bozga, C. Daws, O. Maler, A. Olivero, S. Tripakis, and S. Yovine, "Kronos: a model-checking tool for real-time systems," in *Proceedings of the 10th International Conference on Computer Aided Verification (CAV '98)*, vol. 1427, pp. 546–550, Springer Verlag LNCS, Vancouver, BC, Canada, July 1998.

6

Research on Extended Kalman Filter and Particle Filter Combinational Algorithm in UWB and Foot-Mounted IMU Fusion Positioning

Xin Li [ID],[1] Yan Wang [ID],[1] and Dawei Liu[2]

[1]*School of Computer Science and Technology, China University of Mining and Technology, Xuzhou 221116, China*
[2]*Air Force Logistics College, Xuzhou 221008, China*

Correspondence should be addressed to Yan Wang; wycumtxz@126.com

Academic Editor: Jesus Fontecha

As UWB high-precision positioning in NLOS environment has become one of the hot topics in the research of indoor positioning, this paper firstly presents a method for the smoothing of original range data based on the Kalman filter by the analysis of the range error caused by UWB signals in LOS and NLOS environment. Then, it studies a UWB and foot-mounted IMU fusion positioning method with the integration of particle filter with extended Kalman filter. This method adopts EKF algorithm in the kinematic equation of particle filters algorithm to calculate the position of each particle, which is like the way of running N (number of particles) extended Kalman filters, and overcomes the disadvantages of the inconformity between kinematic equation and observation equation as well as the problem of sample degeneration under the nonlinear condition of the standard particle filters algorithm. The comparison with the foot-mounted IMU positioning algorithm, the optimization-based UWB positioning algorithm, the particle filter-based UWB positioning algorithm, and the particle filter-based IMU/UWB fusion positioning algorithm shows that our algorithm works very well in LOS and NLOS environment. Especially in an NLOS environment, our algorithm can better use the foot-mounted IMU positioning trajectory maintained by every particle to weaken the influence of range error caused by signal blockage. It outperforms the other four algorithms described as above in terms of the average and maximum positioning error.

1. Introduction

With the wide application of indoor positioning technologies in some areas such as supermarket shopping, fire emergency navigation, and hospital patient tracking, indoor positioning can be implemented through the following two approaches. One is based on the various wireless network technologies, such as WiFi (wireless fidelity) [1, 2], RFID (radio frequency identification) [3], and UWB (ultra-wideband) [4, 5], which can be used to realize indoor positioning according to the intensity of received signals, the TOA (time of arrival), or TDOA (time difference of arrival). Among all of these technologies, UWB technology can achieve a decimeter-level positioning precision. However, in some special cases, such as emergency rescue, UWB signals might be blocked by people, walls, or the other barriers in the complex indoor

environment. As it might result in the problems of signal multipath effect or intensity attenuation, high-precision positioning can hardly be achieved in NLOS (nonline of sight) environment through the UWB positioning approach.

The other approach is based on the IMU (inertial measurement unit), such as accelerometer, gyroscope, magnetometer, and so on [6], which can be used for positioning according to the integral or the PDR (pedestrian dead reckoning) method. However, this approach has a deficiency, which is an accumulative error. In order to overcome the problem of error accumulation, in [7], the authors proposed ZUPT (zero velocity update) in 2005 to correct the system error and applied it in NavShoe. In 2012, in [8], the authors proposed the implementation of a shoe-mounted ZUPT-aided open-source INS (inertial navigation system) for real-time positioning. At a cost of around USD (United States dollar)

800, this sample system was able to control the navigation error within the range of 0.2%–1% in a short distance (within 100 meters). Moreover, through the analysis of the limitations of ZUPT and the error model [9], they managed to eliminate the drift error according to the optimization algorithm to enhance the algorithm efficiency [10]. In 2013, based on the shoe-mounted INS, a locally distributed system framework was proposed [11], which could increase significantly the autonomous positioning precision by constraining the course angle deviation of INS in accordance with the distance between both feet. In 2014, Nilsson and his team [12] developed a positioning approach based on the IMU arrays to further increase the reliability and precision of autonomous positioning and at the same time open sourced the experimental positioning platform. In 2017, Wagstaff et al. [13] presented a method to improve the accuracy of a foot-mounted, zero velocity-aided inertial navigation system (INS) by varying the estimator parameters based on a real-time classification of the motion type. By combining the motion classifier with a set of optimal detection parameters, we show how we can reduce the INS position error during mixed walking and running motion. In [14], the authors presented an experimental study on the noise performance and the operating clocks-based power consumption of multi-IMU platforms. It is observed that the four-IMU system is best optimized for cost, area, and power.

Although the ZUPT technology to some extent can realize error correction, it still cannot overcome the problem of error accumulation arising in the long-distance positioning for a long time with an IMU. Therefore, the integration of IMU with UWB is a tendency to achieve the high-precision and real-time indoor positioning. With the integration of IMU, not only the following observations such as velocity and direction can be obtained but also the multipath and NLOS effects can be eliminated [15, 16]. In addition, based on the EKF (extended Kalman filter), loose combination can be adopted to track the pedestrian's movement. In [17], the authors realized the UWB/IMU tightly coupled algorithm based on the EKF and made a comparison with the optical tracking system to show the higher precision. Similarly, in [18], the authors implemented the UWB and inertial data fusion algorithm based on a steady state KF with a fixed gain. The main advantage of this method is that it can be implemented efficiently in low-performance WSN nodes with low-power consumption. With the introduction of a tightly coupled algorithm based on UWB/INS, in [19], the authors analyzed the influence of the integrity monitoring algorithms on the positioning performance. In [20], the authors put forward an adaptive fuzzy Kalman filter method. Their experiment turned out that this algorithm outperformed the basic KF algorithm in terms of the positioning result. In [21], the authors designed a tightly coupled GPS (global positioning system)/UWB/INS integrated system based on the adaptive robust Kalman filter. Yet, it is only for outdoor use [22]. It was the first time that the positioning of a flying drone with the integration of vision, IMU, and UWB was proposed to realize the two-dimensional positioning accuracy of 10 cm. However, in the literature [23], visual-inertial SLAM (simultaneous localization and mapping) technology was used for the

positioning of a flying drone. Meanwhile, the adoption of UWB technology for error correction had obtained a full six-DoF pose of the drone. In [24], the authors studied the EKF loosely/tightly coupled UWB/INS integration based on the PDF algorithm, but they utilized the ray-launching simulations to generate UWB data. In [25], the authors presented an improving tightly-coupled navigation model for indoor pedestrian navigation. In the proposed model, a channel filter is used for the estimation of the distance between the reference node (RN) and blind node (BN) measured by the UWB, and then, a 15-element error state vector is used in the filter for fusing foot-mounted IMU and UWB measurements. The real test results show that the proposed model is effective to reduce the error compared with the conventional model, its mean position error has reduced by about 14.81% compared with the UWB only model. In [26], the authors fused an ultra-wideband (UWB) sensor-based positioning solution with an inertial measurement unit (IMU) sensor-based positioning solution to obtain a robust, yet, optimal positioning performance. Sensor fusion is accomplished via an extended Kalman filter (EKF) design which simultaneously estimates the IMU sensors' systematic errors and corrects the positioning errors. Fault detection, identification, and isolation are built into the EKF design to prevent the corrupted UWB sensor measurement data due to obstructions, multipath, and other interferences from degrading the positioning performance. Computer simulation results indicate that more than 100% positioning performance improvement over the UWB sensor-based positioning solution along can be obtained through the proposed sensor fusion solution. In [27], the authors proposed an approach to combine IMU inertial and UWB ranging measurement for relative positioning between multiple mobile users without the knowledge of the infrastructure. They incorporate the UWB and the IMU measurements into a probabilistic-based framework, which allows cooperatively positioning a group of mobile users and recovering from positioning failures.

Most of the above methods have adopted EKF for UWB/INS fusion positioning and optimization. However, the premise for the use of EKF is to assume that both of system errors and observation errors conform to Gaussian distribution. But in the NLOS condition, signal transmission might be affected by barriers due to the blockage or reflection, which would increase the time delay of signal transmission. Under such a circumstance, if the assumption still holds that UWB ranging errors must conform to Gaussian distribution, it would result in great error. In this paper, UWB and IMU fusion positioning has been studied based on the PF (particle filter). This is because that the particle filter can tackle the multimodal distribution of errors. As long as there are sufficient particles available, an approximate globally optimal solution can be obtained effectively. This paper introduces two UWB and IMU fusion algorithms based on the PF and compares them with the other three UWB or IMU-based positioning algorithms for the analysis.

The remainder of the paper is organized as follows: In Section 2, the analysis and pretreatment of UWB data error is illustrated, and Section 3 introduces two UWB and IMU fusion algorithms based on the PF. In order to facilitate the

FIGURE 1: The experimental area in the entrance hall.

comparison and analysis, it also provides the positioning results based on the foot-mounted IMU, the optimization-based algorithm, and the particle filter algorithm based on pure UWB data. Subsequently, several experiments are analyzed in Section 4, and Section 5 concludes the paper.

2. Analysis and Pretreatment of UWB Data Error

In order to verify the UWB data error in LOS and NLOS environment, we carried out a correlation experiment in the entrance hall of School of Computer Science and Technology in China University of Mining and Technology. As shown in Figure 1, the hall is paved with 0.8 * 0.8 m marble tiles so that the calibration of real positions can be made. As indicated in Figure 2, the core chip used in UWB tag/beacon is the DWM1000 chip from DecaWave.

2.1. Analysis of UWB Data Errors in LOS Condition. As shown in Figure 3, it starts from a distance of 2.4 m. Based on a progressive increase of 0.8 m in range, record the distances between the UWB tag and beacon and calculate the relevant errors in Table 1, where it shows that mean error grows with the increase in distance. For example, when the distance is 1.6 m, the mean error is 0.37 m. However, when the distance is increased to 8 m, the mean error reaches 0.56 m. But generally, as indicated in Figure 4, there is a small standard deviation of errors, which also proves that the positioning result is quite stable.

2.2. Analysis of UWB Data Errors in NLOS Condition. As shown in Figure 5, firstly test the influence of the marble column on the UWB signals in the experimental area, where the column is located 1.13 m away from the UWB beacon. It starts from a distance of 3.39 m between the UWB tag and the beacon. Record the distances between the UWB tag and the beacon based on a progressive increase of 0.8 m in distance, and calculate the related errors in Table 2. As shown in Figure 6, there is a significant increase in the mean error with the increase in distance. When the range is 3.39 m, the mean error is 0.62 m. However, the mean error will grow to 3.45 m when the distance is increased to 10.17 m. Please note that when the distance is increased to more than 9.04 m, the laptop that is connected to the UWB beacon can hardly receive any signal after limited

FIGURE 2: UWB tag/beacon.

FIGURE 3: Ranging test in LOS condition.

groups of range data have been acquired. It reveals that, with the increase in distance, column blockage will lead to an increase in signal attenuation. Actually, when the distance between the tag and the beacon is within 4.52 m, there is a steady change in the ranging result with a standard deviation of 0.04. However, when the distance is greater than 4.52 m, the ranging result will become extremely unstable. There will be a big standard deviation of ranges in the presence of column blockage with the maximum error increased to 4.59 m from 2.01 m.

After that, perform an experiment for the influence of pedestrian blockage on the UWB signals. As shown in Figure 7, it starts from a distance of 1.6 m when the pedestrian moves freely between the beacon and the tag. Then, record the distances between the UWB tag and the beacon based on a progressive increase of 0.8 m in range, and calculate the relevant errors in Table 3. With an increase in distance, the standard deviation of ranges will show a trend of increase first and then decrease. For example, when the distance is below 4.8 m, the standard deviation is around 0.2 m to show a stable ranging result with the maximum error within 1.5 m. However, when the distance is between 5.6 m and 7.2 m, the standard deviation rises quickly. As indicated in Figure 8, big amplitude arises on the corresponding three distance curves, revealing that the ranging results are extremely unstable with the maximum error up to 7.78 m. However, when the distance is over 8.0 m, the ranging result becomes stable again with the standard deviation within 0.3 m. In this experiment, the sudden increase in ranging error always occurs at the moment

TABLE 1: Ranging errors in LOS condition based on a progressive increase of 0.8 m in distance.

Real distance	1.6	2.4	3.2	4.0	4.8	5.6	6.4	7.2	8.0
Mean error	0.37	0.42	0.46	0.45	0.46	0.51	0.50	0.50	0.56
Maximum error	0.42	0.47	0.51	0.49	0.50	0.60	0.55	0.55	0.60
Minimum error	0.33	0.36	0.43	0.42	0.42	0.48	0.44	0.45	0.51
Standard deviation	0.021	0.020	0.015	0.017	0.018	0.023	0.020	0.021	0.024

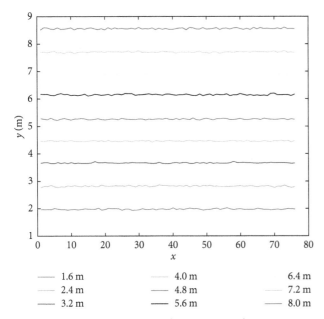

—— 1.6 m	—— 4.0 m	6.4 m
—— 2.4 m	—— 4.8 m	7.2 m
—— 3.2 m	—— 5.6 m	—— 8.0 m

FIGURE 4: Ranging results in LOS condition.

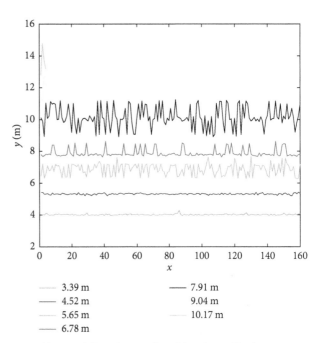

—— 3.39 m	—— 7.91 m
—— 4.52 m	9.04 m
—— 5.65 m	—— 10.17 m
—— 6.78 m	

FIGURE 6: Ranging results with column blockage.

FIGURE 5: Ranging test with column blockage.

FIGURE 7: Ranging test with column blockage.

TABLE 2: Ranging results based on a progressive increase of 1.13 m in distance with column blockage.

Real distance	3.39	4.52	5.65	6.78	7.91	9.04	10.17
Mean error	0.62	0.88	0.94	1.10	2.24	2.89	3.45
Maximum error	0.80	0.91	2.01	1.84	3.32	4.04	4.59
Minimum error	0.56	0.68	0.31	0.86	0.99	2.13	2.60
Standard deviation	0.04	0.04	0.42	0.25	0.64	0.85	0.86

when the pedestrian moves close to the tag or the beacon. This is because that there is a sudden interference from the pedestrian on the UWB signals. But when the distance goes up to a certain degree, signal diffraction occurs to make the signals free from the influence of pedestrian blockage.

2.3. UWB Data Pretreatment. There is an abnormity in the ranging result, when the UWB signals are shielded by the obstruction or the positioning tag is beyond the detection limit of the UWB beacon. Hence, it is necessary to use the KF (Kalman filter) for the smoothing of range data. Algorithm 1 provides the Kalman filtering process based on the data acquired in four beacons when the pedestrian moves around in the experimental area. Use the KF algorithm to filter every column of UWB data. As indicated in Figures 9 and 10, the filtered UWB range data becomes relatively smoother. In this experiment, the pedestrian walks at a speed of 1.5 m/s, and the UWB data have been collected based on a frequency

TABLE 3: Ranging results based on a progressive increase of 0.8 m in distance with pedestrian blockage.

Real distance	1.6	2.4	3.2	4.0	4.8	5.6	6.4	7.2	8.0	8.8
Mean error	0.69	0.80	0.66	0.47	0.53	0.67	0.64	0.96	0.64	0.52
Maximum error	1.14	1.06	1.08	1.09	1.35	7.78	7.29	5.47	1.86	1.12
Minimum error	0.28	0.18	0.18	0.18	0.07	0.09	0.14	0.20	0.14	0.08
Standard deviation	0.27	0.16	0.22	0.15	0.28	0.79	0.70	1.19	0.34	0.25

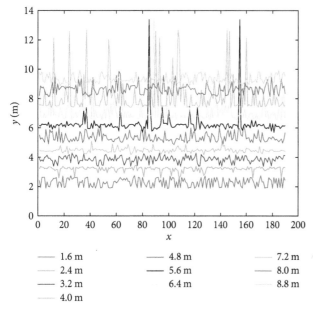

—— 1.6 m	—— 4.8 m	—— 7.2 m
—— 2.4 m	—— 5.6 m	—— 8.0 m
—— 3.2 m	6.4 m	8.8 m
—— 4.0 m		

FIGURE 8: Ranging results with column blockage.

of 3 Hz. Theoretically, the range difference between two adjacent sampling times must be lower than 0.5 m. However, Figure 11 shows that lots of data are above this threshold due to the multipath effect of UWB data and the column blockage. In addition, the range difference is becoming smaller after the Kalman filtering shown in Figure 12. As indicated in Table 4, there is a reduction in the range difference between two adjacent sampling times in four beacons regarding the mean value, the maximum difference, and the variance.

3. IMU/UWB Fusion Positioning and Analysis

This paper presents a UWB and foot-mounted IMU fusion positioning method through the integration of the PF with the EKF. In order to verify the algorithm performance, this paper provides the experimental results obtained according to the foot-mounted IMU-based positioning algorithm, the optimization algorithm-based UWB positioning algorithm, the particle filter-based UWB algorithm, and the particle filter-based IMU/UWB fusion positioning algorithm for the contrast and analysis.

3.1. The Foot-Mounted IMU-Based Positioning Algorithm. Fischer et al. [28] put forward a simple but comparatively more precise positioning algorithm based on the foot-mounted IMU. In summary, their ideas can be concluded into Algorithm 2.

After Line 1 acquires acc_s and $gyro\_s$, get *pitch* and *roll* separately according to the following formula. The value of *yaw* can be obtained with a magnetometer or through the manual setting:

$$pitch = -a\sin\left(\frac{acc_{s(1,1)}}{g}\right),$$

$$roll = a\tan\left(\frac{acc_{s(2,1)}}{acc_{s(3,1)}}\right), \tag{1}$$

$$yaw = init\_heading.$$

Horizontally keep the IMU still for 30~60 seconds to obtain the mean value of the angular velocity noise and take it as the zero bias of $gyro$, $gyro\_bias$. Also, set up an angular velocity skew-symmetric matrix $\mathbf{S}_w$:

$$\mathbf{S}_w = \begin{bmatrix} 0 & -gyro\_s1(3) & gyro\_s1(2) \\ gyro\_s1(3) & 0 & -gyro\_s1(1) \\ -gyro\_s1(2) & gyro\_s1(1) & 0 \end{bmatrix}. \tag{2}$$

The coordinate transformation matrix $\mathbf{C}\_pre$ is provided as below:

$$\mathbf{C}\_pre = \begin{bmatrix} \cos(\theta)*\cos(\psi) & (\sin(\phi)*\sin(\theta)*\cos(\psi)) - (\cos(\phi)*\sin(\psi)) & (\cos(\phi)*\sin(\theta)*\cos(\psi)) + (\sin(\phi)*\sin(\psi)) \\ \cos(\theta)*\sin(\psi) & (\sin(\phi)*\sin(\theta)*\sin(\psi)) + (\cos(\phi)*\cos(\psi)) & (\cos(\phi)*\sin(\theta)*\sin(\psi)) - (\sin(\phi)*\cos(\psi)) \\ -\sin(\theta) & \sin(\phi)*\cos(\theta) & \cos(\phi)*\cos(\theta) \end{bmatrix}. \tag{3}$$

In Line 9, a new coordinate transformation matrix $\mathbf{C}$ will be generated from $\mathbf{C}\_pre$ and $\mathbf{S}_w$, as soon as there are new data arriving. Then, Lines 10~12 will calculate current acceleration, velocity, and position vector in the navigation coordinates. The calculation on these three variables has always been made based on the mean value of the states at the moment and the previous moment. This is because that movement process always occurs between two adjacent data points. After that, construct an observation matrix $\mathbf{H}$ and an observation noise matrix $\mathbf{R}$,

```
Input: UwbData: the original UWB data
Output: UwbDataFilter: UWB data after filtering
 (1) UwbDataFilter = []
 (2) For j in UwbData.cols()
 (3)   EstimateCov = 0.5, MeasureCov = 0.5, Estimate = 0.0
 (4)   For i in UwbData.rows()
 (5)     K = EstimateCov * sqrt(1/(EstimateCov^2 + MeasureCov^2))
 (6)     Estimate = Estimate + K * (UwbData[j, i] − Estimate)
 (7)     EstimateCov = np.sqrt(1 − K) * EstimateCov
 (8)     MeasureCov = np.sqrt(1 − K) * MeasureCov
 (9)     UwbDataFilter[j, i] = Estimate
(10)   End
(11) End
(12) Return UwbDataFilter
```

ALGORITHM 1: *UwbDataFilter* = KF(UwbData).

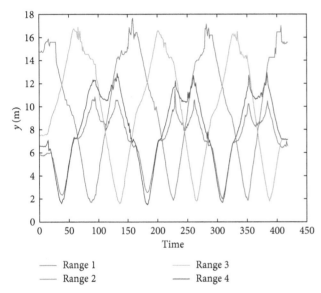

FIGURE 9: Range data obtained in four beacons before the filtering.

where the noise values of zero speed detection are used as the observations:

$$\mathbf{H} = \begin{bmatrix} 0_{3*3} & 0_{3*3} & I_{3*3} \end{bmatrix}, \tag{4}$$

$$\mathbf{R} = \begin{bmatrix} \sigma_v^2 & 0 & 0 \\ 0 & \sigma_v^2 & 0 \\ 0 & 0 & \sigma_v^2 \end{bmatrix}. \tag{5}$$

The acceleration skew-symmetric matrix $\mathbf{S}_a$ is calculated through the following equation:

$$\mathbf{S}_a = \begin{bmatrix} 0 & -acc\_n(3) & acc\_n(2) \\ acc\_n(3) & 0 & -acc\_n(1) \\ -acc\_n(2) & acc\_n(1) & 0 \end{bmatrix}. \tag{6}$$

Calculate the state transfer matrix $\mathbf{F}$ and the system error covariance matrix $\mathbf{Q}$ through the following equations:

$$\mathbf{F} = \begin{bmatrix} I_{3*3} & 0_{3*3} & 0_{3*3} \\ 0_{3*3} & I_{3*3} & dt*I_{3*3} \\ -dt*\mathbf{S}_a & 0_{3*3} & I_{3*3} \end{bmatrix}, \tag{7}$$

$$\mathbf{Q} = \begin{bmatrix} (\sigma_\omega dt)^2 & 0 & 0 & & \\ 0 & (\sigma_\omega dt)^2 & 0 & 0_{3*3} & 0_{3*3} \\ 0 & 0 & (\sigma_\omega dt)^2 & & \\ 0_{3*3} & 0_{3*3} & 0_{3*3} & & \\ & & & (\sigma_a dt)^2 & 0 & 0 \\ 0_{3*3} & 0_{3*3} & 0 & (\sigma_a dt)^2 & 0 \\ & & 0 & 0 & (\sigma_a dt)^2 \end{bmatrix}. \tag{8}$$

With the adoption of the direction error, position error, and velocity error as the state values, calculate the error propagation according to the formula provided in Line 14. In Line 15, if a static state is detected, calculate the Kalman gain K first and then, utilize the velocity vector of the current state to calculate the error vector *delta_x*, which includes the direction error, the position error, and the velocity error. After that, construct an angular error skew-symmetric matrix $\mathbf{S}_e$ according to the value of the direction error:

$$\mathbf{S}_e = \begin{bmatrix} 0 & -attitude\_error(3,1) & attitude\_error(2,1) \\ attitude\_error(3,1) & 0 & -attitude\_error(1,1) \\ -attitude\_error(2,1) & attitude\_error(1,1) & 0 \end{bmatrix}. \tag{9}$$

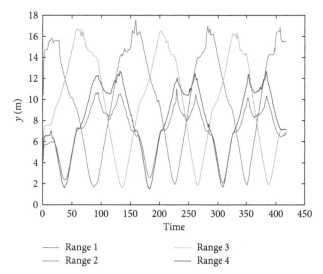

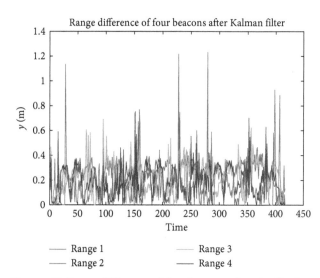

FIGURE 10: Range data obtained in four beacons after the filtering.

FIGURE 12: Range difference of four beacons after the filtering.

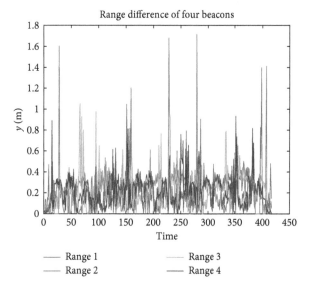

FIGURE 11: Range difference of four beacons before the filtering.

sum up this value with the absolute difference of the corresponding range that has been observed. The result can be taken as the return value of the cost function.

In Line 21, the coordinate transformation matrix **C** is corrected, while Line 22 corrects the current velocity and position value. Line 24 records the directional values in the walking process. Finally, Line 26 returns the positioning result obtained based on the foot-mounted IMU.

3.2. UWB Positioning Based on the Optimization Algorithm.

The UWB data-based positioning result can be calculated through the optimization algorithm with constraints, such as L-BFGS-B. Algorithm 3 gives the computation flow based on the constraint condition that the horizontal position coordinates are within the range of −40 to 40 m. Every time after the range data from the four beacons to the tag is obtained, use the L-BFGS-B algorithm to calculate the positioning result through the minimization of the cost function. In the cost function algorithm (4), first, calculate the range from each beacon to the current pose and then,

3.3. UWB Positioning Based on the PF Algorithm.

In order to track the pedestrian's moving status, this paper adopts the particle filter algorithm for the positioning. The computation flow is provided in Algorithm 5, where the first line is the initialization of the following variables, including the particle number, the initial position, the state noise variance, the evaluation of noise variance, the dynamic array of particle states, the array of particle scores, the weight array, the numerical fusion positioning result, and the array of static beacon coordinates. The state of every particle consists of (x, y), the current position of the particle. Line 2 is the initialization of the particle state. In other words, disperse the particles around the initial position based on the variance *sigma1*. Lines 3–9 are the course of position tracking based on the particle filter. In Line 4, Gaussian noise is added to the position of every particle in the *P_state*. Line 5 calculates the weight of every particle based on the range value acquired from the collected UWB data. As described by Algorithm 6, assume that there are n beacons available. Then, calculate the likelihood function of the kth UWB data observed at the moment according to the variance *sigma2* by taking the range from the current particle to a certain beacon as the mean value. Through the multiplication of n likelihood functions, get the product as the weight of the current particle. Line 7 performs the resampling of particle weight to update the particle state based on the value of the resampling.

3.4. IMU/UWB Fusion Positioning Based on the PF Algorithm.

As indicated by Algorithm 7, this paper adopts the IMU/UWB fusion positioning to reduce the positioning error caused by the deviation of UWB ranges. Lines 1-2 are the initialization of the particle number, the initial position, the variance of sampled noise, the evaluation of noise

TABLE 4: Statistics of range differences in four beacons before and after the filtering.

	The original UWB data				The filtered UWB data			
	Mean difference	Maximum difference	Minimum difference	Difference variance	Mean difference	Maximum difference	Minimum difference	Difference variance
Range 1	0.17	1.67	0	0.15	0.16	1.21	0.008	0.12
Range 2	0.17	1.71	0	0.20	0.16	1.21	0.0006	0.16
Range 3	0.17	1.05	0	0.16	0.16	0.69	0.0008	0.12
Range 4	0.17	0.93	0	0.15	0.16	0.70	0.001	0.12

Input: **ImuData**: the original imu value
Output: **zupt_result**: foot-mounted IMU-based positioning result
(1) Acquire the acceleration acc_s and the gyroscope data $gyro\_s$ from ImuData to initialize yaw, $pitch$, and $roll$
(2) Construct a coordinate transformation matrix $\mathbf{C}\_pre$ and indicate the zero bias of $gyro$ with $gyro\_bias$
(3) Initialize the acceleration, the velocity, and the position vector that are separately indicated by acc_n, vel_n, and pos_n in the navigation coordinate
(4) Initialize the acceleration noise $= 0.01$ m/s, the gyro noise $= 0.01$ rad/s, and the observation noise $= 0.01$ m/s
(5) Construct an observation matrix $\mathbf{H}$ and an observation noise matrix $\mathbf{R}$, and initialize the zero velocity detection threshold $gyro\_threshold = 0.6$ rad/s
(6) **For** t in $ImuData.cols()$
(7) $dt = timestamp(t) - timestamp(t-1)$
(8) $gyro\_s1 = gyro\_s(:,t) - gyro\_bias$
(9) Construct an angular velocity skew-symmetric matrix $\mathbf{S}_w$ and update the coordinate transformation matrix $\mathbf{C} = \mathbf{C}\_pre * (2\mathbf{I}_{3\times3} + \mathbf{S}_w \, dt) \, (2\mathbf{I}_{3\times3} - \mathbf{S}_w \, dt)^{-1}$
(10) $acc\_n(:,t) = 0.5 * (\mathbf{C} + \mathbf{C}\_prev) * acc\_s(:,t)$
(11) $vel\_n(:,t) = vel\_n(:,t-1) + ((acc\_n(:,t) - [0; 0; g]) + (acc\_n(:,t-1) - [0; 0; g])) * dt/2$
(12) $pos\_n(:,t) = pos\_n(:,t-1) + (vel\_n(:,t) + vel\_n(:,t-1)) * dt/2$
(13) Calculate the state transfer matrix $\mathbf{F}$ and the system error covariance matrix $\mathbf{Q}$
(14) $P = \mathbf{F} * P * \mathbf{F}' + \mathbf{Q};$
(15) **If** $norm(gyro\_s1(:,t)) < gyro\_threshold$
(16) $K = (P * (\mathbf{H})')/((\mathbf{H}) * P * (\mathbf{H})' + R)$
(17) $delta\_x = K * vel\_n(:,t)$
(18) $P = (\mathbf{I}_{9\times9} - K * \mathbf{H}) * P$
(19) $attitude\_error = delta\_x(1:3); pos\_error = delta\_x(4:6); vel\_error = delta\_x(7:9)$
(20) Construct an angular error skew-symmetric matrix $\mathbf{S}_e$
(21) $\mathbf{C} = (2 * \mathbf{I}_{3\times3} + \mathbf{S}_e)/(2 * \mathbf{I}_{3\times3} - \mathbf{S}_e * \mathbf{C})$
(22) $vel\_n(:,t) = vel\_n(:,t) - vel\_error; pos\_n(:,t) = pos\_n(:,t) - pos\_error$
(23) **End**
(24) $heading(t) = a\tan 2(\mathbf{C}(2,1), \mathbf{C}(1,1)); \mathbf{C}\_prev = \mathbf{C}$
(25) **End**
(26) **Return** pos_n

ALGORITHM 2: zupt_result = ZuptImu(ImuData).

Input: **UwbData**: the original UWB value;
Output: **uwb_opt_result**: the positioning result obtained through the UWB-based optimization algorithm
(1) $OptResult = []$, $initial\_pose = [0.0, 0.0]$
(2) **For** i in $UwbData.rows()$
(3) $range\_list = UwbData[i,:]$
(4) $res = minimize(CostFunction, initial\_pose, range\_list, bounds = ((-40, 40), (-40, 40)))$
(5) $initial\_pose = res$
(6) $OptResult[i,:] = res$
(7) **End**
(8) **Return** $OptResult$

ALGORITHM 3: uwb_opt_result = OptUwb(UwbData).

Input: **Pose**: current pose; **range_list**: list of ranges from the four beacons to the tag; **beacon_set**: coordinates of four beacons
Output: **Value**: the calculation result based on the cost function
(1) $val = 0.0$
(2) **For** i in $beaconset.rows()$
(3) $val += abs(norm(beaconset[i,:] - pose) - range\_list[i])$
(4) **End**
(5) **Return** val

ALGORITHM 4: Value = CostFunction(Pose, range_list, beacon_set).

Input: **UwbData**: UWB range data;
Output: **pf_result**: the positioning result obtained based on the particle filter algorithm
(1) $particle\_num = 300$; $init\_position = [0.0, 0.0]$; $sigma1 = 0.5$; $sigma2 = 0.5$; $P\_state = [particle\_num, 2]$;
   $score = [particle\_num]$; $weight = [particle\_num]$; $pf\_result = [UwbData.rows(), 2]$; $beacon\_set = [$Coordinates of n beacons$]$
(2) $P\_state = add\_noise(init\_position, sigma1)$
(3) **For** j in $UwbData.rows()$
(4) $P\_state += add\_noise(P\_state, sigma1)$
(5) $score = Evaluation(UwbData[j], P\_state, sigma2, beacon\_set)$
(6) $weight * = score$
(7) $P\_state, weight = Resample(P\_state, weight)$
(8) $pf\_result[j] = get\_result(P\_state, weight)$
(9) **End**
(10) **Return** pf_result

ALGORITHM 5: pf_result = PfUwb(UwbData).

Input: **UwbData**: UWB range data; **P_state**: particle state array; **sigma2**: the deviation value for particle evaluation; **beacon_set**: the coordinates of four beacons.
Output: **score**: the score of every particle after the evaluation
(1) **For** i in $particle\_num$
(2) $score[i] = 1.0$
(3)   **For** k in $beacon\_set.rows()$
(4) $score[i] * = normPdf(range\_list[i], norm(beacon\_set[i,:] - pose), sigma2)$
(5)   **End**
(6) **End**
(7) **Return** $score$

ALGORITHM 6: Score = Evaluation(UwbData[j], P_state, sigma2, beacon_set).

---

variance, the array of particle states, the array of particle scores, the weight array, the numerical fusion positioning result, and the array of static beacon coordinates. Line 3 utilizes the first 5 groups of UWB data to estimate the pedestrian's initial position through the triangle method. Centered by the initial position, particles are dispersed randomly based on the variance $sigma1$ to initialize the array of particle states. Lines 5~12 give the particle filter-based fusion process based on the ImuPath and UWB data. Line 6 acquires the increment of the trajectory between two adjacent UWB data points based on the ZuptImu algorithm. Line 7 takes the increment of the trajectory as the mean value to sum up the value of the P_state array according to the variance $sigma1$. Line 8 makes an evaluation on the particle state based on the currently acquired UWB data according to the principle similar to that used in Algorithm 6. Lines 9~10

indicate the resampling of particles and weight updating with the particle state to be updated based on the resampling value. Line 11 provides the final fused positioning result through the weighted calculation on the weight of the current particle.

3.5. IMU/UWB Positioning Based on the Integration of PF and EKF Algorithm. As indicated by Algorithm 8, this paper presents a positioning method through the integration of the particle filter with the Kalman filter. Lines 1-2 are the initialization of the following variables, including the IMU data counts, the UWB data counts, the number of particles, the initial position, the initial direction, the variance of state noise, the evaluation of noise variance, the pose of the particle, the score array, the weight array, the dynamic array

Input: **UwbData**: UWB range data; **ImuPath**: path calculated based on the ZuptImu algorithm with the fragmentation stored based on the timestamp of the UWB data.
Output: **pf_imu_uwb_result**: Uwb and Imu fusion positioning result
(1) *particle_num* = 300; *init_position* = [0, 0]; *sigma1* = 0.5; *sigma2* = 0.5; *P_state* = [*particle_num*, 2]; *score* = [*particle_num*]
(2) *weight* = [*particle_num*]; *FusingResult* = [*UwbData*.rows(),2]; *beaconset* = [the coordinates of *n* beacons]
(3) *init_position* = trianglepose(*BeaconSet*, *UwbData*[1:5,])
(4) *P_state* = add_noise(*init_position*, *sigma1*)
(5) **For** *j* in *UwbData*.rows()
(6)   *delta_pose* = *ImuPath*[*j*,:] − *ImuPath*[*j* − 1,:]
(7)   *P_state* = add_odo_noise(*P_state*, *delta_pose*, *sigma1*)
(8)   *score* = Evaluation(*UwbData*[*j*], *P_state*, *sigma2*)
(9)   *weight* ∗ = *score*
(10)   *P_state*, *weight* = Resample(*P_state*, *weight*)
(11)   *FusingResult*[*j*] = get_result(*P_state*, *weight*)
(12) **End**
(13) **Return** *FusingResult*

ALGORITHM 7: pf_imu_uwb_result = PfImuUwb(UwbData, ImuPath).

Lines 10–15 indicate that if IMU data is read first, then calculate every particle based on the foot-mounted IMU through the EKF. In Line 12, the *add_Noise* function represents the pose estimation after the Gaussian noise is added to the acceleration and gyrodata each time.

Lines 16–27 indicate that when UWB data is received, pose data will be updated based on the new observations. Lines 16–19 show the calculation on the weight of the current particle according to the newly acquired range value. Lines 20–23 indicate the updating of the pose array based on the new weight, and Line 24 records the UWB and foot-mounted IMU fused positioning results into the *FusingResult* array. Line 25 indicates the resampling of particle weight to update the particle state according to the value of resampling.

Input: **UwbData**: UWB range data; **ImuData**: IMU data; **ZuptData**: the zero velocity state of IMU data at a specific moment
Output: **pf_ekf_result**: the positioning result with the integration of Uwb and Imu
(1) *imu_index* = 0; *uwb_index* = 0; *particle_num* = 300; *init_postion* = [0, 0]; *init_heading* = 0; *sigma1* = 0.5; *sigma2* = 0.5; *pose* = [0, 0];
(2) *weight* = [*particle_num*]; *score* = [*particle_num*]; *Ekf_List* = [*particle_num*]; *beacon_set* = [the coordinates of *n* beacons];
(3) **For** *i* in *particle_num*
(4)   *Ekf_List*[*i*] = init_Nav_Eq(*ImuData*[1:20], *init_postion*, *init_heading*)
(5) **End**
(6) **while** *true*
(7)   **If** *imu_index* == *ImuData*.rows() || *uwb_index* == *UwbData*.rows()
(8)   **break**;
(9)   **End**
(10)   **If** time(*UwbData*(*uwb_index*)) > time(*ImuData*(*imu_index*))
(11)   **For** *i* in *particle_num*
(12)   *pose*[*i*] = *Ekf_List*[*i*].get_Position(add_Noise(*ImuData*, *ZuptData*, *sigma1*))
(13)   **End**
(14)   *imu_index*++
(15)   **Else**
(16)   **For** *i* in *particle_num*
(17)   *score*[*i*] = *Ekf_List*[*i*].Evaluation(*beaconset*, *UwbData*(*uwb_index*), *sigma2*)
(18)   *weight*[*i*] ∗ = *score*[*i*]
(19)   **End**
(20)   *pose_fusing* = [0,0,0]; *weight*[*i*] = *weight*[*i*]/Sum(*weight*[*i*])
(21)   **For** *i* in *particle_num*
(22)   *pose_fusing* += *pose*[*i*] ∗ *weight*[*i*]
(23)   **End**
(24)   *FusingResult*[*uwb_index*] = *pose_fusing*
(25)   *Ekf_List*, *weight* = Resample(*Ekf_List*, *weight*)
(26)   *uwb_index*++;
(27)   **End**
(28) **End**
(29) **Return** *FusingResult*

ALGORITHM 8: pf_ekf_result = PfEkfImuUwb(UwbData, ImuData, ZuptData).

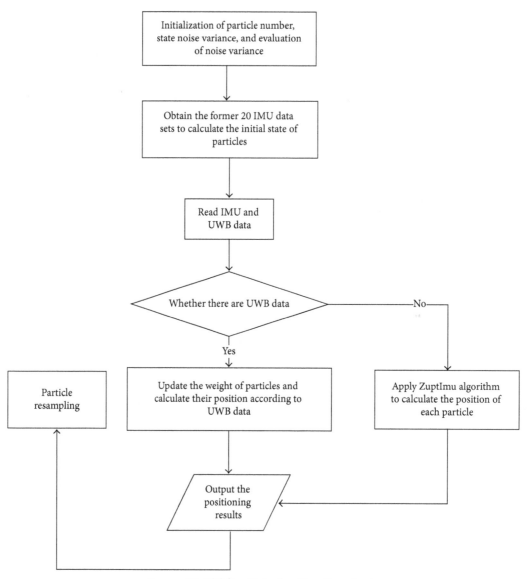

FIGURE 13: PfEkfImuUwb algorithm flow chart.

FIGURE 14: Foot-mounted IMU.

of particle states, and the array of static beacon coordinates. Every particle consists of the computational nodes on the path of the ZUPT-based foot-mounted IMU. That is to say, every particle can make the real-time calculation on the

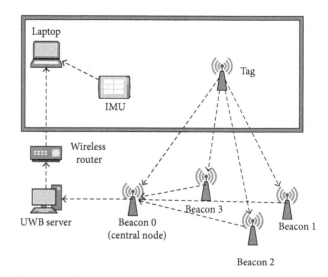

FIGURE 15: Data communication among experimental facilities.

FIGURE 16: Experiment with a laptop in the hand.

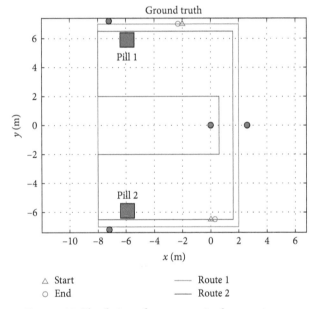

FIGURE 17: The design of two routes in the experiment.

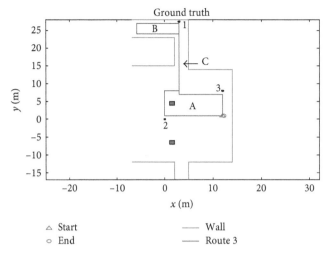

FIGURE 18: Route 3 in the experiment.

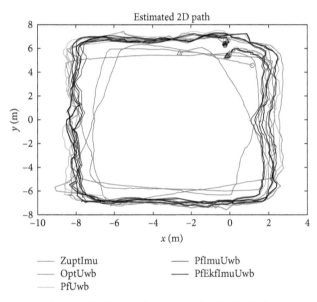

FIGURE 19: The positioning results for Route 1.

movement path according to the EKF algorithm. Lines 3–5 are the initialization of the particle state, or in other words, the estimation on the pose based on the first 20 groups of IMU data at the initial moment. Lines 7–9 indicate that the program quits if the IMU data or UWB data have been read.

The algorithm flow chart is shown in Figure 13.

4. Experiment

The experimental field was established in the entrance hall of School of Computer Science and Technology in China University of Mining and Technology, where the tester had been mounted with an IMU device X-IMU produced by a UK company X-IO on the foot, with the output frequency

of 128 Hz, as shown in Figure 14. The data communication between the UWB positioning system and IMU is shown in Figure 15, in which there are four positioning beacons (Beacon 0~3) and one positioning tag, they are connected via wireless links, and their ranging data are transmitted to the UWB server by Beacon 0. After the data are preprocessed by the UWB server, they are transmitted to laptop via WiFi. Meanwhile, the IMU data on the foot are also transmitted to laptop via Bluetooth. So, the UWB data and IMU data can be transmitted to laptop for time synchronization. In Figure 15, experimental facilities in the solid box are all carried by testers, in which the tag is installed on a helmet, the IMU is mounted on a foot, and the laptop is held by a tester, as indicated in Figure 16, who walks at a constant velocity.

Three routes have been designed in the experiment. As indicated in Figure 17, there are Route 1, a rectangular route with fewer turns, and Route 2, a polygonal route with more turns. Route 3, as shown in Figure 18, may lose some UWB data during the walking process. There are four beacons

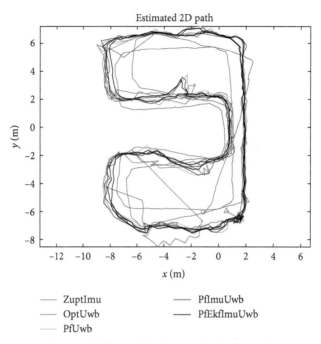

FIGURE 20: The positioning results for Route 2.

TABLE 5: Error comparison and analysis based on the five algorithms.

| | Route 1 | | | Route 2 | | |
	Mean error	Maximum error	Minimum error	Mean error	Maximum error	Minimum error
ZuptImu	0.987	3.405	0.012	0.789	2.825	0.006
OptUwb	0.596	2.011	0.003	0.751	6.678	0.001
PfUwb	0.624	2.067	0.009	0.527	1.499	0.003
PfImuUwb	0.637	2.087	0.001	0.531	1.462	0.001
PfEkfImuUwb	0.685	2.576	0.003	0.505	1.356	0.009

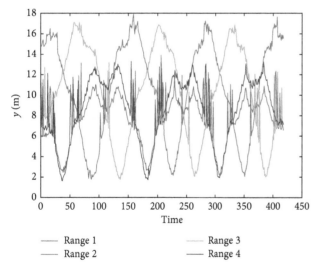

FIGURE 21: The UWB ranges in Route 1 under the interference.

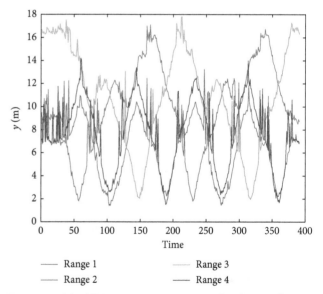

FIGURE 22: The UWB ranges in Route 2 under the interference.

established in Figure 17, three beacons established in Figure 18, and their positions are indicated by red dots. Also, in Figures 17 and 18, there were two red blocks representing both of the marble columns in the hall as shown in Figure 1.

Three scenarios have been set up in the experiment. The first is the UWB and IMU fused positioning test without any interference from the pedestrian. The second is the UWB

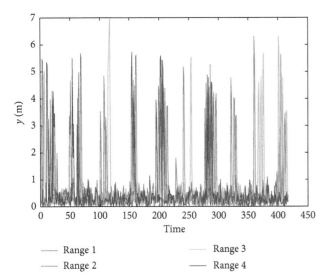

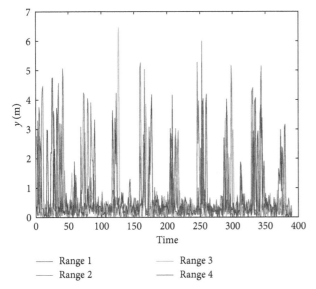

FIGURE 23: The difference in the adjacent ranges from four beacons in Route 1.

FIGURE 24: The difference in the adjacent ranges from four beacons in Route 2.

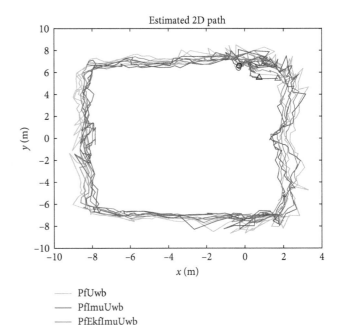

FIGURE 25: The positioning results in Route 1.

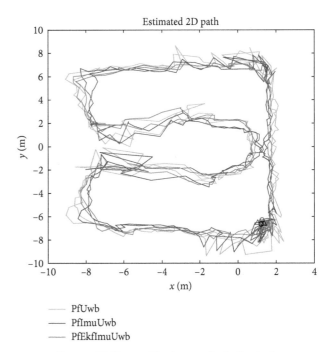

FIGURE 26: The positioning results in Route 2.

and IMU fused positioning test in the context of intentional interference from multiple persons. The third is the experiment performed under the situation of losing some UWB data. The performance and the reliability of the algorithm can be assessed through the calculation of the positioning errors in the various schemes.

4.1. Analysis on the Positioning Paths without Pedestrian Blockage.

In the entrance hall of School of Computer Science and Technology, walk along the route marked in red for three circles and along the route marked in blue for two circles as indicated in Figure 17, where the start and end of each route has been marked out. Figures 19 and 20 demonstrate the positioning paths based on the various algorithms with the positioning errors provided in Table 5 after the calculation.

Based on the original IMU data, Scheme I denoted as ZuptImu is indicated by the pink path in Figures 19 and 20. Due to the accumulative error in IMU data, the positioning result obtained based on this scheme will deviate from the real path with the mean error up to 0.987 m and the maximum error at 3.405 m.

Based on the optimization algorithm, Scheme II denoted as OptUwb is indicated by the red path in Figures 19 and 20. As most of the path agrees well with the real trajectory based on this algorithm, big error will arise in the positioning result when signals are blocked. Or in other words, the optimization

TABLE 6: Error comparison and analysis based on the three approaches.

	Route 1			Route 2		
	Mean error	Maximum error	Minimum error	Mean error	Maximum error	Minimum error
PfUwb	0.696	2.981	0.001	0.587	2.299	0.017
PfImuUwb	0.735	2.896	0.003	0.571	1.816	0.007
PfEkfImuUwb	0.624	2.576	0.003	0.527	1.524	0.008

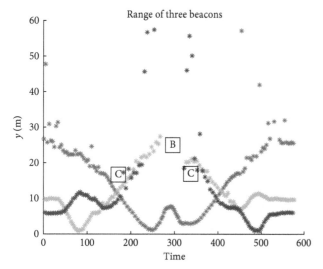

FIGURE 27: The UWB ranges in Route 3.

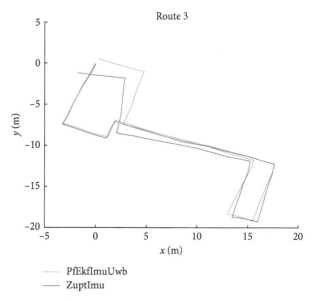

FIGURE 28: The positioning results in Route 3 based on the two algorithms.

algorithm fails to converge to the correct result. For example, in Route 2, the maximum positioning error reaches 6.678 m, which indicates that the trajectory has deviated from the real path.

Based on the UWB signals, Scheme III is denoted as PfUwb with the use of the particle filter for the positioning. Based on this scheme, the current positioning result can be corrected according to the range value acquired from the four beacons. With this algorithm, the mean error in Route 1, and Route 2 is separately 0.624 m and 0.527 m. It proves that the positioning result is quite stable.

As indicated in Figures 19 and 20, Schemes IV and V, respectively, denoted as PfImuUwb and PfEkfImuUwb are indicated by the blue and black paths. Both of the schemes can guarantee a stable positioning result in Route 1 and Route 2 with the mean positioning error approximate to that in Scheme III.

4.2. Analysis on the Positioning Path in NLOS Condition.

Section 4.1 reveals that all of these three schemes, including PfUwb, PfImuUwb, and PfEkfImuUwb can steadily implement the calculation on the path when there is no interference from the pedestrian. In order to verify the stability of these three algorithms without pedestrian interference, we had three pedestrians to move about in the experimental area. Figures 21 and 22 show the UWB ranges in Route 1 and Route 2 with pedestrian interference. Figures 23 and 24 demonstrate the difference in adjacent ranges from four beacons, revealing that pedestrian blockage will lead to lots of errors in range data. Figures 25 and 26 show the different positioning trajectories obtained based on the various algorithms with the positioning errors provided in Table 6.

In the context of signal interference, great deviation will arise with the PfUwb algorithm. Especially when the pedestrian that causes signal interference is close to the beacon, the sudden change in signal transmission will lead to a big leap in the positioning result. As indicated by the green path in Figures 25 and 26, the mean error is up to 0.696 m in Route 1 with a maximum error of 2.981 m. In Route 2, the mean error reaches 0.587 m and the maximum error is 2.299 m.

As indicted by the blue path in Figures 25 and 26, the PfImuUwb algorithm can alleviate to some extent the positioning error arising in the PfUwb algorithm with the aid of the IMU positioning result. However, this algorithm has very limited deviation correction ability. Therefore, in most cases, the positioning result obtained through this algorithm will be affected by the UWB signals to make the positioning result similar to that through the PfUwb algorithm. For example, the maximum positioning error in Route 1 reaches 2.896 m through this method, and significant distortion can be found on part of the path.

As the PfEkfImuUwb algorithm can utilize every particle to maintain the IMU-based EKF positioning and tracking, the positioning result is equivalent to the integration of several results obtained in multiple positioning paths to weaken the influence from abnormal UWB signals. On this regard, the positioning result through this method is comparatively smoother. Meanwhile, the mean error in Route 1 and 2 is separately 0.624 and 0.527 m, which also proves that this algorithm can guarantee a stable positioning result.

4.3. Analysis on the Positioning Path in NLOS Condition. Route 3 starts from Zone A, passes Point C, and reaches Zone B. With a clockwise walk, it finally goes back to the starting point. As shown in Figure 27, there are two areas where the loss of UWB data occurs during the walking process. One is that when the tester gets to Point C, the data exception of Beacon 3 occurs due to the occlusion of walls, and these abnormal data can be masked by the filtering algorithm. The other is that when the tester walks into Zone B, the ranging signals of Beacons 2 and 3 cannot be received due to the occlusion of walls as well as the increase of distance.

The positioning results of the ZuptImu algorithm and PfEkfImuUwb algorithm are shown in Figure 28. The main disadvantage of the ZuptImu algorithm is that there is a big deviation in direction computation. After turning several corners, the deviation is bigger and bigger, and the final closing error reaches 2.3 m, with an obvious mismatch between positioning trajectories and actual trajectories. But, in the PfEkfImuUwb algorithm, each particle maintains EKF positioning and tracking based on the IMU, which guarantees a small motion deviation of each particle. Under the situation of losing some UWB data, the motion angle at the point of error particles is corrected through the constraint of only one UWB beacon in Zone B and the ranging constraint of two UWB beacons when passing through the straight line of Point C, with a almost overlap between positioning trajectories and actual trajectories.

5. Conclusions

This paper presents a UWB and foot-mounted IMU fusion positioning method through the integration of PF with EKF. Although this algorithm can achieve good positioning result in the context of pedestrian blockage, it needs to be further improved and perfected in terms of the followings: on the one hand, as path calculation based on the EKF is maintained by every particle, it undoubtedly will increase the computational burden, which can be solved through the parallel algorithm; on the other hand, some variables that are deduced from the IMU algorithm, such as velocity, acceleration, angle, and angular velocity, can be added into the particle state, and since they are equivalent to the addition of a constraint of uniform change, a better effect can be achieved theoretically. This positioning is based on the detection method of ArUCO beacons, and its accuracy can reach about 7 cm after optimization of the backend system [29]. At present, this algorithm is still under study.

Authors' Contributions

The corresponding author Yan Wang proposed the research and was involved in the writing of the manuscript. Xin Li designed the experiment, performed the data analysis, and drafted the manuscript. Dawei Liu was responsible for data collection and was involved in data processing.

Acknowledgments

This work was supported by the National Natural Science Foundation of China under Grant no. 41674030, China Postdoctoral Science Foundation under Grant no. 2016M601909, and grant of China Scholarship Council.

References

[1] X. Li, J. Wang, C. Liu, L. Zhang, and Z. Li, "Integrated WiFi/PDR/smartphone using an adaptive system noise extended Kalman filter algorithm for indoor localization," *ISPRS International Journal of Geo-Information*, vol. 5, no. 2, p. 8, 2016.

[2] X. Li, J. Wang, and C. Liu, "A Bluetooth/PDR integration algorithm for an indoor positioning system," *Sensors*, vol. 15, no. 10, pp. 24862–24885, 2015.

[3] R. Mautz, "The challenges of indoor environments and specification on some alternative positioning systems," in *Proceedings of the Workshop on Positioning, Navigation and Communication (WPNC 2009)*, pp. 29–36, Hannover, Germany, March 2009.

[4] J. Chóliz, A. Hernández, and A. Valdovinos, "A framework for UWB-based communication and location tracking systems for wireless sensor networks," *Sensors*, vol. 11, no. 9, pp. 9045–9068, 2011.

[5] F. Zampella, A. D. Angelis, I. Skog, and D. Zachariah, "A constraint approach for UWB and PDR fusion," in *Proceedings of the International Conference on Indoor Positioning and Indoor Navigation*, vol. 41, pp. 1–9, Sydney, NSW, Australia, November 2012.

[6] D. Gusenbauer, C. Isert, and J. Krosche, "Self-contained indoor positioning on off-the-shelf mobile devices," in *Proceedings of the International Conference on Indoor Positioning and Indoor Navigation (IPIN 2010)*, pp. 1–9, Zurich, Switzerland, 2010.

[7] E. Foxlin, "Pedestrian tracking with shoe-mounted inertial sensors," *IEEE Computer Graphics and Applications*, vol. 25, no. 6, pp. 38–46, 2005.

[8] J. O. Nilsson, I. Skog, P. Handel, and K. V. S. Hari, "Foot-mounted INS for everybody—an open-source embedded implementation," in *Proceedings of the IEEE/ION Plans 2012*, vol. 1, pp. 140–145, Myrtle Beach, SC, USA, April 2012.

[9] J. O. Nilsson, I. Skog, and P. Händel, "A note on the limitations of ZUPTs and the implications on sensor error modeling," *Signal Processing*, 2012.

[10] J. O. Nilsson and P. Händel, "Standing still with inertial navigation," in *Proceedings of the International Conference on Indoor Positioning and Indoor Navigation (IPIN)*, Montbéliard, France, October 2013.

[11] J. O. Nilsson, D. Zachariah, I. Skog, and P. Händel, "Cooperative localization by dual foot-mounted inertial sensors and inter-agent ranging," *EURASIP Journal on Advances in Signal Processing*, vol. 2013, no. 1, pp. 164–181, 2013.

[12] I. Skog, J. O. Nilsson, and P. Händel, "Pedestrian tracking using an IMU array," in *Proceedings of the IEEE International Conference on Electronics, Computing and Communication Technologies*, pp. 1–4, IEEE, Bangalore, India, January 2014.

[13] B. Wagstaff, V. Peretroukhin, and J. Kelly, "Improving foot-mounted inertial navigation through real-time motion

classification," in *Proceedings of the International Conference on Indoor Positioning and Indoor Navigation (IPIN 2017)*, Sapporo, Japan, September 2017.

[14] B. Subhojyoti, K. Amit, and P. Gupta, "On the noise and power performance of a shoe-mounted multi-IMU inertial positioning system," in *Proceedings of the International Conference on Indoor Positioning and Indoor Navigation (IPIN 2017)*, Sapporo, Japan, September 2017.

[15] S. Sczyslo, J. Schroeder, S. Galler, and T. Kaiser, "Hybrid localization using UWB and inertial sensors," in *Proceedings of the IEEE International Conference on Ultra-Wideband (ICUWB 2008)*, vol. 3, pp. 89–92, Hanover, Germany, September 2008.

[16] S. Pittet, V. Renaudin, and B. Merminod, "UWB and MEMS based indoor navigation," *Journal of Navigation*, vol. 61, no. 3, pp. 369–384, 2008.

[17] J. D. Hol, F. Dijkstra, H. Luinge, and T. B. Schon, "Tightly coupled UWB/IMU pose estimation," in *Proceedings of the IEEE International Conference on Ultra-Wideband (ICUWB 2008)*, pp. 688–692, Vancouver, BC, Canada, September 2009.

[18] A. Savioli, E. Goldoni, P. Savazzi, and P. Gamba, "Low complexity indoor localization in wireless sensor networks by UWB and inertial data fusion," *Computer Science*, vol. 52, no. 4, pp. 723–732, 2013.

[19] C. Ascher, L. Zwirello, T. Zwick, and G. Trommer, "Integrity monitoring for UWB/INS tightly coupled pedestrian indoor scenarios," in *Proceedings of the International Conference on Indoor Positioning and Indoor Navigation (IPIN 2011)*, pp. 1–6, Guimarães, Portugal, September 2011.

[20] Q. Fan, Y. Wu, J. Hui, L. Wu, Z. Yu, and L. Zhou, "Integrated navigation fusion strategy of INS/UWB for indoor carrier attitude angle and position synchronous tracking," *Scientific World Journal*, vol. 2014, Article ID 215303, 13 pages, 2014.

[21] J. Wang, Y. Gao, Z. Li, X. Meng, and C. M. Hancock, "A tightly-coupled GPS/INS/UWB cooperative positioning sensors system supported by V2I communication," *Sensors*, vol. 16, no. 7, p. 944, 2016.

[22] A. Benini, A. Mancini, and S. Longhi, "An IMU/UWB/vision-based extended Kalman filter for mini-UAV localization in indoor environment using 802.15.4a wireless sensor network," *Journal of Intelligent and Robotic Systems*, vol. 70, no. 1–4, pp. 461–476, 2013.

[23] H. E. Nyqvist, M. A. Skoglund, G. Hendeby, and F. Gustafsson, "Pose estimation using monocular vision and inertial sensors aided with ultra wide band," in *Proceedings of the International Conference on Indoor Positioning and Indoor Navigation (IPIN 2015)*, pp. 1–10, Banff, AB, Canada, October 2015.

[24] L. Zwirello, C. Ascher, G. F. Trommer, and T. Zwick, "Study on UWB/INS integration techniques," in *Proceedings of the Workshop on Positioning Navigation and Communication (WPNC 2011)*, pp. 13–17, Dresden, Germany, April 2011.

[25] Y. Xu, X. Chen, J. Cheng, Q. Zhao, and Y. Wang, "Improving tightly-coupled model for indoor pedestrian navigation using foot-mounted IMU and UWB measurements," in *Proceedings of the Instrumentation and Measurement Technology Conference (I2MTC 2016)*, IEEE, Taipei, Taiwan, May 2016.

[26] Y. Leehter, W. Yeong, and W. Andy, "An integrated IMU and UWB sensor based indoor positioning system," in *Proceedings of the International Conference on Indoor Positioning and Indoor Navigation (IPIN)*, Sapporo, Japan, September 2017.

[27] R. Liu, C. Yuen, T. N. Do, D. Jiao, X. Liu, and U. X. Tan, "Cooperative relative positioning of mobile users by fusing IMU inertial and UWB ranging information," in *Proceedings of the IEEE International Conference on Robotics and Automation (ICRA 2017)*, pp. 5623–5629, IEEE, Piscataway, NJ, USA, 2017.

[28] C. Fischer, P. T. Sukumar, and M. Hazas, "Tutorial: implementing a pedestrian tracker using inertial sensors," *IEEE Pervasive Computing*, vol. 12, no. 2, pp. 17–27, 2013.

[29] J. Bacik, F. Durovsky, P. Fedor, and D. Perdukova, "Autonomous flying with quadrocopter using fuzzy control and ArUco markers," *Intelligent Service Robotics*, vol. 10, no. 3, pp. 185–194, 2017.

7

UAV-Assisted Data Dissemination in Delay-Constrained VANETs

Xiying Fan [ID],[1,2] **Chuanhe Huang** [ID],[1,2] **Bin Fu**,[3] **Shaojie Wen** [ID],[1,2] **and Xi Chen**[1,2]

[1]*School of Computer Science, Wuhan University, Wuhan, China*
[2]*Collaborative Innovation Center of Geospatial Technology, Wuhan, China*
[3]*Department of Computer Science, The University of Texas Rio Grande Valley, Edinburg, TX, USA*

Correspondence should be addressed to Chuanhe Huang; huangch@whu.edu.cn

Academic Editor: Nicola Bicocchi

Due to the high mobility of vehicles, the frequent path failures caused by dynamic network topology, and a variety of obstructions, efficient data dissemination with delay constraint in vehicular ad hoc networks (VANETs) is a challenging issue. To address these problems, a novel mobile relaying technique by employing unmanned aerial vehicles (UAVs) is considered to facilitate data dissemination in vehicular environments where the communication infrastructures are not available or the network connectivity is poor. This paper studies and formulates the throughput maximization problem in UAV-assisted VANETs, which aims to achieve high throughput while guarantee the delay constraint of data flows to the vehicles in the area. To maximize the network throughput, the maximization problem tries to find an optimal delivery strategy for data dissemination by optimizing the transmission rate. To solve the problem, the knapsack problem can be reduced to the maximization problem, which is proved NP-hard. A polynomial time approximation scheme is proposed to achieve an approximate solution. Detailed theoretical analysis including time complexity and approximation ratio of the proposed algorithm is presented. Simulation results demonstrate the effectiveness of the proposed algorithm.

1. Introduction

As important components of Intelligent Transportation System (ITS), vehicular ad hoc networks (VANETs) are large-scale mobile ad hoc networks composed of vehicles with communication functions and roadside infrastructures, which aim to provide services for autonomous driving and high-speed information sharing [1, 2]. In VANETs, drivers mainly obtain real-time road conditions and safety information sent by other vehicles through wireless communication technology. In this way, traffic accidents and road congestion can be effectively avoided while travel time and energy consumption can be reduced. Meanwhile, VANETs can provide information services, such as news and entertainment, which can add fun to the boring journey.

However, VANETs have some unique characteristics that other ad hoc networks do not share, such as high vehicle mobility, dynamic network topology, and intermittent network connectivity. These features bring a variety of challenges to data dissemination. To deal with the issues,

unmanned aerial vehicles (UAVs) can be utilized to cooperate with VANETs. Compared to traditional terrestrial wireless communications, UAV-enabled communications are significantly less affected by channel impairments such as shadowing and fading and in general possess more reliable air-to-ground channels due to higher possibility of having line-of-sight (LoS) links with ground users [3]. Additionally, in the areas where the infrastructures are difficult or too costly to install and maintain to provide ideal network coverage, UAVs can serve as a viable option, as they can collect information from an area of interest and transmit the information to ground VANETs [4, 5]. They can also act as relays to ground networks when direct multihop communications are not available.

Considering the advantages of UAVs, a cooperative hybrid network framework is proposed, integrating UAVs with ground vehicles for data dissemination in VANETs. In the studied scenario, a vehicle carries a message and aims to transmit the message to a target area where exist a number of vehicles and UAVs. To complete the transmission, the message can be either transmitted over vehicle-to-vehicle (V2V) links, vehicle-to-infrastructure (V2I), or air-to-ground (A2G) communication links. To improve the performance of data dissemination, it should transmit data as much as possible in a specific period, which means to maximize the network throughput. Transmission rate and transmission delay of the links over which data is transmitted are utilized to reflect the throughput. Therefore, to achieve the maximum throughput is equivalent to maximize the sum of transmission rate of selected links on data delivery path. As there may exist more than one path from source to destination, the study aims to select a path with the maximum throughput while satisfying a predefined delay threshold.

Graph theory is applied to abstract the network as a connected graph, then the well-known 0/1 knapsack problem can be reduced to the throughput maximization problem. Due to the property of transmission in VANETs, the problem is regarded as the graph knapsack problem which is one of the classical NP-complete problems [6]. Then, a polynomial time approximation algorithm for the graph knapsack problem is derived based on the approximation scheme for subset sum problem [7]. Since the throughput maximization problem can be reduced from the graph knapsack problem, the proposed approximation algorithm can be applied to the maximization problem and to obtain an end-to-end path with the maximum throughput.

The main contributions of this paper are described as below.

(i) A throughput maximization problem in delay-constrained UAV-assisted VANETs is formulated, which considers the tradeoff of data transmission rate and transmission delay. Then, a multiedge graph knapsack problem is constructed based on 0/1 knapsack problem and reduced to the throughput maximization problem, which is proved to be NP-hard.

(ii) A polynomial time approximation scheme is developed for the multiedge graph knapsack problem to obtain the approximate solution. In the proposed scheme, the edges and vertices are assigned with values to indicate their weight. To select a path with the maximum weight, a trim procedure is applied to remove the unnecessary values. Theoretical analysis proves that the algorithm runs in polynomial time with a bound which is polynomial in the size of the input and $1/\epsilon$, where ϵ denotes the approximation parameter. Additionally, the approximation ratio caused by trimming the unnecessary edges in path selection is also derived as $1 + \epsilon$. The results can be applied to general graph knapsack problem.

(iii) An efficient data dissemination algorithm based on the approximate scheme for the graph knapsack problem is proposed to solve the throughput maximization problem. The values of edges in the graph knapsack problem correspond to the transmission rate and delay of the links. Considering the approximation in the knapsack problem, the proposed algorithm for the maximization problem has a quadratic approximation, which is the combination of the approximation to obtain the optimal transmission rate of links and the approximation to trim the unnecessary edges when selecting the path with the maximum throughput. The time complexity and approximation ratio of the proposed algorithm are also given.

The remainder of the paper is organized as follows. Section 2 overviews the related work. Section 3 describes system model and problem formulation. Section 4 develops a polynomial time approximation scheme for the graph knapsack problem, based on which an algorithm for the throughput maximization problem is proposed. Performance evaluation is presented in Section 5. Section 6 concludes the paper. Finally, Section 7 discusses the tradeoff between the benefit and cost of employing UAVs and gives the direction of future work.

2. Related Work

Lots of research has been done to achieve data dissemination with high efficiency in vehicular networks, most of which is devoted to analyzing the performance of delay, throughput and utility of data dissemination [8]. In this section, data dissemination in ground VANETs and UAV-assisted VANETs is mainly discussed.

2.1. Data Dissemination in Ground VANETs. Tan et al. [9] proposed an analytical model to characterize the downlink average throughput and distribution achieved for each vehicle during the sojourn time by the Markov reward model. Zhang et al. [10] proposed an analytical model to facilitate the real-time data delivery as well as delay-tolerant data delivery, in which the theoretical per-vehicle throughput was derived. Lin et al. [11] developed an analytical model that accurately characterized the maximum throughput rate performance achievable under a prescribed outage probability constraint. As the first study on reliable transmission for bulk or stream-like data in DTNs (delay tolerant networks) [12], Zeng et al. proposed a dynamic segmented network coding scheme to efficiently exploit the transmission opportunity. Xing et al. [13] formulated the multimedia scheduling problem to maximize the utility and designed a heuristic algorithm. As continuous research, the authors [14] investigated multimedia dissemination for large-scale VANETs considering the tradeoff of delivery delay, the quality of service (QoS) of delivered data, and the storage cost.

As an emergent paradigm, some research work has applied SDN (Software Defined Network) to support applications in DTNs while reduce the operating costs [15] as it separates the control and data communication layers to simplify the network management. Liu et al. [16] described the application of SDN concept in VANETs and studied data scheduling problem. Nobre et al. [17] defined an architecture that adapted SDN to battlefield networking (BN), which integrated BN and SDN into dynamic and heterogeneous network-centric environments. Zacarias et al. [18] combined SDN and DTN concepts to address the needs of tactical-operational networks, which could support the diverse range of strict requirements for applications.

2.2. UAV-Assisted Wireless Communications. To provide wireless communications to a given geographical area, Mozaffari et al. [19] analyzed the deployment of an UAV as a flying base station and derived an analytical framework for the coverage and rate analysis for the device-to-device communication network. Then they investigated the optimal 3D deployment of multiple UAVs [20] to maximize the downlink coverage performance with a minimum transmit power. Orfanus et al. [21] utilized the self-organizing paradigm to design efficient UAV relay networks, to provide robust connections to the devices on the military field. Oubbati et al. [22] proposed a UAV-assisted routing protocol to assist data dissemination and improve the reliability of data delivery by filling the communication gap. Wang et al. [23] studied hybrid VANETs that utilized on-vehicle drones and proposed a distributed location-based routing protocol. Xiao et al. [24] employed UAVs to improve network performance against smart jammers and formulated the interaction between UAVs and jammers as an anti-jamming UAV relay game. Seliem et al. [25] proposed a mathematical framework to obtain the minimum drone density, which was equivalent to the maximum separation distance between two adjacent drones, to limit the worst delay of vehicle-drone packet transmissions. Shilin et al. [26] considered a drone-aided communication network model in an isolated VANET segment to enhance network connectivity. Fawaz et al. [27] developed a mathematical model that utilized drones to evaluate the impact of non-cooperative vehicles on forwarding path availability.

Most of the related work did not consider maximizing the network throughput taking consideration of delay constraint in UAV-aided vehicular networks, which motivates this research.

3. System Model and Problem Formulation

In this section, the system model is presented while the maximization problem is formulated.

3.1. System Model. To improve the reliability and efficiency of data dissemination in VANETs, UAVs are employed to form a cooperative air-to-ground network. By exploiting the UAV-aided VANETs, UAVs can help ground vehicles explore the area of interest and enhance network connectivity.

As stated in [22], most urban applications that use UAVs like small Quad-Copters do not fly at high altitudes [28]. Thus, this study assumes that UAVs have a low and constant altitude during the flight in order to communicate with vehicles on the ground. IEEE 802.11p MAC protocol is adopted for both V2V and A2G communications. UAVs in the network use a large transmission range (i.e., up to 1000 m [29]) and have a global view of the network. Vehicles and UAVs are equipped with GPS and digital maps to obtain

their geographical positions. UAVs can also act as relay nodes to forward data packets when direct multihop V2V links are not available.

The cooperative network architecture of the UAV-assisted VANETs is depicted as Figure 1, which is composed of the UAV network and the ground vehicular network. The scenario includes A2G and V2V communication links, which is a hybrid mode that allows the network to apply both A2G and V2V communications for data dissemination in VANETs.

The network can be abstracted as an edge-weighted graph $G(V, E)$ (see Definition 1), where V is a set of vehicles and UAVs, E is a set of edges to indicate the communication links for data dissemination. The weight of each edge is represented by the transmission condition of the corresponding link.

Definition 1. Given a weighted network graph $G(V, E)$, where V is the set of vertices and E is the set of edges. Let (w_e, d_e) denotes the value of edge e, where w_e indicates the transmission rate of e, and d_e indicates the transmission delay. The tuple (W_i, D_i) denotes the value of node $i \in G \cdot V$, where W_i and D_i indicates the total transmission rate and transmission delay from source node s to i, respectively.

Note that $G(V, E)$ only considers the edges over which two nodes can communicate, which means no silent edges are included.

3.2. Problem Formulation. Assume a packet with size K carried by vehicle s needs to be transmitted to a specific area. There may exist more than one end-to-end path from source node s to other nodes in V, which can be denoted by P_s and p indicates a path in P_s. It is important to note that the paths may only exist among vehicles through V2V links or they may contain a hybrid of A2G and V2V communications. For simplicity, the A2G and V2V links are considered as common links with different properties hereinafter. The differences of the links are reflected by their transmission rate and delay.

As the network throughput can be mapped by the transmission rate of the end-to-end path, this study discusses how to optimize the transmission rate of each individual path to achieve the maximum throughput. To guarantee the real-time transmission, the end-to-end delay is limited to a predefined threshold. A continuous convex function $f(r_l) = \log r_l$ with the transmission rate as parameter is utilized to depict the throughput, where l denotes a link on path p and r_l denotes the transmission rate of link l. The reason to consider $f(r_l)$ instead of r_l is that the logarithmic utility function $\log r_l$ can better reflect the transmission rate of the delivery path and guarantee the maximum transmission rate. Meanwhile, the logarithm is concave and, hence, has diminishing returns. Here, it seeks a utility for that naturally achieves the maximum throughput and some level of fairness among the links.

To optimize the transmission rate of links and improve the network throughput, the throughput maximization problem can be formulated as below:

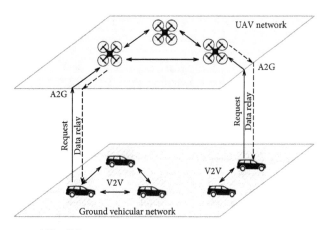

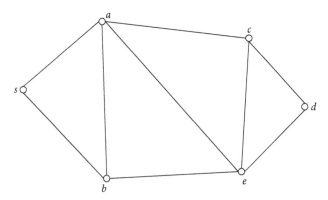

FIGURE 2: An example of an undirected graph $G(V, E)$ with six vertices, each edge connecting the nodes is assigned with weight (w_e, d_e), which helps illustrate how to find a path from s to d.

FIGURE 1: An overview of the cooperative air-to-ground network architecture.

$$\max \quad \sum_{l \in L} f(r_l) \cdot x_l,$$

$$\text{s.t.} \quad \sum_{l \in L} d_l \cdot x_l \leq \delta, \tag{1}$$

where $r_l \in (1, c_l]$, c_l indicates the maximum capacity of link l, d_l is the transmission delay of link l, and δ is the predefined delay threshold. If link l is selected, x_l is equal to 1, otherwise, x_l is 0. Transmission delay d_l can be calculated by the following equation according to the channel model [30]:

$$d_l = K / (c_l - r_l). \tag{2}$$

The problem can be stated as follows. Given a delay threshold δ and n pairs of positive values (r_l, d_l) to indicate the transmission rate and transmission delay of link l, it aims to select a delivery path which contains a few links to maximize the transmission rate while satisfying the delay constraint. The well-known 0/1 knapsack problem can be reduced to the throughput maximization problem. Then, a polynomial time approximation scheme is proposed to solve the problem.

3.3. An Example. An example is given to illustrate how to derive an approximation solution for the maximization problem. An undirected graph $G(V, E)$ with six vertices is shown in Figure 2. In graph G, each edge has a pair of values (w_e, d_e) and each node has (W_i, D_i) as its weight. The goal is to find a path from source node s to node d, such that the path has the maximum transmission rate W_d while the transmission delay D_d does not exceed δ. The procedure to obtain the path utilizing the approximation method is described.

First, the weight of each edge is given as (w_{sa}, d_{sa}), (w_{sb}, d_{sb}), (w_{ab}, d_{ab}), (w_{ac}, d_{ac}), (w_{be}, d_{be}), (w_{ae}, d_{ae}), (w_{ce}, d_{ce}), (w_{cd}, d_{cd}), and (w_{ed}, d_{ed}). The initial value of node s is $(0, 0)$ and other nodes is $(0, \infty)$. The values of the nodes are recorded, and a list of values for each node will be generated. There might be quite a few values if there are

a large number of nodes. To eliminate redundant values, a trim procedure will be executed if two values in L are close to each other since there is no need to keep both of them. More accurately, a trimming parameter α is utilized such that $0 < \alpha < 1$. When trimming a list by α, remove as many elements as possible, in such a way if L' is the result of trimming L, then for every element y that was removed from L, there is still an element z still in L' that approximates y, that is,

$$\frac{y}{1 + \alpha} \leq z \leq y. \tag{3}$$

Through the trim procedure, the approximate values (W_d, D_d) of node d can be obtained; thus, an approximate path from s to d will be achieved.

Before explaining the details of the proposed scheme, a list of variables that will be used throughout this research is provided as Table 1.

4. Proposed Solution

In this section, the knapsack problem is reduced to the throughput maximization problem first. Then, a polynomial time approximation scheme is proposed to solve the graph knapsack problem, which can return an approximate solution. Finally, a throughput maximization algorithm is presented based on the approximate scheme for graph knapsack. The approximation ratio and the running time of the proposed algorithm are also analyzed.

4.1. Problem Reduction. To maximize the network throughput, this study optimizes the transmission rate r_l while the sum of d_l does not exceed δ is satisfied. The relation between the transmission rate r_l and transmission delay d_l is presented as $d_l = K / (c_l - r_l)$, subject to $\sum_{l \in L} d_l \leq \delta$. From the equation, it can be seen that the transmission delay d_l will increase when r_l increases. There should exist an optimal transmission rate r_l^* with corresponding d_l^*, so the selected path could achieve the maximum throughput while the total delay is within delay constraint. To reduce the complexity of obtaining the optimal values, the approximate values for r_l, d_l are derived.

TABLE 1: Variables used in this paper.

Variable	Definition
$Neighbor[i]$	List of neighbors of node i
δ	Predefined delay threshold
ϵ	A positive real number used for approximation
l	Link between two nodes
L	Set of links on the path for data delivery
r_l	Transmission rate of link l
x_l	$X_l = 1$ means links l is selected otherwise not
s	Source node that carries the information
u, v	Nodes active in the network
W_v	Total weighted transmission rate at node v
D_v	Total transmission delay at node v
w_e	Weighted transmission rate of link between u and v
d_e	Transmission delay of link e between u and v
w_l	Weighted transmission rate of link l, calculated by $f(r_l)$
d_l	Transmission delay of link l
Y	Set of pair values (W_v, D_v) of vehicle v
Y'	Trimmed list of (W_v, D_v) of vehicle v
P^*	Optimal solution of the maximization problem
P	Approximate solution of the maximization problem

According to the equation $d_l = K/(c_l - r_l)$, $r_l = c_l - K/d_l$ holds. When $r_l \longrightarrow 1$, d_l has the minimum value, that is $d_l \longrightarrow K/(c_l - 1)$. Hence, the range of d_l is $(K/(c_l - 1), \delta]$, while the range of r_l is $(1, c_l]$. To achieve the approximate values of r_l and d_l, let r_l increase $1 + \epsilon$ each time until it reaches the largest value, where $0 < \epsilon < 1$ is the parameter used for approximation. Then a list of values for r_l is obtained, which is shown as $\{1, 1 + \epsilon, (1 + \epsilon)^2, \cdots, (1 + \epsilon)^t\}$. According to $(1 + c)^t \le c_l$, it has $t \le \log_{1+\epsilon} c_l$, and t is the largest integer satisfying the inequality, which means $(1 + \epsilon)^{t+1} > c_l$. d_l may have different values according to different r_l calculated by $d_l = K/(c_l - r_l)$, satisfying the condition that $d_l \in (K/(c_l - 1), \delta]$. Therefore, t pairs of (r_l, d_l) can be derived. After $f(r_l)$, t pairs of corresponding values (w_l, d_l) are generated. Consequently, the approximation ratio to obtain the approximate value of r_l is $1 + \epsilon$.

Different pairs of values for each link can be treated as different weights of corresponding edges between two nodes. Accordingly, there may exist multiple edges between two nodes. Then Definition 2 is described.

Definition 2. Given a weighted graph $G(V, E)$, there may exist multiple tuples of values for each edge, which can be treated that there are multiple edges with different values between the corresponding nodes. Accordingly, $G(V, E)$ becomes a multiedge-weighted graph.

As each link is represented by an edge of $G(V, E)$, the graph knapsack problem can be reduced to the throughput maximization problem. The study aims to select a set of links over which the maximum throughput can be achieved while the delay constraint is satisfied.

4.2. Approximation Scheme for the Graph Knapsack Problem. As there may exist more than one path from source node s to node i, node i could have different pairs of weighted values (W_i, D_i). Let Y denote the set of values. Assume Y is sorted

into monotonically increasing order of W_i. A procedure *trim*() (see Algorithm 1) is designed to remove unnecessary values of node i, based on the idea of approximation. The procedure scans the elements of Y in monotonically increasing order. An element is appended onto the returned list Y' only if it is the first element of Y or if it cannot be represented by the most recent values placed into Y'. The output of the procedure *trim*() described as Algorithm 1 is a trimmed, sorted list.

Given the trim procedure, a polynomial time approximation scheme can be constructed for the graph knapsack problem, which is described as Algorithm 2. The approximation procedure takes as input a set of values for node u, $Q = \{(W_{u_1}, D_{u_1}), (W_{u_2}, D_{u_2}), \cdots, (W_{u_n}, D_{u_n})\}$ (in arbitrary order), the delay threshold δ, and the approximation parameter ϵ. Algorithm 2 calls Algorithm 1 to trim the input list. An approximate solution denoted by P within a $1 + \epsilon$ factor of the optimal solution will be returned by the scheme. Lemma 1 is developed to prove that the proposed scheme runs in polynomial time. Meanwhile, Theorem 1 is derived to show that there is a polynomial time approximation algorithm for the multiedge graph knapsack problem.

Lemma 1. *The algorithm for the multiedge graph knapsack problem runs in $O(3n^2 m \ln W^*/\epsilon)$ time, where m, n denotes the number of edges and vertices of graph G, respectively.*

Proof. It will show that the running time of the proposed scheme is polynomial in both $1/\epsilon$ and the size of the input. The first part of the algorithm runs in time $O(nm)$, since the initialization in line 1 takes $\Theta(n)$ time, each of the $|V| - 1$ passing over the edges takes $\Theta(m)$ time, where $n = |V|, m = |E|$.

Now the running time of trim process will be analyzed. Assume W^* is the optimal weighted transmission rate of link l and $y.W < y.W^*$. After trimming, successive elements y and Y' of Y have the relationship $Y'.W/y.W > 1 + \epsilon/2n$; that is, they differ by a factor of at least $1 + \epsilon/2n$. Thus, each list contains possibly the value 1 and up to $\log_{1+\epsilon/2n} W^*$ values. It can be deduced that the number of elements in each list Y is at most

$$
\lfloor \log_{1+\epsilon/2n} W^* \rfloor + 1 = \frac{\ln W^*}{\ln(1 + \epsilon/2n)} + 1
$$

$$
\le \frac{2n(1 + \epsilon/2n)\ln W^*}{\epsilon} + 1 \quad (4)
$$

$$
< \frac{3n \ln W^*}{\epsilon} + 1.
$$

In summary, the overall running time of the algorithm is $O(3n^2 m \ln W^*/\epsilon)$. This bound is polynomial in the size of the input n and $1/\epsilon$. $\square$

Theorem 1. *The proposed algorithm for the multiedge graph knapsack problem is a polynomial time approximation scheme with an approximation ratio $1 + \epsilon$.*

Input: Y: a list of $(W_i, D_i), \forall i \in V$;
     δ: the predefined delay threshold;
     ϵ: a real number
Output: Y': a trimmed list of Y

(1) Let Y' be empty;
(2) Remove every tuple (W, D) in Y with $D > \delta$;
(3) Partition the tuples of Y into $A_1, \cdots, A_t$ such that for every two tuples (W, D), $(W', D',)$ in the same A_s, they satisfy $W \leq (1 + \epsilon/2n)W'$ and $W' \leq (1 + \epsilon/2n)W$;
(4) **for** $i = 1$ to t **do**
(5)   Select one tuple (W, D) from A_i with the least D;
(6)   Append (W, D) to Y';
(7) **end for**
(8) **return** Y'

ALGORITHM 1: Trim (Y, δ, ϵ).

Input: Q: a list of (W_u, D_u) for node u, every $u \in G.V$;
     δ: the predefined delay threshold
Output: approximation solution P

(1) INITIALIZE $G(V, E)$, set up the value of (w_e, d_e) for corresponding link;
(2) Let $Y_u = \varnothing$ for every $u \in G.V$;
(3) **for** $i = 1$ to $|G \cdot V| - 1$ **do**
(4)   **for** each edge $e = (u, v) \in G \cdot E$ **do**
(5)   **for** each $(W_u, D_u) \in Y_u$ **do**
(6)   Calculate $(W_v, D_v) = (W_u, D_u) + (w_e, d_e)$;
(7)   Add (W_v, D_v) to Y_v;
(8)   **end for**
(9)   Trim(Y_v);
(10)   **end for**
(11) **end for**
(12) **return** P, which contains the set of links selected

ALGORITHM 2: Polynomial time approximation scheme for graph knapsack.

Proof. Let P^* denote the optimal solution of the problem. From the proposed scheme, it is easily seen that $P \leq P^*$. It needs to show that $P^*/P \leq 1 + \epsilon$.

After trimming, successive tuples y and y' of Y' have the relationship $y \cdot W/y \cdot W > 1 + \epsilon/2n$, where n indicates the number of nodes in G. Scan all the edges, find the path from source node s to destination node u, and let $v_1 v_2 \cdots v_t$ denote the path, $s = v_1, u = v_t$. Due to the trim process executed at the receiver node of each edge, there exists a $(1 + \epsilon/2n)$ factor approximation.

As to v_i, there are $i - 1$ edges between s to v_i; therefore, the approximation ratio should be $(1 + \epsilon/2n)^{i-1}$. Then, v_i reaches v_{i+1} through edge $v_i v_{i+1}$, the approximation ratio at v_i after trimming should be $(1 + \epsilon/2n)^{i-1} \cdot (1 + \epsilon/2n)$, which is equal to $(1 + \epsilon/2n)^i$. Since there are totally n nodes, the number of edges on the path is at most $n - 1$. From the induction of the above procedure, an overall approximation ratio can be expressed as $(1 + \epsilon/2n)^{n-1}$, which can be presented as below:

$$P^*/P \leq (1 + \epsilon/2n)^{n-1}. \tag{5}$$

Now, it needs to show that $P^*/P \leq 1 + \epsilon$, by proving

$$(1 + \epsilon/2n)^{n-1} \leq 1 + \epsilon. \tag{6}$$

Since $\lim_{n \to \infty} (1 + x/n)^n = e^x$, the equation $\lim_{n \to \infty} (1 + \epsilon/2n)^{n-1} = e^{\epsilon/2}$ holds. Since $d/dn(1 + \epsilon/2n)^{n-1} > 0$, function $(1 + \epsilon/2n)^{n-1}$ is monotonically increasing, which means the function increases with n as it approaches the limit $e^{\epsilon/2}$. Thus, the following inequality stands:

$$(1 + \epsilon/2n)^{n-1} \leq e^{\epsilon/2} \leq 1 + \epsilon/2 + (\epsilon/2)^2 \leq 1 + \epsilon. \tag{7}$$

Combine with $P^*/P \leq (1 + \epsilon/2n)^{n-1}$, it has

$$P^*/P \leq (1 + \epsilon), \tag{8}$$

and the analysis of the approximation ratio is completes.

Combined with Lemma 1, it proves that the proposed scheme is a polynomial time approximation scheme. $\square$

4.3. Proposed Algorithm for Throughput Maximization Problem. As the graph knapsack problem is reduced to the throughput maximization problem, a throughput maximization algorithm is proposed based on the approximation scheme for the graph knapsack problem in this section.

Given n items, the ith item is worth w_i and d_i pounds in weight. The 0/1 knapsack problem aims to find a subset of items that the total value is maximum while the total weight is limited to a value. Assume that d_i is at most δ and the items are indexed in monotonically increasing order of their values, that is, $w_1 \leq w_2 \leq \cdots \leq w_n$. Theorem 2 is derived to show that the throughput maximization problem with delay constraint is NP-hard.

Theorem 2. *The throughput maximization problem is NP-hard.*

Proof. Reduce the 0/1 knapsack problem to the throughput maximization problem. Consider Q is a list of n items, denoted by $v_1, v_2, \cdots, v_n$, with corresponding values $\{(w_1, d_1), (w_2, d_2), \cdots, (w_n, d_n)\}$, where (w_i, d_i) indicates the value of the ith item.

Construct a graph $G(V, E)$ with $V = \{s, t, v_1, \cdots, v_n\}$, where s and t denote the source node and destination node, respectively. For every node $v_i \in V$, there is a pair of values (w_i, d_i) for every edge (v, v_i) that goes from v to v_i for any $v \neq v_i$ and $v \neq t$. There is a pair of values $(0, 0)$, for every edge (v, t) that goes from v to t for any $v \neq s$ and $v \neq t$. The knapsack problem aims to select a subset $U \subseteq \{v_1, v_2, \cdots, v_n\}$ of items such that $\sum_{v_i \in U} w_i$ is maximized and the total weight $\sum_{v_i \in U} d_i \leq \delta$. A subset U is a feasible solution for the knapsack problem if and only if there is a path that goes from s to the vertex v_i with $v_i \in U$ and then to t. It is easy to see that the time of construction is in polynomial time.

Select items satisfying the required conditions and add them to the knapsack, which is also the way to select the path for the problem. Therefore, if there exists a solution for the knapsack problem, the maximization problem can be solved. Vice Versa, existence of a solution to the maximization problem means there is a solution to the knapsack problem. Thus, the maximization problem is NP-hard.

After reducing the graph knapsack problem to the throughput maximization problem, a Throughput Maximization algoRithm (TMR) is proposed based on Algorithm 2, shown as Algorithm 3. Assume source node s intends to disseminate information to a specific area, an approximate delivery path is desired to achieve the maximum throughput and satisfy the delay constraint.

To be more clearer, a detailed description on how the proposed TMR works on the throughput maximization problem to solve the path-finding issue is presented as follows.

In the initialization process, let $N[v]$ represent a set of v's neighbors. Starting from source node s, execute the following steps to each edge $e \in G.E$.

(1) To find the neighbor vehicles $N[v]$ for node v, exchange information and obtain the corresponding values $(W_{N[i]}, D_{N[i]})$ of the neighbors.

(2) To obtain the links connecting node v with its neighbors, calculate the channel capacity c_l

according to the channel condition. Then, get the transmission rate and delay of the corresponding link. Calculate $(W_v, D_v) \in Y_v$ for v by adding the value of its neighbor u, (W_u, D_u) with (w_e, d_e) of the corresponding link e. Therefore, a list Y_v for v can be achieved.

(3) According to the previous step, several pairs of values (W_v, D_v) may exist for node v. First, remove the values with delay that are larger than the delay threshold δ. Then, if there are values with the same delay, keep those with larger transmission rate. Also, remove the values with larger delay and smaller transmission rate. In the following case, such as $Y_v(i) \in Y_v$ with values of $(W_v(i), D_v(i))$, $Y_v(j) \in Y_v$ with $(W_v(j), D_v(j))$, if $W_v(i) > W_v(j)$ and $D_v(i) > D_v(j)$, which means item $Y_v(i)$ is with larger transmission rate but also with larger delay compared with $Y_v(j)$, a trim procedure will be executed to determine whether to remove an element. If $W_v(i) > (1 + \epsilon/2n) \cdot W_v(j)$, append $Y_v(i)$ onto list Y'; otherwise, remove $Y_v(i)$ and append $Y_v(j)$ onto list Y'.

After the iterative operations, paths containing a set of selected links are obtained. If there are more than one path from the source to the destination, choose the one with the largest transmission rate, which is the approximate solution intended to achieve for data delivery.

Theorem 3 is presented to show that the proposed TMR is a polynomial approximation algorithm with an approximation ratio of $1 + \gamma$, where $0 < \gamma < 1$.

Theorem 3. *The throughput maximization algorithm can achieve an approximation ratio $1 + \gamma$ within running time $O(n^2 m \ln C / \gamma)$, where m denotes the number of edges, $C = \sum_{i=1}^{m} c_i$, c_i indicates the transmission capacity of link i.*

Proof. The input of the proposed maximization algorithm is w_l, d_l. As stated in 4.1, the value of w_l, d_l is within a $1 + \epsilon$ factor approximation of the optimal value. Considering the approximation in the multiedge knapsack problem, the throughput maximization problem should have a quadratic approximation. According to theorem 1, the comprehensive approximation ratio is

$$(1 + \epsilon)^2 = 1 + 2\epsilon + \epsilon^2. \tag{9}$$

Assume $\gamma = \epsilon/3$, the inequality $(1 + \epsilon)^2 = 1 + 2\epsilon + \epsilon^2 < 1 + \gamma$ holds.

Hence, an approximation solution within a $1 + \gamma$ factor of the optimal solution can be achieved.

As $r_l \in (1, c_l]$, assume there are m edges on the path. Let $C = \sum_{i \in m} c_i$, then $\forall y \in (m, C)$. From Lemma 1, it has

$$\lfloor \log_{1 + \epsilon/2n} C \rfloor + 1 < \frac{3n \ln C}{\epsilon} + 1$$

$$= \frac{n \ln C}{\gamma} + 1. \tag{10}$$

Therefore, the total running time is $O(n^2 m \ln C / \gamma)$. $\square$

Input: A list V containing all the vehicles and UAVs in the network
Output: The selected path

(1) Let $N[v]$ denote the neighbor of node v;
(2) $a = 0$, which denotes the number of executions;
(3) **for** each node v **do**
(4)   Send a request to its neighbors;
(5)   Receive the channel information (CI) from its neighbors;
(6)   Calculate c_l according to CI;
(7)   Calculate the approximate values of r_l and d_l according to $d_l = K/(c_l - r_l)$;
(8)   Obtain the transmission rate and delay of the link between each neighbor and node v, denoted as (w_e, d_e);
(9) **end for**
(10) **repeat**
(11) Calculate the transmission rate and delay for each node v in the network, by adding the neighbor's corresponding values to
       (w_e, d_e), denoted as W_v, D_v;
(12) Apply the trim procedure to remove unnecessary values of node v;
(13) $a = a + 1$;
(14) **until** $(a = |G \cdot V| - 1)$
(15) **return** A path containing a set of selected links

ALGORITHM 3: Throughput Maximization Algorithm.

5. Performance Evaluation

In this section, simulation settings and results are presented and analyzed.

5.1. Simulation Settings. To evaluate the performance of the proposed algorithm, TMR is implemented and compared with other algorithms. In the simulations, the following default settings are used.

The simulations select a 2000 m × 2000 m rectangle street area on the map of Los Angeles and extract the area using openstreetmap [31], the satellite map of which is presented as Figure 3(a). Then, Simulation of Urban Mobility (SUMO) [32] is used to convert the extracted area to the road topology layout, shown in Figure 3(b). The realistic mobility trace of vehicles is generated by the open-source microscopic space-continuous and time-discrete vehicular traffic generator package SUMO. SUMO uses a collision-free car-following model to determine the speeds and the positions of the vehicles. The simulations deploy a number of UAVs which can cooperatively form a full coverage of the simulated area. The speed of UAVs varies from 0 to 15 m/s, and the UAVs maintain a constant altitude that does not exceed 200 m during the flight. The random walk mobility model is applied for the UAVs covering the area. Table 2 gives a list of simulation parameters.

The simulations implement the proposed algorithm and two other algorithms which are UVAR [33] and VBN [11], respectively. Extensive simulations are conducted to thoroughly investigate the efficiency of the proposed algorithm in aspect of delivery ratio, throughput, and number of hops when the number of vehicles varies and the deployed UAVs are set to 20. Additionally, the performance of the proposed algorithm with different UAV densities is evaluated when the number of vehicles is set to 300. A comparison between the proposed solution and optimal solution in terms of throughput is also presented.

5.2. Impact of Number of Vehicles on Delivery Ratio. Delivery ratio is defined as the percentage of packets that are successfully delivered, that is, the ratio of the total number of data packets received by the target destinations to the total number of data packets generated from the sources. A higher delivery ratio means better performance.

In Figure 4, the delivery ratio under different number of vehicles for the compared algorithms is compared. As shown in the figure, the evaluated schemes achieve higher delivery ratio when there are more vehicles in the network. Besides, it can be seen that the proposed scheme and UVAR have better delivery ratio, due to the advantage of UAVs that applied to maintain better network connectivity and guarantee a significant accuracy of path selection. VBN mainly chooses the delivery paths based on cooperation among RSUs and vehicles, which cannot be accurate all the time and may select the paths that are not appropriate for data transmission, resulting in lower delivery ratio.

5.3. Impact of Number of Vehicles on Throughput. An important performance indicator of the algorithms is the throughput of the path from the source to the destination nodes.

In Figure 5, the throughput is plotted versus the number of vehicle nodes. It is observed that the proposed algorithm TMR outperforms the other two algorithms and has the highest network throughput. When the number of vehicles is 50, the throughput of TMR is 1.48 Mbps, higher than that of UVAR and VBN. It also shows that all the compared algorithms achieve higher network throughput as the vehicle density level increases. As the number of vehicles increases to 400, the corresponding throughput of the three schemes increases to 3.45, 3.3, and 3.1 Mbps.

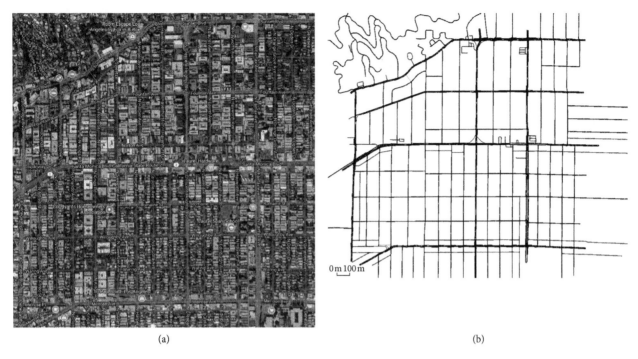

(a)

(b)

FIGURE 3: Selected area of Los Angeles, CA, USA.

TABLE 2: Simulation setup.

Parameters	Settings
Simulation area	2000 m × 2000 m
MAC protocol	IEEE 802.11p
Communication range of vehicles	300 m
Communication range of UAVs	1000 m
Vehicle velocity	0–13 m/s
UAV velocity	0–15 m/s
Number of vehicles	50–400
Number of UAVs	10–40

5.4. Impact of Number of Vehicles on Number of Hops.
Number of hops can be obtained from the total number of hops performed by disseminating the message to the target area.

The trend of required hops with the increasing number of vehicle nodes is shown in Figure 6. The proposed algorithms TMR and UVAR perform fewer hops than VBN. This is because that with the help of employed UAVs, available delivery paths from the source node to the destination area can be quickly found by avoiding unnecessary transmissions among vehicle nodes, such that the number of hops consumed is smaller. As the network size becomes larger, the number of hops increases for all the compared schemes. This is mainly because that as the number of vehicles in the target area increases, more hops are needed to deliver the message to the vehicles and complete the dissemination.

5.5. Evaluating the Throughput Maximization Algorithm.
Figure 7 shows the impact of the number of UAVs on data delivery ratio and delivery delay. As more UAVs

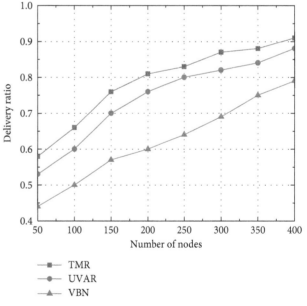

FIGURE 4: Data delivery ratio comparison when the number of nodes increases from 50 to 400.

participate in data transmission, the delivery ratio increases while the delivery delay tends to decrease. This is because that the UAVs can serve relay nodes in data dissemination when there are no available vehicles to carry and forward the data. When more UAVs participate in the communications, the vehicle nodes could select a better UAV relay with a higher probability, which results in the changing data delivery delay and delivery ratio shown in the figure.

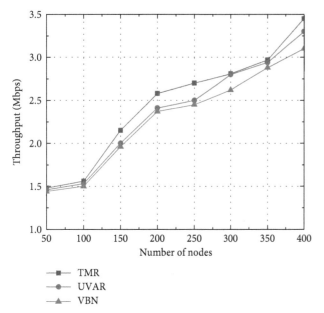

FIGURE 5: Throughput comparison when the number of nodes increases from 50 to 400.

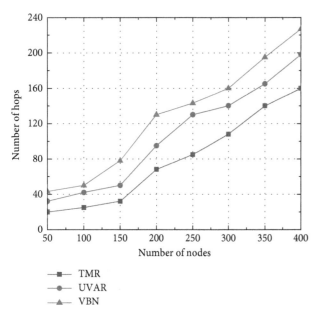

FIGURE 6: Consumed hops comparison when the number of nodes increases from 50 to 400.

To show the difference between the proposed solution and the best possible solution, Figure 8 compares the proposed and optimal solutions in terms of throughput, considering the UAV-assisted vehicular environment. Comparing the simulation throughput with the optimal throughput in Figure 8, it can be seen that the simulation result consists with the optimal throughput to a great extent. Meanwhile, the approximation ratio is smaller than 1.1, which also verifies the effectiveness of the proposed algorithm. Observing the changing trends of the throughput, it is easy to find that the system throughput improves when the number of nodes in the network increases.

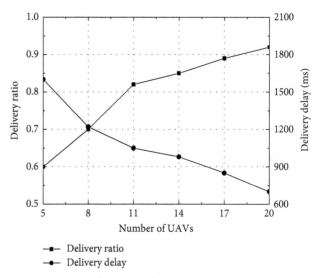

FIGURE 7: Impact of number of UAVs on data delivery efficiency, illustrated by delivery ratio and delivery delay.

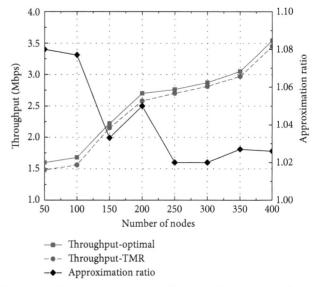

FIGURE 8: Throughput comparison between the proposed solution and the optimal solution, the approximation ratio reflects the throughput difference and the effectiveness of the proposed solution.

6. Conclusion

In this paper, efficient data dissemination in cooperative UAV-assisted VANETs is investigated. To optimize the network throughput, this study formulates a network throughput maximization problem to find the best delivery strategy and select the optimal paths for data delivery, with consideration of the transmission rate of links and the delay constraint for data dissemination. Then reduce the graph knapsack problem to the throughput maximization problem, and a polynomial time approximation scheme is proposed to solve the graph knapsack problem. As to the maximization problem, a throughput maximization algorithm is developed based on the approximation scheme.

Theoretical analysis including the approximation ratio and running time of the proposed solution is provided. Finally, simulations are conducted to evaluate the performance of the proposed algorithm.

7. Discussions and Future Work

While the utilization of UAVs brings significant advantages, it also faces the cost problem. UAV communications are subjected to the additional energy consumption to fly at high altitudes, which is more significant than the communication energy consumption due to signal processing. Nevertheless, the limited on-board energy due to high propulsion energy consumption of UAVs poses critical limits on their communication performance and endurance.

It can be seen that there exists a fundamental tradeoff between the achievable utility benefit and system cost in UAV-assisted communication networks. Using UAVs can increase the network throughput and improve quality of service, which is important to users, especially to the applications with high quality of service requirements. Although the use of UAVs increases the system cost, UAVs have significant advantages over common roadside infrastructure. The tradeoff between the benefit and cost can be achieved by energy-efficient design to enhance the performance of UAV-assisted communication, which is a promising future work direction, such that the deployment and trajectory of UAVs can be carefully designed to save the energy consumption and improve the quality of communications (improved transmission rate and transmission delay).

Despite the contributions presented in this work, many challenges remain to be solved by academia and industry. Future work will focus on the frequent handover problem and interference caused by the high mobility of UAVs and vehicles. Also, energy efficiency of the UAVs remains a relevant topic to be explored to achieve fully utilization of UAVs and improve data dissemination in cooperative network. Further, the integration of the proposed architecture with the concept of SDN and the development of envisaged applications which can adapt to more complicated scenarios might be considered as another future work direction.

Acknowledgments

This work is supported by the the National Natural Science Foundation of China (nos. 61772385, 61373040, and 61572370).

References

[1] R. Ghebleh, "A comparative classification of information dissemination approaches in vehicular ad hoc networks from distinctive viewpoints: a survey," *Computer Networks*, vol. 131, pp. 15–37, 2018.

[2] F. Cunha, L. Villas, A. Boukerche et al., "Data communication in VANETs: protocols, applications and challenges," *Ad Hoc Networks*, vol. 44, pp. 90–103, 2016.

[3] Q. Wu and R. Zhang, "Common throughput maximization in UAV-enabled ofdma systems with delay consideration," 2018, https://arxiv.org/abs/1801.00444.

[4] N. Zhang, S. Zhang, P. Yang et al., "Software defined space-air-ground integrated vehicular networks: challenges and solutions," *IEEE Communication Magazine*, vol. 55, no. 7, pp. 101–109, 2017.

[5] Y. Zhou, N. Cheng, N. Lu, X. Shen, and Sherman, "Multi-UAV-aided networks: aerial-ground cooperative vehicular networking architecture," *IEEE Vehicular Technology Magazine*, vol. 10, no. 4, pp. 36–44, 2015.

[6] M. R. Karp, *Reducibility among Combinatorial Problems*, Springer, New York, NY, USA, 1972.

[7] H. T. Cormen, E. C. Leiserson, L. R. Rivest, and C. Stein, *Introduction to Algorithms*, MIT Press, Cambridge, UK, 3rd edition, 2009.

[8] N. Benamar, K. D. Singh, M. Benamar et al., "Routing protocols in vehicular delay tolerant networks: a comprehensive survey," *Computer Communications*, vol. 48, no. 8, pp. 141–158, 2014.

[9] W. L. Tan, W. C. Lau, O. C. Yue et al., "Analytical models and performance evaluation of drive-thru Internet systems," *IEEE Journal on Selected Areas in Communications*, vol. 29, no. 1, pp. 207–222, 2011.

[10] B. Zhang, X. Jia, K. Yang et al., "Design of analytical model and algorithm for optimal roadside AP placement in VANETs," *IEEE Transactions on Vehicular Technology*, vol. 65, no. 9, pp. 7708–7718, 2016.

[11] Y. Lin and I. Rubin, *Throughput maximization under guaranteed dissemination coverage for VANET systems Information Theory and Applications Workshop*, pp. 313–318, ITA, UK, 2015.

[12] D. Zeng, S. Guo, and J. Hu, "Reliable bulk-data dissemination in delay tolerant networks," *IEEE Transactions on Parallel and Distributed Systems*, vol. 25, no. 8, pp. 2180–2189, 2014.

[13] M. Xing, J. He, and L. Cai, "Maximum-utility scheduling for multimedia transmission in drive-thru Internet," *IEEE Transactions on Vehicular Technology*, vol. 65, no. 4, pp. 2649–2658, 2016.

[14] M. Xing, J. He, and L. Cai, "Utility maximization for multimedia data dissemination in large-scale VANETs," *IEEE Transactions on Mobile Computing*, vol. 16, no. 4, pp. 1188–1198, 2017.

[15] K. Zheng, L. Hou, H. Meng et al., "Soft-defined heterogeneous vehicular network: architecture and challenges," *IEEE Network*, vol. 30, no. 4, pp. 72–80, 2016.

[16] K. Liu, J. K. Y. Ng, V. C. S. Lee et al., "Cooperative data scheduling in hybrid vehicular ad hoc networks: VANET as a software defined network," *IEEE/ACM Transactions on Networking*, vol. 24, no. 3, pp. 1759–1773, 2016.

[17] J. Nobre, D. Rosario, C. Both et al., "Toward software-defined battlefield networking," *IEEE Communications Magazine*, vol. 54, no. 10, pp. 152–157, 2016.

[18] I. Zacarias, L. P. Gaspary, A. Kohl et al., "Combining software-defined and delay-tolerant approaches in last-mile tactical edge networking," *IEEE Communications Magazine*, vol. 55, no. 10, pp. 22–29, 2017.

[19] M. Mozaffari, W. Saad, M. Bennis, and M. Debbah, "Unmanned aerial vehicle with underlaid device-to-device communications: performance and tradeoffs," *IEEE Transactions on Wireless Communications*, vol. 15, no. 6, pp. 3949–3963, 2016.

[20] M. Mozaffari, W. Saad, M. Bennis, and M. Debbah, "Efficient

deployment of multiple unmanned aerial vehicles for optimal wireless coverage," *IEEE Communications Letters*, vol. 20, no. 8, pp. 1647–1650, 2016.

[21] D. Orfanus, P. E. Freitas de, and F. Eliassen, "Self-organization as a supporting paradigm for military UAV relay networks," *IEEE Communications Letters*, vol. 20, no. 4, pp. 804–807, 2016.

[22] O. S. Oubbati, A. Lakas, F. Zhou, M. Gunes, N. Lagraa, and M. Yagoubi, "Intelligent UAV-assisted routing protocol for urban VANETs," *Computer Communications*, vol. 107, pp. 93–111, 2017.

[23] X. Wang, L. Fu, Y. Zhang et al., "VDNet: an infrastructure-less UAV-assisted sparse VANET system with vehicle location prediction," *Wireless Communications & Mobile Computing*, vol. 16, no. 17, pp. 2991–3003, 2016.

[24] L. Xiao, X. Lu, D. Xu et al., "UAV relay in VANETs against smart jamming with reinforcement learning," *IEEE Transactions on Vehicular Technology*, vol. 99, p. 1, 2018.

[25] H. Seliem, M. Ahmed, R. Shahidi et al., "Delay analysis for drone-based vehicular ad-hoc networks," in *Proceedings of 28th IEEE Annual International Symposium on Personal, Indoor, and Mobile Radio Communications*, IEEE, Montreal, QC, Canada, October 2017.

[26] P. Shilin, R. Kirichek, A. Paramonov et al., "Connectivity of VANET segments using UAVs," in *Internet of Things, Smart Spaces, and Next Generation Networks and Systems*, Springer International Publishing, New York, NY, USA, 2016.

[27] W. Fawaz, "Effect of non-cooperative vehicles on path connectivity in vehicular networks: a theoretical analysis and UAV-based remedy," *Vehicular Communications*, vol. 11, pp. 12–19, 2018.

[28] L. Gupta, R. Jain, and G. Vaszkun, "Survey of important issues in UAV communication networks," *IEEE Communications Surveys & Tutorials*, vol. 18, no. 2, pp. 1123–1152, 2016.

[29] W. Fisher, *Development of DSRC/Wave Standards*, IEEE, Annapolis, MD, USA, 2007.

[30] S. Guo, C. Dang, and Y. Yang, "Joint optimal data rate and power allocation in lossy mobile ad hoc networks with delay-constrained traffics," *IEEE Transactions on Computers*, vol. 64, no. 3, pp. 747–762, 2015.

[31] M. Haklay and P. Weber, "OpenStreetMap: user-generated street maps," *IEEE Pervasive Computing*, vol. 7, pp. 12–18, 2008.

[32] SUMO-simulation of urban mobility, http://sumo.sourceforge.net.

[33] O. S. Oubbati, A. Lakas, N. Lagraa, and M. Yagoubi, "UVAR: an intersection UAV-assisted VANET routing protocol," in *Proceedings of the IEEE Wireless Communications and Networking Conference (WCNC)*, IEEE, Doha, Qatar, April 2016.

[34] X. Fan, C. Huang, X. Chen, S. Wen, and B. Fu, "Delay-constrained throughput maximization in UAV-assisted VANETs wireless algorithms, systems, and applications (WASA)," in *Lecture Notes in Computer Science*, Vol. 10874, Springer, Cham, Switzerland, 2018.

8

User-Adaptive Key Click Vibration on Virtual Keyboard

Seokhee Jeon,[1] Hongchae Lee,[1] Jiyoung Jung,[1] and Jin Ryong Kim ⓘ[2]

[1]*Department of Computer Engineering, Kyung Hee University, Yongin 17104, Republic of Korea*
[2]*Smart UI/UX Device Laboratory, ETRI, Daejeon 34129, Republic of Korea*

Correspondence should be addressed to Jin Ryong Kim; jessekim@etri.re.kr

Academic Editor: Byeong-Seok Shin

This study focuses on design of user-adaptive tactile keyboard on mobile device. We are particularly interested in its feasibility of user-adaptive keyboard in mobile environment. Study 1 investigates how tactile feedback intensity of the virtual keyboard in mobile devices affects typing speed and user preference. We report how different levels of feedback intensity affect user preferences in terms of typing speed and accuracy in different user groups with different typing performance. Study 2 investigates different tactile feedback modes (i.e., whether feedback intensity is linearly increased, linearly decreased, or constant from the centroid of the key, and whether tactile feedback is delivered when a key is pressed, released, or both pressed and released). We finally design and implement user-adaptive tactile keyboards on mobile device to explore the design space of our keyboards. We close by discussing the benefits of our design along with its future work.

1. Introduction

Touchscreen-based mobile devices are pervasive, and a number of smartphone users prefer typing on a virtual keyboard in mobile phone as they spend more time on it [1]. Mobile-based text entry system using virtual keyboards has many benefits. Its portability allows people not to carry on their own keyboards, and it does not require significant learning effort from users since it sometimes adapts the training-free QWERTY layout. Due to this, it is even possible to perform eyes-free typing on mobile device [2]. Such benefits of text entry with mobile device can bring another opportunity in other computing areas such as virtual reality (VR).

A good example that effectively takes these benefits is VR. Typing in VR is becoming more important as it gradually replaces conventional communication media. In this sense, typing is required in a broad range of areas in VR such as education, media consumption, gaming, and training. However, typing in VR is still difficult due to many technical challenges such as poor hand and finger tracking, lack of tactile feedback, limitation of vertical field of view, and so on. Since user's typing fingers are invisible in HMD-based VR, there is a need for alternative ways to achieve typing in VR. One good approach would be text entry on virtual keyboard using mobile device in VR. Since text entry on mobile device can provide portability, convenient, and even eyes-free typing, it can be a good match with VR as today's VR focuses more on the interactive aspects of VR. In fact, a number of studies [3–5] showed the great benefits and promising results for using mobile device as text input.

In this study, we focus more on how to improve the typing performance and final user experience to the level of conventional mechanical keyboard typing scenario. Assuming that one of the important missing cues in current virtual keyboard typing is the absence of proper tactile feedback, we investigate the effect of various factors in the key click vibration on the final user experience. Two tactile cues during typing are essential for typing experience: (1) mechanical click feeling to provide a user confirmation of pressing and (2) the feeling of valley among keys to help a user navigating the fingers to locate a specific key. In virtual keyboard, the former is partly provided by playing vibrotactile feedback when clicking, but there are huge variations in individual preference on this virtual click feedback due to low fidelity of the feedback. In addition, the latter cue, while this significantly affects the typing accuracy, is not usually provided in virtual keyboard typing scenario.

To overcome these shortcomings, we proposed two new techniques to improve the quality of user's typing experience. First, in order to cope with various demands of the users about the virtual key click, the relationship between the user's expertise on typing and user's preference on the strength of the virtual click feedback is experimentally investigated. This idea is based on our initial hypothesis that expert users do not rely on the feedback since they already have confidence on their typing skill, while novice users like to have feedback to strengthen their internal confidence. Using this relationship, we designed an adaptive key click strength control scheme. This scheme predicts user's typing expertise in real time based on his or her typing speed and adaptively controls the strength of the feedback to provide optimum experience.

Our second approach is an initial investigation on providing cue equivalent to the valley among keys using vibrotactile feedback. To achieve this goal, vibrotactile feedback is designed in a way that its strength is systematically altered when a user clicks the edge of the key so that he or she can notice the edge of keys. Two different algorithms are provided, and the effect of these is investigated through user experiment.

We finally integrate all the proposed techniques in one virtual tactile keyboard and test the significance of the approaches by observing user's workload using a TLX questionnaire.

The main contributions of the paper are as follows:

(1) New algorithm to adaptively control the amplitude of key click vibration through user preference estimation based on his/her typing expertise

(2) New key click vibration rendering technique to deliver virtual keys' edge information

(3) Experimental validations on the new above algorithms

2. Related Works

There exists a broad range of research works focusing on user experience improvement in virtual keyboard using sensory modality. Auditory feedback [6–8] can be a good solution to improve the user experience on virtual keyboards, but it has clear limitation in noisy environment. Visual feedback is the most commonly used modality for user experience and usability of virtual keyboard [9–27]. Sears [24] investigated the palm-style QWERTY keyboard, and they changed their keyboard size and location to investigate the user performance. Even though the size does not affect the user's typing performance, they revealed that there is a difference in user performance between numeric type keyboard and QWERTY keyboard. Nakagawa and Uwano proved the relationship between location of keyboard and user's performance such as typing error rate and typing speed. This clearly showed that the location of keyboard is also important factor for the user experience [21]. Mackenzie and Zhang [18] designed the new soft keyboard and their keyboard design improved the user's text entry speed. Kim et al. [19] developed one key keyboard that can

be worn on the wrist. Their keyboard could increase the input speed of users and accelerate their text entry learning ability.

Tactile feedback is another sensory modality that is available on virtual keyboards [8, 28–36]. Brewster et al. [28] found that typing with tactile feedback improves typing performance on mobile device. Users were able to enter more texts, make fewer errors, and correct more errors with tactile feedback. They demonstrated that tactile feedback is important role in touchscreen devices. Rabin and Gordon [29] studied the role of tactile feedback. They analysed kinematics of the right index finger with and without tactile feedback with special gloves and sensors. Their results suggested that tactile cues can provide information about the start location of the finger in which it is necessary to perform finger movement more accurately. Hoffmann et al. [30] developed a new tactile device for text entry. They used tactile feedback as a detector to prevent errors during the typing. Basic concept is that if a user types an incorrect key, the resistive force of the key becomes stronger so that the key makes the user to press the incorrect key harder. Kaaresoja et al. [32] used tactile feedback to mobile touchscreen and they demonstrated that tactile feedback is helpful in both usability and user experience. Nishino et al. [33] used tactile feedback as a communication modality. In their study, they used various types of vibration pattern such as strength, length, and effect. As a result, they found out the guidelines for building a practical system for tactile communication. Lylykangas et al. [37] focused on tactile feedback output delay and duration time. In their study, they found out the optimal duration and delay of tactile feedback when button is pressed. More recently, there exist a number of studies that are focused on perception [38] and performance [39–41] using tactile feedback in mobile and flat keyboard environments.

3. Study 1: Effects of Tactile Feedback Intensity

Study 1 investigates how tactile feedback intensity of the virtual keyboard in the mobile device affects typing speed and how strong user prefers based on their typing speed. We classified four levels of different feedback intensity and conducted a user study to observe the typing speed along with a follow-up questionnaire.

3.1. Apparatus. We built an Android-based typing program application with Mackenzie and Soukoreff phrase set [42]. In this program, a phrase is displayed on the top of the screen and the virtual keyboard is located under the phrase display area. The typing application stores typing speed, total elapsed time, and experimental condition. The typing application runs on Samsung Galaxy Note 3 due to its large touchscreen for typing phrases.

The vibration in the experiment is generated by the phone's internal linear resonant actuator (LRA) with 200 Hz resonant frequency and 4.043 m/s^2 maximum acceleration.

3.2. Participants. A total of eighteen university students participated in this study (mean age: 28.34, SD = 3.93). All participants were paid for their participation. They reported no disabilities. They also reported that they had prior experience of typing in English and familiar with virtual keyboards.

3.3. Experimental Design. We have four levels of tactile feedback intensity: *None, Low, Mid,* and *High.* Each has different intensity of tactile feedback signal of 0, 3.430, 3.798, and 4.043 m/s², respectively. The signal frequency was set to 200 Hz for all four levels of signal. We chose four levels of signal intensity because of *JND* of tactile feedback since participants may not be able to distinguish among the levels if the number of level exceeds four. Number of level less than four may not give us enough data to analyse the experiment.

In each level, twenty phrases were randomly assigned. The order of intensity level was selected by Latin square to reduce any ordering effect [43].

3.4. Procedure. A number of preliminary typing trials were conducted prior to the main experiment in order to allow participants to be accustomed to the typing application. In the main experiment, we asked participants to take a seat and naturally hold the touchscreen phone with their two hands (Figure 1). We then asked participants to type as fast and as accurately as they can based on the given phrase using their two thumbs [44]. We kept silence in the room during the experiment to avoid any noise effect. For each session, we recorded typing speed in words per minute (WPM), total elapsed time, and intensity level.

After completing each condition, we asked participants to rate the application for the following three questions on a 7-point Likert-type scale from 1 (strongly disagree) to 7 (strongly agree): *Typing speed*—this tactile feedback is helpful for increasing typing speed; *Typing accuracy*—this tactile feedback is helpful for reducing typing errors; *Preference*—I prefer this feedback. We also had a debrief session to ask more questions about their typing experiences with different intensity levels after the main experiment.

3.5. Data Analysis. We measured several performance metrics for this experiment. We measured typing speed in words per minute (WPM), keystroke per character (KSPC) for measuring typing efficiency, and minimum string distance (MSD) [44] for typing accuracy.

3.6. Results: Feedback Intensity versus Typing Speed. Figure 2 shows typing speed with different intensity levels. It is clearly observed that the WPM increases when the intensity level increases. Although one-way repeated-measure ANOVA shows that there is only a weak difference in typing speed ($F = 2.1$, $p = 0.098$), the trend is well observed.

3.7. Results: Feedback Intensity versus User Type. We further divided the participants into three groups based on their typing speeds—that is, *Beginner, Intermediate,* and *Expert.* Participants having average WPM value lower than 25 were categorized in *Beginner* group, and participants who scored average WPM higher than 30 were classified into *Experts.* Rest of the participants were classified as *Intermediates.* These thresholds are decided based on [44] and our observation on the distribution of the WPM. As a result, *Beginner* group and *Expert* group had 5 participants each, and *Intermediate* group had 8 participants.

As can be seen in Figures 3–5, people in each group prefer different intensity levels. *Beginner* group reported that they prefer *Mid* level of feedback intensity, whereas *Intermediate* group reported that they prefer both *Mid* and *High* levels of feedback intensity. *Expert* group reported that they prefer *Low* level of feedback intensity.

For different expertise groups and different measurements, one-way repeated-measure ANOVA tests were conducted. For all cases, no significant effect was observed. However, weak evidences (0.05 < *p* value < 0.1) of the effect of feedback intensity on user preferences were captured through a post hoc test (Bonferroni test) in some conditions as shown in the figures as blue lines.

3.8. Discussion. In this study, we first found that increasing feedback intensity is likely to lead to higher typing speed. Although the results were only weakly supported (0.05 < *p* value < 0.1), we clearly observed a linear trend of performance improvement with increased feedback intensity. We then observed the preference of different user group. Interestingly, people in *Beginner* and *Intermediate* groups preferred comparably higher intensity levels (*Mid* for *Beginner* and *Mid*/*High* for *Intermediate*) than those in *Expert* group (*Low* for *Expert*). It seems that *High* level was too strong for people in *Beginner.* However, they clearly preferred relatively higher intensity as they think that it helped them in increasing the typing speed and accuracy.

For people in *Expert* group, low tactile feedback intensity showed promising results as compared to other expertise groups. It is probably due to the fact that people in *Expert* group are already good at typing on virtual keyboards so that they neither need any strong key click feedback nor no feedback. This can be weak evidence that our initial hypothesis is on the relationship between typing expertise and the preference on feedback strength.

4. Study 2: Effects of Vibrotactile Edge

The goal of this study is to investigate the effect of vibrotactile feedback that encodes information about the keyboard edge. The idea of the encoding is that the distance between the finger and the intended keyboard is linearly mapped to the strength of the click vibration. Two different mapping functions were used: linearly increasing as distance increases, and the other way around. Additionally, we also investigated the effect of the moment when the encoded feedback is provided, at the moment of key

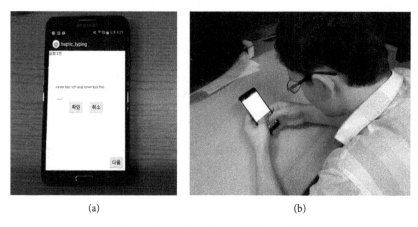

(a)   (b)

FIGURE 1: Typing program application (a) and a subject holding a phone during the experiment (b).

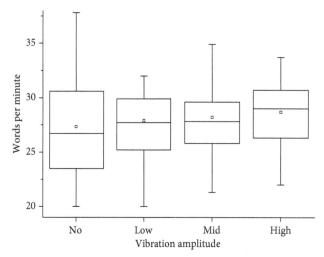

FIGURE 2: Typing speed in words per minute (WPM) for each tactile feedback intensity.

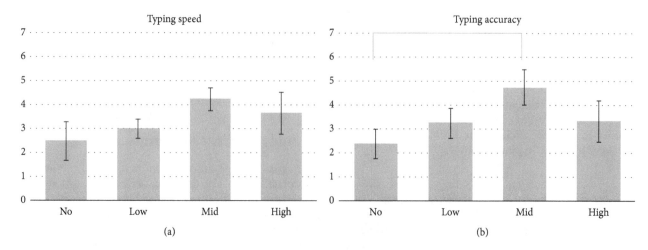

FIGURE 3: Average score for user preference on typing speed (a) and accuracy (b) for each intensity level in *Beginner* group.

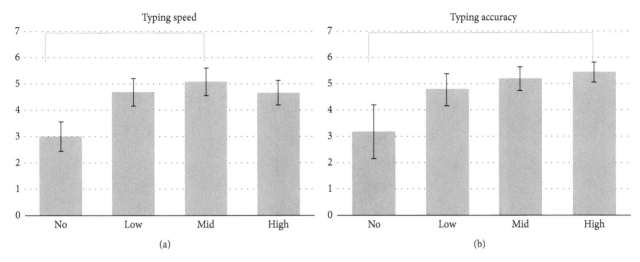

FIGURE 4: Average score for user preference on typing speed (a) and accuracy (b) for each intensity level in *Intermediate* group.

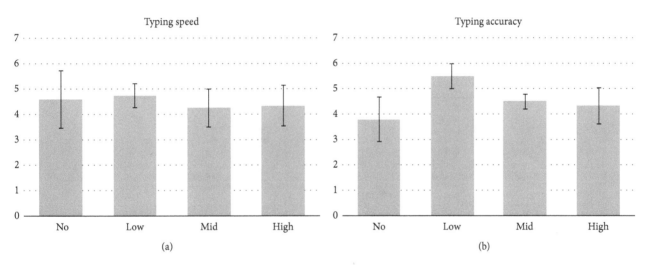

FIGURE 5: Average score for user preference on typing speed (a) and accuracy (b) for each intensity level in *Expert* group.

pressing or at the moment of key releasing. In this experiment, we experimentally find the best suitable combination among various combinations of the two mapping functions and two moments of feedback by observing the users' preference on their typing speed, error, and users' confidence.

4.1. Apparatus. We built another Android-based typing program apparatus with customized virtual keyboard (Figure 6). In this keyboard, we provide different intensity of tactile feedback signal based on the location of user's key press on the key. The intensity of signal is determined by the experimental conditions (see the later subsection for further explanation). This keyboard further supports temporal feedback conditions in which tactile feedback is delivered based on key's touch state. For example, tactile feedback is delivered when key is pressed, released, or both pressed and released. The typing application stores user's typing speed in WPM, number of key pressed, and the position of finger touched on the key.

4.2. Participants. A total of ten university students who did not participate in Study 1 participated in this study (mean = 25.34, SD = 1.92). Participants were paid for their participation. They reported that they are healthy and have no disabilities. They also reported that they had prior experience of typing in English and familiar with virtual keyboards.

4.3. Experimental Conditions. Table 1 shows the combinations of spatial-temporal feedback conditions—we call this tactile feedback mode.

As shown in Figures 7 and 8, and Table 1, we have three spatial tactile feedback conditions for this experiment: *Linear Feedback*, *Reversed Feedback*, and *Constant Feedback*. In *Linear Feedback*, tactile feedback intensity is linearly increased in regard to the distance of the touched location from the centroid of the key. Stronger tactile feedback is delivered when touched point is relatively farther from the centroid of the key. In *Reversed Feedback*, tactile feedback intensity is linearly decreased in regard to the touched point

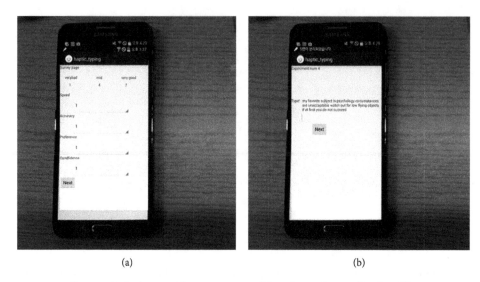

(a)   (b)

FIGURE 6: A phone with a survey page (a) and a typing application (b).

TABLE 1: Tactile feedback mode based on spatial-temporal feedback conditions.

	Linear	Reversed	Constant
Attached	AL	AR	AC
Detached	DL	DR	DC
Both ways	BL	BR	BC

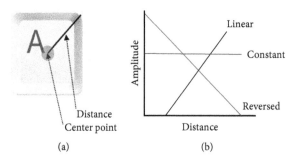

(a)   (b)

FIGURE 7: A key showing the distance from the centroid to its corner (a) and a chart showing three tactile feedback conditions—*Linear*, *Reversed*, and *Constant* based on the distance from the centroid (b).

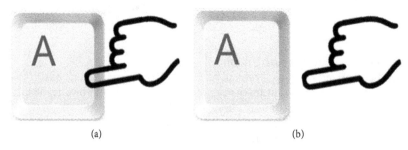

(a)   (b)

FIGURE 8: A finger is attached to a key (a) and detached from a key (b).

from the centroid of the key. In *Constant Feedback*, the constant intensity of tactile feedback is delivered no matter which area is touched within a key. We also provide no feedback when the centroid of key is touched, providing

confidence to the users that they touched the centroid of the key—we call this *Dead Zone*.

We also have three temporal tactile feedback conditions: *Pressed*, *Released*, and *Pressed-Released*. In *Pressed*, tactile feedback is delivered when key is being pressed (touched). In *Released*, tactile feedback is delivered when key is being released. In *Pressed-Released*, tactile feedback is delivered when key is being pressed and being released.

4.4. Data Analysis.

We measured several performance metrics for this experiment. We measured typing speed in words per minute (WPM), keystroke per character (KSPC) for measuring typing efficiency, and minimum string distance (MSD) [44] for typing accuracy. We also measured key distance ratio for typing accuracy. Given that $P(x, y)$ is user's touch point on the key, we divided the touched key into four regions using two diagonal lines to obtain a distance ratio from the centroid to user's touch point $P(x, y)$ (Figure 9).

If $P(x, y)$ is in *Surface* 1 or 3, we calculate the ratio by the following equation:

$$\text{Ratio}R = \frac{((P(y) - (\text{height}/2) \times 100)}{(\text{height}/2)}, \quad \text{if } P(y) > \frac{\text{height}}{2},$$

$$\text{Ratio}R = \frac{(P(y) \times 100)}{\text{height}/2}, \quad \text{if } P(y) < \frac{\text{height}}{2}. \tag{1}$$

And if $P(x, y)$ is in *Surface* 2 or 4, then:

$$\text{Ratio}R = \frac{((P(x) - (\text{width}/2)) \times 100)}{\text{width}/2}, \quad \text{if } P(x) > \frac{\text{width}}{2},$$

$$\text{Ratio}R = \frac{(P(x) \times 100)}{(\text{width}/2)}, \quad \text{if } P(y) < \frac{\text{width}}{2}, \tag{2}$$

where height and width are the sizes of actual virtual key.

4.5. Procedure.

We conducted a typing test experiment to measure the typing performance on a virtual keyboard apparatus as shown in Figure 6. Similar to Study 1, we asked participants to type as fast and accurately as possible using their two thumbs while holding the touchscreen phone.

Combinations of three feedback intensity conditions (i.e., *Linear*, *Reversed*, and *Constant*) and three feedback delivery time conditions (i.e., *Pressed*, *Released*, and *Pressed-Released*) were used for this experiment. No feedback was also used for the comparison. Each session was composed of 10 lines of phrases and the session is repeated three times, yielding thirty tasks in total. Latin square was used to reduce the ordering effect.

After completing each session, we asked participants to rate their typing experiences based on the following questions on a 7-point Likert-type scale from 1 (strongly disagree) to 7 (strongly agree): *Typing speed*—this tactile feedback is helpful for increasing typing speed; *Typing accuracy*—this tactile feedback is helpful for reducing typing errors; *Preference*—I prefer this feedback; and *Confidence*—this tactile feedback gives me the confidence of key click. We added this *Confidence* question to find out

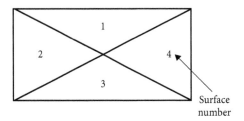

FIGURE 9: A key with four regions and their surface numbers. The key is divided using two diagonal lines.

whether degradation of confidence is caused by spatial or temporal feedback.

4.6. Results: Typing Speed and KSPC.

Figure 10(a) shows typing speed in WPM. As observed in this figure, the feedback condition of the "Attached" and "Constant" feedback reached the highest. Figure 10(b) shows typing efficiency in KSPC. We observed that the lowest KSPC (meaning highest efficiency) was observed in the "Attached" and "Linear" condition and second lowest KSPC was observed in "Attached" and "Constant" condition. A two-way repeated-measure ANOVA confirmed that feedback condition was not a significant factor for typing speed ($F = 0.276$, $p = 0.981$) or typing efficiency ($F = 0.45$, $p = 1.00$).

4.7. Results: Different Ratio.

We also measured the calculated ratio used in the data analysis section. This value can be an indicator of how close user's key press is to the centroid of a key. Figure 11 shows the ratio for each experimental condition. As illustrated in this figure, the lowest ratio was observed in AC (Attached and Constant) condition, meaning highest typing accuracy.

4.8. Results: User Preference.

Figure 12 shows the results of user preference on *Accuracy, Speed, Comfort,* and *Confidence* for temporal feedback conditions (*Attached, Detached, Both,* or *None*), respectively. It is clearly showed that participants preferred tactile feedback that is provided when key is attached (pressed) in terms of accuracy, but preferred feedback that is provided when key is detached (released) in terms of confidence.

Figure 13 shows the results of user preferences for spatial feedback conditions (*No, Linear, Reversed,* or *Constant*), respectively. It is clearly shown that participants preferred *Reversed Feedback* condition to provide key click confidence. However, participants preferred *Constant Feedback* condition in terms of typing speed.

A two-way repeated-measure ANOVA showed that there was a significant effect of both temporal conditions and spatial conditions for *Confidence* measurement ($F(2, 90) = 3.326$, $p = 0.05$ for temporal condition and $F(2, 90) = 4.122$, $p = 0.02$ for spatial conditions), but not for other measurements. Post hoc Tukey tests on both spatial and temporal feedback conditions showed the significantly

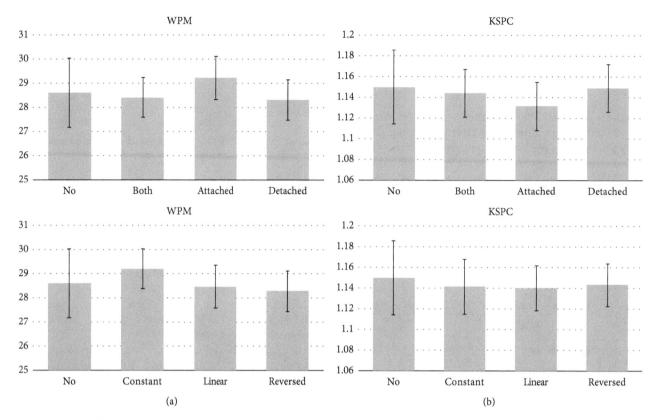

FIGURE 10: Simple Effects. Typing speed in words per minute (a) and typing efficiency in KSPC (b) for each condition. Upper images are for temporal encoding factors, and the lower images are for edge encoding factors.

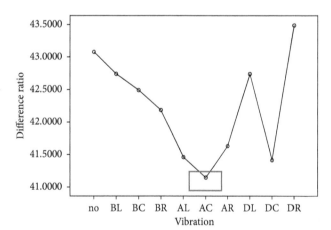

FIGURE 11: Ratio for each condition.

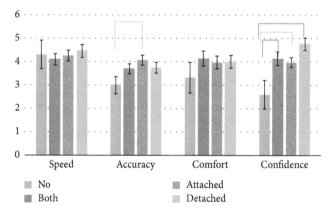

FIGURE 12: User preference on *Accuracy* (a) and *Confidence* (b) for temporal feedback conditions.

different pairs (see red connecting lines in Figures 12 and 13 for *p* value less than 0.05 and blue connecting lines for *p*value less than 0.1).

4.9. Discussion. In this study, we confirmed that the spatial and temporal modifications on the vibration feedback did not give us a physical performance enhancement as shown in Figure 10. In addition, spatial modification of the feedback did not show significant effect on the user's preference compared to the *Constant Feedback* case, while the feedback itself was

clearly advantageous on giving a user confidence. Temporal modification on the feedback also did not have clear effect on user's preference. This result indicates that the proposed temporal and spatial modification techniques are not very effective on physical performance and on user preference.

We presume that these results are due to the fact that participants are naïve to the information embedded on the feedback, so they did not successfully utilize the feedback. This can lead to the need of more intensive experiment that involves prolonged usage of the interface.

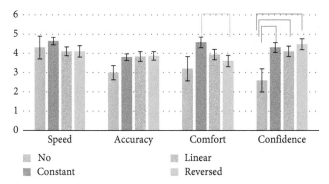

FIGURE 13: User preference on *Confidence* (a) and *Typing Speed* (b) for spatial feedback conditions.

In next section, in order to partially avoid this problem, we examined the user's workload when using a virtual keyboard with new feedback techniques. This time, participants were aware of the meaning of difference in the feedback.

5. User Evaluation: Work Load Analysis

Based on the findings from our first study, we learned that there exist different preferences of tactile feedback for different groups of users. We first discovered that people with higher typing performance (i.e., people who type fast with mobile keyboard) prefer comparably reduced intensity of tactile feedback, whereas beginner and intermediate-level users prefer comparably higher intensity of tactile feedback. From the second study, although there was no statistical meaning, we also had evidences that people prefer *Attached* with *Constant* tactile feedback to increase typing speed and provide confidence for their key click confirmation. Based on these findings, we propose a user-adaptive tactile keyboard on mobile device and compare it with the existing keyboard to explore the feasibility of our work (Figure 14).

5.1. Development of User-Adaptive Tactile Keyboards. We developed two different versions of user-adaptive tactile keyboards for this study. First one is feedback mode change-based keyboard. Basically, it measures the user's typing speed in real time and adaptively changes its feedback mode based on user's typing performance—we call this keyboard *Feedback Mode Change Keyboard*. For example, if typing speed is slow, the mode becomes DR (Detached-Reversed—meaning feedback is delivered when key is detached, and tactile feedback intensity is linearly decreased in regard to the touched point from the centroid of the key (Table 1)). If typing speed is increased up to intermediate level, the mode becomes BL (Attached/Detached-Linear—meaning feedback is delivered when key is attached and also detached, and tactile feedback intensity is linearly increased in regard to the touched point from the centroid of the key). If typing speed is further increased up to expert level, the mode is changed to AC (Attached-Constant—meaning feedback is delivered when key is attached, and tactile feedback intensity is constantly provided).

FIGURE 14: Adaptive keyboard design.

We also developed another keyboard that changes its tactile feedback intensity based on user's typing speed but does not consider the feedback mode—we call this keyboard *Feedback Intensity Change Keyboard*. In this keyboard, we set a number of intensity levels and only allowed the level to change one level at a time. This is due to the fact that people felt uncomfortable when the tactile feedback intensity changed dramatically during our pilot study.

For the baseline of this experiment, we also used an ordinary virtual keyboard in which tactile feedback intensity is fixed and constant at all times. We set the size of all keys to the same size and added tactile feedback for space bar and delete key. We stored user's typing speed and all the characters of the keys that they typed.

5.2. Participants. A total of ten university students who did not participate in Study 1 or Study 2 participated in this study. Participants were paid for their participation. They reported that they are healthy and have no disabilities. They also reported that they had prior experience of typing in English and familiar with virtual keyboards.

5.3. Procedure. Similar to studies 1 and 2, we asked participants to take a seat and naturally hold the touchscreen phone with their two hands for thumb typing (Figure 15). However, instead of asking them to type as fast and as accurately as they can, we asked them to type as comfortable as they can just like they perform the typing task during ordinary days. Since the goal of this study is to find the feasibility of our user-adaptive tactile keyboard in daily life setting, we focus more on comfort use of user-adaptive tactile keyboard than typing performance. For this reason, we provided 10 lines of multiple sentences to simulate real-world scenarios of daily text entry on mobile phone. Three sessions with three different keyboard types (*Feedback Mode Change Keyboard*, *Feedback Intensity Change Keyboard*, and baseline) were provided for each user.

After each session, we asked participants to evaluate the workload by providing a questionnaire based on NASA-TLX [45]. A total of six questions were asked: mental demand, physical demand, temporal demand, overall performance, frustration level, and effort (Figure 16).

(a)

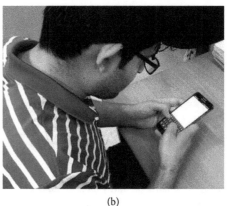
(b)

FIGURE 15: Typing application (a) and a user conducting a typing task (b).

Mental demand	How much focus on or effect while typing?
Physical demand	How much physical activity was required? (finger or arm?)
Temporal demand	How much time did you spend to finish the task?
Performance	How successful were you in performing the task?
Effort	How annoyed were you due to vibration?
Frustration	How hard did you have to type the key affected by vibration?

FIGURE 16: Six questions based on NASA-TLX.

In order to calculate the final workload of the conditions, we used analytic hierarchy process (AHP) to objectify the subjective response from the participants by assigning a weight to subjective response from NASA-TLX [45].

5.4. Results and Discussion: NASA-TLX. Based on the weight from AHP, we calculated the workload with each keyboard condition. The workload for *Feedback Mode Change Keyboard* and *Feedback Intensity Change Keyboard* is 31.44 and 38.72, respectively. The workload for baseline was 36.98. Compared to the baseline, the *Feedback Mode Change Keyboard* reduced the workload by 17%. This is notable since the feedback modes did not have statistical effect in study 2. From this, we can speculate that the feedback modes have positive effect to reduce user's mental load, although the users do not have preference.

6. General Discussion

This work focuses on feasibility of user-adaptive tactile keyboard on mobile touchscreen. We noticed that there exists a number of tactile feedback that mobile device can provide. We also noticed that not every user likes the same and simple tactile feedback. We hypothesized that there exists a relationship between feedback intensity and users, and we further hypothesized that these users can be grouped by a factor—such as typing speed. We also believed that we

can build a user-adaptive tactile keyboard for better usability and performance, and this can be extended to virtual and augmented reality.

We first observed how mobile users behave based on tactile feedback intensity and what intensity level that different user group prefers. We then studied if different tactile feedback mode affects the user preference based on user's typing speed. Interestingly, we discovered that users preferred tactile feedback that is provided when key is attached (pressed) in terms of accuracy, but preferred feedback that is provided when key is detached (released) in terms of confidence. We also discovered that people in *Beginner* and *Intermediate* groups preferred comparably higher levels of feedback intensity (*Mid* for *Beginner* and *Mid/High* for *Intermediate*) than those in *Expert* group (*Low* for *Expert*).

Based on our findings, we built two different versions of user-adaptive tactile keyboard on mobile phone. We conducted a user study to investigate the feasibility of the keyboards by analysing the workload. As results, *Feedback Mode Change Keyboard* reduced the workload by 17 percent. We believe that this achievement will shed light on the development of user-adaptive tactile keyboard on mobile platform. Our future work will extend the present study by considering the use of adaptive keyboard in VR setting as typing is one of most challengeable tasks in VR and user-adaptive keyboard can be a good solution to address this issue.

7. Conclusions

This work investigates the effects of user-adaptive tactile keyboard on mobile touchscreen. We performed two studies to investigate the relationship between tactile feedback intensity and user preference. We then implemented user-adaptive tactile keyboards on mobile platform to verify their feasibility. We performed a user study to evaluate the workload of our proposed keyboard and showed the improvement in workload.

Acknowledgments

This work was supported by the NRF of Korea through the Basic Research Program (NRF-2017R1D1A1B03031272) and by the MSIP through IITP (no. 2017-0-00179) (HD Haptic Technology for Hyper Reality Contents).

References

[1] Pew Research Center, *Smartphone Ownership and Internet Usage Continues to Climb in Emerging Economies*, Pew Research Center, Washington, DC, USA, 2016.

[2] S. Zhu, T. Luo, X. Bi, and S. Zhai, "Typing on an invisible keyboard," in *Proceeding of SIGCHI Conference on Human Factors in Computing Systems (CHI 2018)*, pp. 1–13, ACM, Montreal, QC, Canada, April 2018.

[3] J. Gugenheimer, D. Dobbelstein, C. Winkler, G. Haas, and E. Rukzio, "FaceTouch: enabling touch interaction in display fixed UIs for mobile virtual reality," in *Proceeding of Annual Symposium on User Interface Software and Technology (UIST 2016)*, pp. 49–60, ACM, Tokyo, Japan, October 2016.

[4] Y. Kim and G. J. Kim, "HoVR-Type: smartphone as a typing interface in VR using hovering," in *Proceeding of ACM Conference on Virtual Reality Software and Technology*, pp. 333–334, ACM, Munich, Germany, November 2016.

[5] J. Gugenheimer, "Nomadic virtual reality: exploring new interaction concepts for mobile virtual reality head mounted displays," in *Proceeding of Annual Symposium on User Interface Software and Technology (UIST 2016 Adjunct)*, pp. 9–12, ACM, Tokyo, Japan, October 2016.

[6] E. Hoggan, A. Crossan, S. Brewster, and T. Kaaresoja, "Audio or tactile feedback: which modality and when?," in *Proceeding of the SIGCHI Conference on Human Factors in Computing Systems (CHI 2009)*, pp. 2253–2256, ACM, New York, NY, USA, April 2009.

[7] H. Korhonen, "Audio feedback on a mobile phone for number dialling," in *Proceeding of the 7th International Conference on Human Computer Interaction with Mobile Devices & Services*, pp. 345–346, ACM, Salzburg, Austria, September 2005.

[8] T. Noguchi, Y. Fukushima, and I. Yairi, "Evaluating information support system for visually impaired people with mobile touch screens and vibration," in *Proceeding of the 13th International ACM SIGACCESS Conference on Computers and Accessibility*, pp. 243-244, ACM, Dundee, Scotland, UK, October 2011.

[9] E. Matias, I. S. MacKenzie, and W. Buxton, "One-handed touch typing on a QWERTY keyboard," *Human-Computer Interaction*, vol. 11, no. 1, pp. 1–27, 1996.

[10] N. Green, J. Kruger, C. Faldu, and R. Amant, "A reduced QWERTY keyboard for mobile text entry," in *Proceeding of CHI Extended Abstracts on Human Factors in Computing Systems*, pp. 1429–1432, ACM, Vienna, Austria, April 2004.

[11] S. Hwang and G. Lee, "Qwerty-like 3×4 keypad layouts for mobile phone," in *Proceeding of CHI Extended Abstracts on Human Factors in Computing Systems*, pp. 1479–1482, ACM, Portland, Oregon, April 2005.

[12] A. Chang and C. O'Sullivan, "Audio-haptic feedback in mobile phones," in *Proceeding of CHI Extended Abstracts on Human Factors in Computing Systems*, pp. 1264–1267, ACM, Portland, Oregon, April 2005.

[13] H. Chen, J. Santos, M. Graves, K. Kim, and H. Tan, "Tactor localization at the wrist," in *Proceeding of Haptics: Perception, Devices and Scenarios*, pp. 209–218, Springer, Madrid, Spain, June 2008.

[14] K. B. Perry and J. P. Hourcade, "Evaluating one handed thumb tapping on mobile touchscreen devices," in *Proceeding of Graphics Interface 2008*, pp. 57–64, Canadian Information Processing Society, Windsor, Canada, May 2008.

[15] M. Silfverberg, I. S. MacKenzie, and P. Korhonen, "Predicting text entry speed on mobile phones," in *Proceeding of SIGCHI Conference on Human Factors in Computing Systems (CHI 2000)*, pp. 9–16, ACM, Hague, The Netherlands, April 2000.

[16] Y.-S. Jeong, T.-K. Yeom, J. S. Park, and J. H. Park, "Efficient model of Korean graphemes based on a smartphone keyboard," *Electronic Commerce Research*, vol. 13, no. 3, pp. 357–377, 2013.

[17] K. Lyons, D. Plaisted, and T. Starner, "Expert chording text entry on the twiddler one-handed keyboard," in *Proceeding of Eighth International Symposium on Wearable Computers, 2004. ISWC 2004*, vol. 1, pp. 94–101, IEEE, Arlington, VA, October 2004.

[18] I. S. MacKenzie and S. X. Zhang, "The design and evaluation of a high performance soft keyboard," in *Proceedings of the SIGCHI Conference on Human Factors in Computing Systems*, pp. 25–31, ACM, Pittsburgh, Pennsylvania, USA, May, 1999.

[19] S. Kim, M. Sohn, J. Pak, and W. Lee, "One-key keyboard: a very small QWERTY keyboard supporting text entry for wearable computing," in *Proceeding of 18th Australia Conference on Computer-Human Interaction: Design: Activities, Artefacts and Environments*, pp. 305–308, ACM, Sydney, Australia, November 2006.

[20] A. Karlson, "Interface design for single-handed use of small devices," in *Proceeding of Annual ACM Symposium on User Interface Software and Technology (UIST 2008)*, pp. 27–30, ACM, Monterey, CA, USA, October 2008.

[21] T. Nakagawa and H. Uwano, "Usability differential in positions of software keyboard on smartphone," in *Proceeding of IEEE Global Conference on Consumer Electronics (GCCE 2012)*, pp. 304–308, Tokyo, Japan, October 2012.

[22] L. Findlater, J. O. Wobbrock, and D. Wigdor, "Typing on flat glass: examining ten-finger expert typing patterns on touch surfaces," in *Proceeding of SIGCHI Conference on Human Factors in Computing Systems*, pp. 2453–2462, ACM, Vancouver, Canada, 2011.

[23] T.-B. Ryu, "Performance analysis of text entry with preferred one hand using smart phone touch-keyboard," *Journal of the Ergonomics Society of Korea*, vol. 30, no. 1, pp. 259–264, 2011.

[24] A. Sears, "Improving touchscreen keyboards: design issues

and a comparison with other devices," *Interacting with Computers*, vol. 3, no. 3, pp. 253–269, 1991.

[25] A. Sears, D. Revis, J. Swatski, R. Crittenden, and B. Shneiderman, "Investigating touchscreen typing: the effect of keyboard size on typing speed," *Behaviour and Information Technology*, vol. 12, no. 1, pp. 17–22, 1993.

[26] T. Page, "Usability of text input interfaces in smartphones," *Journal of Design Research*, vol. 11, no. 1, pp. 39–56, 2013.

[27] A. Sears and Y. Zha, "Data entry for mobile devices using soft keyboards: understanding the effects of keyboard size and user tasks," *International Journal of Human-Computer Interaction*, vol. 16, no. 2, pp. 163–184, 2003.

[28] S. Brewster, F. Chohan, and L. Brown, "Tactile feedback for mobile interactions," in *Proceeding of SIGCHI Conference on Human Factors in Computing Systems (CHI 2007)*, pp. 159–162, ACM, San Jose, CA, USA, April 2007.

[29] E. Rabin and A. M. Gordon, "Tactile feedback contributes to consistency of finger movements during typing," *Experimental Brain Research*, vol. 155, no. 3, pp. 362–369, 2004.

[30] A. Hoffmann, D. Spelmezan, and J. Borchers, "Typeright: a keyboard with tactile error prevention," in *Proceedings of SIGCHI Conference On Human Factors In Computing Systems*, pp. 2265–2268, ACM, Boston, MA, USA, April 2009.

[31] D. Rudchenko, T. Paek, and E. Badger, "Text revolution: a game that improves text entry on mobile touchscreen keyboards," in *Proceeding of Pervasive Computing*, pp. 206–213, Springer, San Francisco, CA, USA, June 2011.

[32] T. Kaaresoja, L. M. Brown, and J. Linjama, "Snap-crackle-pop: tactile feedback for mobile touch screens," in *Proceedings of Eurohaptics*, pp. 565-566, Paris, France, July 2006.

[33] H. Nishino, R. Goto, T. Kagawa et al., "A touch screen interface design with tactile feedback," in *Proceeding of IEEE International Conference on Complex, Intelligent and Software Intensive Systems (CISIS 2011)*, pp. 53–60, Seoul, South Korea, June 2011.

[34] J. Luk, J. Pasquero, S. Little, K. MacLean, V. Levesque, and V. Hayward, "A role for haptics in mobile interaction: initial design using a handheld tactile display prototype," in *Proceeding of SIGCHI Conference on Human Factors in Computing Systems*, pp. 171–180, ACM, Montréal, Québec, Canada, April 2006.

[35] J. R. Kim and H. Z. Tan, "A study of touch typing performance with keyclick feedback," in *Proceeding of IEEE Haptics Symposium (HAPTICS 2014)*, pp. 227-233, Houston, TX, USA, February 2014.

[36] S. Brewster and A. King, "An investigation into the use of tactons to present progress information," in *Proceeding of Human-Computer Interaction (INTERACT 2005)*, pp. 6–17, Springer, Rome, Italy, September 2005.

[37] J. Lylykangas, V. Surakka, K. Salminen et al., "Designing tactile feedback for piezo buttons," in *Proceeding of SIGCHI Conference on Human Factors in Computing Systems (CHI 2011)*, pp. 3281–3284, ACM, Boston, MA, USA, 2011.

[38] K. Kim, "Perception-based tactile soft keyboard for the touchscreen of tablets," *Mobile Information Systems*, vol. 2018, Article ID 4237346, 9 pages, 2018.

[39] B. Han, K. Kim, K. Yatani, and H. Z. Tan, "Text entry performance evaluation of haptic soft QWERTY keyboard on a tablet device," in *Proceeding of the EuroHaptics 2014*, pp. 325–332, Springer, Versailles, France, June 2014.

[40] J. R. Kim and H. Z. Tan, "Haptic feedback intensity affects touch typing performance on a flat keyboard," in *Proceeding of Euro-Haptics 2014*, pp. 369–375, Springer, Versailles, France, June 2014.

[41] J. R. Kim and H. Z. Tan, "Effect of information content in sensory feedback on typing performance using a flat keyboard," in *Proceeding of IEEE World Haptics Conference 2015*, pp. 228–234, Chicago, Illinois, June 2015.

[42] I. S. MacKenzie and R. W. Soukoreff, "Phrase sets for evaluating text entry techniques," in *Proceeding of Extended Abstracts of the SIGCHI Conference on Human Factors in Computing Systems*, pp. 754-755, ACM, 2003.

[43] D. A. Grant, "The Latin square principle in the design and analysis of psychological experiments," *Psychological Bulletin*, vol. 45, no. 5, p. 427, 1948.

[44] R. W. Soukoreff and I. S. MacKenzie, "Metrics for text entry research: an evaluation of MSD and KSPC, and a new unified error metric," in *Proceeding of SIGCHI Conference on Human Factors in Computing Systems (CHI 2003)*, pp. 113–120, ACM, Lauderdale, Florida, USA, April 2003.

[45] S. G. Hart and L. E. Stavenland, "Development of NASA-TLX (Task Load Index): results of empirical and theoretical research," in *Human Mental Workload*, P. A. Hancock and N. Meshkati, Eds., NASA Ames Research Center, Mountain View, CA, USA, 1988.

9

Multimodal Affective Computing to Enhance the User Experience of Educational Software Applications

Jose Maria Garcia-Garcia (ID),[1] **Víctor M. R. Penichet** (ID),[1] **María Dolores Lozano** (ID),[1] **Juan Enrique Garrido** (ID),[2] **and Effie Lai-Chong Law**[3]

[1]*Research Institute of Informatics, University of Castilla-La Mancha, Albacete, Spain*
[2]*Escuela Politécnica Superior, University of Lleida, Lleida, Spain*
[3]*Department of Informatics, University of Leicester, Leicester, UK*

Correspondence should be addressed to María Dolores Lozano; maria.lozano@uclm.es

Academic Editor: Salvatore Carta

Affective computing is becoming more and more important as it enables to extend the possibilities of computing technologies by incorporating emotions. In fact, the detection of users' emotions has become one of the most important aspects regarding Affective Computing. In this paper, we present an educational software application that incorporates affective computing by detecting the users' emotional states to adapt its behaviour to the emotions sensed. This way, we aim at increasing users' engagement to keep them motivated for longer periods of time, thus improving their learning progress. To prove this, the application has been assessed with real users. The performance of a set of users using the proposed system has been compared with a control group that used the same system without implementing emotion detection. The outcomes of this evaluation have shown that our proposed system, incorporating affective computing, produced better results than the one used by the control group.

1. Introduction

In 1997, Rosalind W. Picard [1] defined Affective Computing as "computing that relates to, arises from, or influences emotions or other affective phenomena." Since then, a general concern about the consideration of the emotional states of users for different purposes has arisen in different research fields (phycology [2, 3], marketing, computing, etc.).

Concretely, the underlying idea of Affective Computing is that computers that interact with humans need the ability to at least recognize affect [4]. Indeed, affective computing is a new field, with recent results in areas such as learning [5], information retrieval, communications [6], entertainment, design, health, marketing, decision-making, and human interaction where affective computing may be applied [7]. Different studies have proved the influence of emotions in consumers' behaviour [8] and decision-making activities [9].

In computer science research, we could study emotions from different perspectives. Picard mentioned that if we want computers to be genuinely intelligent and to interact naturally with us, we must give computers the ability to recognize, understand, even to have and express emotions. In another different research work, Rosalind pointed out some inspiring challenges [10]: sensing and recognition, modelling, expression, ethics, and utility of considering affect in HCI. Studying such challenges still makes sense since there are gaps to be explored behind them. In human-computer interaction, emotion helps regulate and bias processes in a helpful way.

In this paper, we focus our research in the use of emotions to dynamically modify the behaviour of an educational software application according to the user feelings, as described in Section 3. This way, if the user is tired or stressed, the application will decrease its pace and, in some cases, the level of difficulty. On the other hand, if the user is getting bored, the application will increase the pace and the difficulty level so as to motivate the user to continue using the application.

Finally, we have assessed the application to prove that including emotion detection in the implementation of educational software applications considerably improves users' performance.

The rest of the paper is organized in the following sections: In Section 2, some background concepts and related works are presented. In Section 3, we describe the educational software application we have developed enhanced with affective computing-related technologies. Section 4 shows the evaluation process carried out to prove the benefits of the system developed. Finally, Section 5 presents some conclusions and final remarks.

2. Background Concepts and Related Works

In this section, a summary of the background concepts of affective computing and related technologies is put forward. We provide a comparison among the different ways of detecting emotions together with the technologies developed in this field.

2.1. Affective Computing. Rosalind Picard used the term "affective computing" for the first time in 1995 [11]. This technical report established the first ideas on this field. The aim was not to answer questions such as "what are emotions?," "what causes them?," or "why do we have them?," but to provide a definition of some terms in the field of affective computing.

As stated before, the term "affective computing" was finally set in 1997 as "computing that relates to, arises from, or deliberately influences emotion or other affective phenomena" [1]. More recently, we can find the definition of Affective computing as the study and development of systems and devices that can recognize, interpret, process, and simulate human affects [4]. In other words, any form of computing that has something to do with emotions. Due to the strong relation with emotions, their correct detection is the cornerstone of Affective Computing. Even though each type of technology works in a specific way, all of them share a common core in the way they work, since an emotion detector is, fundamentally, an automatic classifier.

The creation of an automatic classifier involves collecting information, extracting the features which are important for our purpose, and finally training the model so it can recognize and classify certain patterns [12]. Later, we can use the model to classify new data. For example, if we want to build a model to extract emotions of happiness and sadness from facial expressions, we have to feed the model with pictures of people smiling, tagged with "happiness" and pictures of people frowning, tagged with "sadness." After that, when it receives a picture of a person smiling, it identifies the shown emotion as "happiness," while pictures of people frowning will return "sadness" as a result.

Humans express their feelings through several channels: facial expressions, voices, body gestures and movements, and so on. Even our bodies experiment visible physical reactions to emotions (breath and heart rate, pupil's size, etc.).

Because of the high potential of knowing how the user is feeling, this kind of technology (emotion detection) has experienced an outburst in the business sector. Many technology companies have recently emerged, focused exclusively on developing technologies capable of detecting emotions from specific input. In the following sections, we present a brief review of each kind of affective information channel, along with some existing technologies capable of detecting this kind of information.

2.2. Emotion Detection Technologies. This section presents a summary of the different technologies used to detect emotions considering the various channels from which affective information can be obtained: emotion from speech, emotion from text, emotion from facial expressions, emotion from body gestures and movements, and emotion from physiological states [13].

2.2.1. Emotion from Speech. The voice is one of the channels used to gather emotional information from the user of a system. When a person starts talking, they generate information in two different channels: primary and secondary [14].

The primary channel is linked to the syntactic-semantic part of the locution (what the person is literally saying), while the secondary channel is linked to paralinguistic information of the speaker (tone, emotional state, and gestures). For example, someone says "That's so funny" (primary channel) with a serious tone (secondary channel). By looking at the information of the primary channel, the message received is that the speaker thinks that something is funny and by looking at the information received by the secondary channel, the real meaning of the message is worked out: the speaker is lying or being sarcastic.

Four technologies in this category can be highlighted: *Beyond Verbal* [15], *Vokaturi* [16], *EmoVoice* [17] and *Good Vibrations* [18]. Table 1 shows the results of the comparative study performed on the four analyzed technologies.

2.2.2. Emotion from Facial Expressions. As in the case of speech, facial expressions reflect the emotions that a person can be feeling. Eyebrows, lips, nose, mouth, and face muscles: they all reveal the emotions we are feeling. Even when a person tries to fake some emotion, still their own face is telling the truth. The technologies used in this field of emotion detection work in an analogous way to the ones used with speech: detecting a face, identifying the crucial points in the face which reveal the emotion expressed, and processing their positions to decide what emotion is being detected.

Some of the technologies used to detect emotions from facial expressions are *Emotion API* (Microsoft Cognitive Services) [19], *Affectiva* [20], *nViso* [21], and *Kairos* [22]. Table 2 shows a comparative study.

As far as the results are concerned, every tested technology showed considerable accuracy. However, several conditions (reflection on glasses and bad lightning) mask important facial gestures, generating wrong results. For

TABLE 1: Comparison of emotion detection technologies from speech.

Name	API/SDK	Requires Internet	Information returned	Difficulty of use	Free software
Beyond verbal	API	Yes	Temper, arousal, valence, and mood (up to 432 emotions)	Low	No
Votakuri	SDK	No	Happiness, neutrality, sadness, anger, and fear	Medium	Yes
EmoVoice	SDK	No	Determined by developer	High	Yes
Good vibrations	SDK	—	Happy level, relaxed level, angry level, scared level, and bored level	Medium	No

TABLE 2: Comparison of emotion detection technologies from facial expressions.

Name	API/SDK	Requires Internet	Information returned	Difficulty of use	Free software
Emotion API	API/SDK	Yes	Happiness, sadness, fear, anger, surprise, neutral, disgust, and contempt	Low	Yes (limited)
Affectiva	API/SDK	Yes	Joy, sadness, disgust, contempt, anger, fear, and surprise[1]	Low	Yes, with some restriction
nViso	API/SDK	No	Happiness, sadness, fear, anger, surprise, disgust, and neutral	—	No
Kairos	API/SDK	Yes	Anger, disgust, fear, joy, sadness, and surprise[2]	Low	Yes, only for personal use

[1]Besides, it also detects different facial expressions, gender, age, ethnicity, valence, and engagement. [2]Besides, it also detects user head position, gender, age, glasses, facial expressions, and eye tracking.

example, an expression of pain, in a situation in which eyes and/or brows cannot be seen, can be detected as a smile by these technologies (because of the stretching, open mouth).

As far as time is concerned, *Emotion API* and *Affectiva* show similar times to scan an image, while *Kairos* takes much longer to produce a result. Besides, the amount of values returned by *Affectiva* provides much more information to the developers, and it is easier to interpret the emotion that the user is showing than when we just have the weight of six emotions, for example. It is also remarkable the availability of *Affectiva*, which provides free services to those dedicated to research and education or producing less than $1,000,000 yearly.

2.2.3. Emotion from Text. There are certain situations in which the communication between two people, or between a person and a machine, does not have the visual component inherent to face-to-face communication. In a world dominated by telecommunications, words are powerful allies to discover how a person may be feeling. Although emotion detection from text (also referred as sentiment analysis) must face more obstacles than the previous technologies (spelling errors, languages, and slang), it is another source of affective information to be considered. Since emotion detection from texts analyzes the words contained on a message, the process to analyze a text takes some more steps than the analysis of a face or a voice. There is still a model that needs to be trained, but now text must be processed in order to use it to train a model [23]. This processing involves tasks of tokenization, parsing and part-of-speech tagging, lemmatization, and stemming, among others. Four technologies of this category are *Tone Analyzer* [24], *Receptiviti* [25], *BiText* [26], and *Synesketch* [27].

Due to the big presence of social media and writing communication in the current society, this field is, along

with emotion detection from facial expressions, one of the most attractive fields to companies: posts from social media, messages sent to "Complaints" section, and so on. Companies which can know how their customers are feeling have an advantage over companies which cannot. Table 3 shows a comparative study of some of the key aspects of each technology. It is remarkable that as far as text is concerned, most of the companies offer a demo or trial version on their websites, while companies working on face or voice recognition are less transparent in this aspect. Regarding their accuracy, the four technologies have yielded good values. On the one hand, *BiText* has proved to be the simplest one, as it only informs if the emotion detected is good or bad. This way, the error threshold is wider and provides less wrong results. On the other hand, *Tone Analyzer* has proved to be less clear on its conclusions when the text does not contain some specific key words.

As far as the completeness of results is concerned, *Receptiviti* has been the one giving more information, revealing not only affective information but also personality-related information. The main drawback is that all these technologies (except *Synesketch*) are pay services and may not be accessible to everyone. Since *Synesketch* is not as powerful as the rest, it will require an extra effort to be used.

2.2.4. Emotion from Body Gestures and Movement. Even though people do not use body gestures and movement to communicate information in an active way, their body is constantly conveying affective information: tapping with the foot, crossing the arms, tilting the head, changing our position a lot of times while seated, and so on. Body language reveals what a person is feeling in the same way our voice does.

However, this field is quite new, and there is not a clear understanding about how to create systems able to detect emotions relating to body language. Most researchers have

TABLE 3: Comparison of emotion detection technologies from text.

Name	API/SDK	Requires Internet	Information returned	Difficulty of use	Free software
Tone analyzer	API	Yes	Emotional, social, and language tone	Low	No
Receptiviti	API	Yes	See [29]	Low	No
BiText	API	Yes	Valence (positive/negative)	Low	No
Synesketch	SDK	No	Six basic emotions	Medium	Yes

focused on facial expressions (over 95 per cent of the studies carried out on emotions detection have used faces as stimuli), almost ignoring the rest of channels through which people reveal affective information [28].

Despite the newness of this field, there are several proposals focused on recognizing emotional states from body gestures, and these results are used for other purposes. Experimental psychology has already demonstrated how certain types of movements are related to specific emotions [29]. For example, people experimenting fear will turn their bodies away from the point which is causing that feeling; people experimenting happiness, surprise, or anger will turn their bodies towards the point causing that feeling.

Since there are no technologies available for emotion detection from body gestures, there is not any consensus about the data we need to detect emotions in this way. Usually, experiments on this kind of emotion detection use frameworks (as for instance, *SSI*) or technologies to detect the body of the user (as for instance, Kinect), so the researches are responsible for elaborating their own models and schemes for the emotion detection. These models are usually built around the joints of the body (hands, knees, neck, head, elbows, and so on) and the angle between the body parts that they interconnect [30], but in the end, it is up to the researchers.

2.2.5. Emotion from Physiological States. Physiologically speaking, emotions originate on the limbic system. Within this system, the amygdala generates emotional impulses which create the physiological reactions associated with emotions: electric activity on face muscles, electrodermal activity (also called galvanic skin response), pupil dilatation, breath and heart rate, blood pressure, brain electric activity, and so on. Emotions leave a trace on the body, and this can be measured with the right tools.

Nevertheless, the information coming directly from the body is harder to classify, at least with the category system used in other emotion detection technologies. When working with physiological signals, the best option is to adopt a classification system based on a dimensional approach [25]. An emotion is not just "happiness" or "sadness" anymore, but a state determined by various dimensions, like valence and arousal. It is because of this that the use of physiological signals is usually reserved for research and studies, for example, related to autism. There are no emotion detection services available for this kind of detection based on physiological states, although there are plenty of sensors to read these signals.

In a recent survey on mobile affective computing [31], authors make a thorough review of the current literature on affect recognition through smartphone modalities and show the current research trends towards mobile affective computing. Indeed, the special capacities of mobile devices open new research challenges in the field of affective computing that we aim to address in the mobile version of the system proposed.

Finally, we can also find available libraries to be used in different IDEs (integrated development environments) supporting different programming languages. For instance, NLTK, in python [32] can be used to analyze natural language for sentiment analysis. Scikit-learn [33], also in python, provides efficient tools for data mining and data analysis with machine learning techniques. Lastly, OpenCV (Open Source Computer Vision Library) [34] supports C++, Python, and Java interfaces in most operating systems. It is designed for computer vision and allows the detection of elements caught by the camera in real time to analyze the facial points detected according, for instance, to the Facial Action Coding System (FACS) proposed by Ekman and Rosenberg [35]. The data gathered could be subsequently processed with the scikit-learn tool.

```
aff_information = get_affective_information()
#aff_information = {"face": [...], "voice": [...],
"mimic": [...]}
stress_flags = {"face": 0.0, "voice": 0.0, "mimic": 0.0}
#values from 0 to 1 indicating stress levels detected
for er_channel, measures in aff_information:

    measure_stress(er_channel, measures, stress_flags)
if (stress_flags["face"] > 0.6 and

    stress_flags["voice"] < 0.3 and
    stress_flags["mimic"] < 0.1):
    #reaction to affective state A

if (stress_flags["face"] < 0.1 and

    stress_flags["voice"] < 0.1 and
    stress_flags["mimic"] < 0.5):
    #reaction to affective state B

...
```

3. Modifying the Behaviour of an Educational Software Application Based on Emotion Recognition

Human interaction is, by definition, multimodal [36]. Unless the communication is done through phone or text, people can see the face of the people they are talking to, listen to their voices, see their body, and so on. Humans are, at this point, the best emotion detectors as we combine information from several channels to estimate a result. This is how multimodal systems work.

It is important to remark that a multimodal system is not just a system which takes, for example, affective information from the face and from the voice and calculates the average of each value. The hard part of implementing one of these systems is to combine the affective information correctly. For example, a multimodal system combining text and facial expressions that detects a serious face and the message "it is very funny" will return "sarcasm/lack of interest," while the result of combining these results in an incorrect way will return "happy/neutral." It is proven that by combining information from several channels, the accuracy of the classification improves significantly.

For example, let us imagine we need to assess the stress levels of a person considering the affective information gathered through three different channels: affective information extracted from facial expressions, voice, and body language. Since we have more than one channel, we can support each measure taken from each channel with values detected in the others.

This way, it is possible not only to confirm with a high level of certainty the occurrence of an affective state, but also to detect situations that could not be sensed without performing multimodal emotion detection, as sarcasm.

The following code snippet shows an easy example of affective information combination. The mere fact of considering a measure in the context of more affective information gives us a whole new dimension of information.

To this end, we have developed an initial prototype in order to study how using multimodal emotion detection systems on educational software applications could enhance the user experience and performance. The proposed prototype, named emoCook, has been developed as a game to teach English to 9–11-year-old children. Information about this prototype can be found at [37]. At present, the prototype is only available in Spanish as it is initially addressed to Spanish-speaking children in the process of learning English.

The architecture of this application is shown in Figure 1. During the gameplay, the user is transmitting affective information (Figure 1-1) through their face, their voice, their behaviour, and so on. The prototype is receiving this information (Figure 1-3) and sending it to several third-party emotion detection services (Figure 1-4). After retrieving this information (Figure 1-5), we put it in context to extract conclusions from it about the user's performance (Figure 1-6). Based on these results, the pace and difficulty level of the game changes (Figure 1-7), adapting it to the user's affective state (Figure 1-2).

The theme of the game was focused on cooking issues to practice vocabulary and expressions related to this topic. It is organized in different recipes, from the easiest to hardest. Each recipe is an independent level and is divided into two parts. The first part is a platform game in which the player must gather all the ingredients needed to cook the recipe (Figure 2). The ingredients are falling from the sky all the time, along with other food we do not need for the recipe. If the player catches any food that is not in the ingredients list,

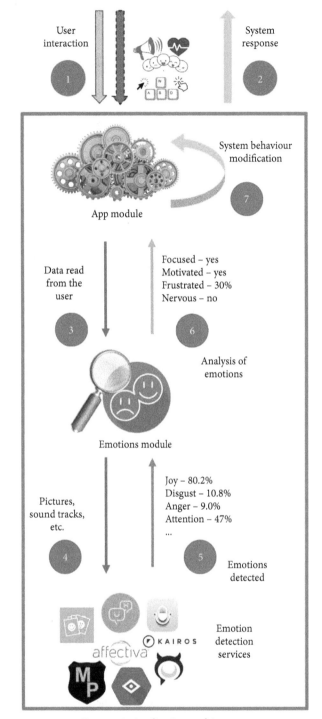

FIGURE 1: Application architecture.

it is considered as a mistake. The maximum number of mistakes allowed per level is five.

After finishing this first part, the system shows a set of sentences (more or less complex) including vocabulary related to the recipe that the player has to read out loud to practice speaking and pronunciation. If the user fails thrice to read a sentence, the system will move to the next one, or finish the exercise if it is the last sentence.

This prototype has been implemented with three emotion detection technologies, which monitor the player's

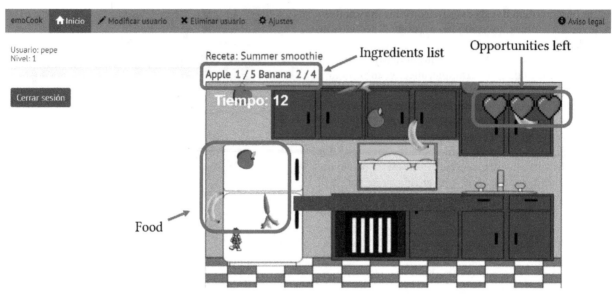

FIGURE 2: emoCook prototype.

affective state, and the results obtained are used to change the difficulty level and the pace of the game. Each time the player finishes a level, the affective data are analyzed, and according to the results, the difficulty of the next level is set. The technologies integrated in the system are the following:

(i) *Affectiva*. It uses the camera feed to read the facial expression of the player.

(ii) *Beyond Verbal*. It gathers the audio collected during the speech exercise to identify the affective state of the player attending to their speech features.

(iii) *Keylogger*. The game keeps a record of the keys pressed by the users, considering different factors: when they press a correct key, when they do not, when they press it too fast, and so on.

Because of changes on Beyond Verbal API, affective data from the speech could not be collected, so in the end, only data from the facial expression (using Affectiva) and from the behaviour when pressing keys (using Keylogger) were used. Affectiva is a third-party service, while Keylogger was developed within the prototype.

A mobile version of the system is also available, and it can be used through a browser running on a mobile device [37]. This way, the game can be controlled both with the arrow keys in a keyboard and by touching on a tactile screen. Touching on the left-hand side of the screen makes the character move to the left. Touching on the right-hand side of the screen makes the character move to the right and touching twice very quickly in any part of the screen makes the character jump upwards.

Figure 3 shows a screenshot of the mobile version of the application running in the Firefox browser in a mobile device. The possibility of using the system through a mobile device opens new ways of detecting emotions that we aim to explore in further research. For instance, we could use sensors such as the accelerometer or gyroscope to gather affective information. Initial trials have been performed with

FIGURE 3: Mobile version of emoCook system.

the API offered in [38] with promising results that will be further explored.

4. Evaluation of the System

In order to prove the initial hypothesis, the system has been assessed with real users by applying the method described in this section.

4.1. Participants and Context. We recruited sixteen children aged between 10 and 11 years old belonging to the same primary school and with a similar level of English knowledge to avoid differences in the education level that could affect the evaluation results. Their parents had been previously informed and authorised their participation in this evaluation. The setup of the experiment consisted of two laptops, one in front of the other so that participants could not see each other. Both laptops were equipped with mouse and webcam and Windows 10 as operating system and were connected to the same Wi-Fi network. The prototype was accessed through the browser Google Chrome in both laptops. This setup was prepared in a room the English teachers of the primary school provided us within the school premises.

4.2. Evaluation Metrics. The system was measured considering three types of metrics: effectiveness, efficiency, and satisfaction, that is, the users' subjective reactions when using the system. Effectiveness was measured by considering task completion percentage, error frequency, and frequency of assistance offered to the child. Efficiency was measured by calculating the time needed to complete an activity, specifically, the mean time taken to achieve the activity. Besides, some other aspects were also considered such as the number of attempts needed to successfully complete a level, number of keystrokes, and the number of times a key was pressed too fast as an indicative signal of nervousness.

Finally, satisfaction was measured with the System Usability Scale (SUS) slightly adapted for teenagers and kids [39]. This questionnaire is composed of ten items related to the system usage. The users had to indicate the degree of agreement or disagreement on a 5-point scale.

4.3. Experimental Design. After several considerations regarding the evaluation process for games used in learning environments [40], the following features were established:

(i) *Research Design.* The sample of participants was divided into two groups of the same size, being one of them the control group. This control group tested the application implemented without emotion detection and hence without modifying the behaviour of the application in real time according to the child's emotions. This one was called the System 2 group. The other group tested the prototype implemented with emotion detection which adapted its behaviour, by modifying the pace of the game and difficulty level, according to the emotions detected on the user, in such a way that if the user becomes bored, the system increases the pace of the game and difficulty level and on the contrary, if the user becomes stressed or nervous, the system decreases the speed of the game and difficulty level. This one was called the System 1 group. By doing this, it can be shown how using emotion detection to dynamically vary the difficulty level of an educational software application influences the performance and user experience of the students.

(ii) *Intervention.* The test was conducted in the premises of the primary school in a quiet room where just the participants (two at a time) using System 1 and System 2 and the evaluators were present. We prepared two laptops of similar characteristics, one of them running System 1 with the version of the application implemented with emotion recognition and the other laptop running System 2 with the version of the application without emotion recognition.

The whole evaluation process was divided into two parts:

(i) *Introduction to the Test.* At the beginning of the evaluation, the procedure was explained to the sixteen children at a time, and the game instructions for the different levels were given.

(ii) *Performing the Test.* Kids were called in pairs to the room where the laptops running System 1 and System 2 were prepared. None of the children knew what system they were going to play with. At the end of the evaluation sessions, the sixteen children completed the SUS questionnaire. Researchers were present all the time, ready to assist the participants and clarify doubts when necessary. When a participant finished the test, they returned to their classroom and called the next child to go in the evaluation room.

To keep the results of each participant fully independent, the sixteen users were introduced on the database of the prototype with the key "evalX," being "X" a number. Users with an odd "X" used System 1, while those with an even "X" were assigned to System 2 (control group).

The task that the participants had to perform was to play the seven levels of the prototype, including each level a platform game and a reading out loud exercise. The data collected during the evaluation sessions were subsequently analyzsed, and the outcomes are described next.

4.4. Evaluation Outcomes and Discussion. Although participants with System 1 needed, on average, a bit more time per level to finish (76.18 seconds against 72.7), we could appreciate an improvement on the performance of the participants using System 1, as most of them made less than 5 mistakes on the last level, while only one of the control group users of System 2 had less than 5 mistakes.

Figure 4 shows the evolution of the average number of mistakes, which increases in the control group (System 2) from level 4 onwards. Since the game adapts its difficulty (in System 1), after detecting a peak of mistakes in the fourth level (as a sign of stress, detected as a combination of negative feelings found in the facial expression and the way the participant used the keyboard), the difficulty level was reduced. This adaptation made the next levels easier to play for participants using System 1, what was reflected in less mental effort. Since participants using System 2 did not have this feature, their average performance got worse.

On average, participants using System 1 needed 1.33 attempts to finish each level, while participants using System 2 needed 1.59, almost 60% more. Also, the ratio of mistakes to total keystrokes was also higher in the case of System 2 users (19% against the 12% from users of System 1). Likewise, System 2 users asked for help more often (13 times) than System 1 users (10 times). In future experimental activities, the sample size would be increased in order to obtain more valuable data.

The evaluation was carried out as a between-subjects design with *emotion recognition* as the independent variable (using or not using emotion recognition features) and

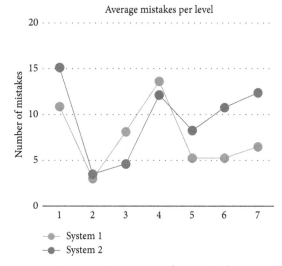

FIGURE 4: Average mistakes per level.

TABLE 4: Satisfaction results for System 1.

Participant	SUS score
1	90
3	90
5	90
7	100
9	92.5
11	87.5
13	82.5
15	80
Mean	89.06

TABLE 5: Satisfaction results for System 2.

Participant	SUS score
2	92.5
4	75
6	87.5
8	85
10	75
12	92.5
14	72.5
16	90
Mean	83.75

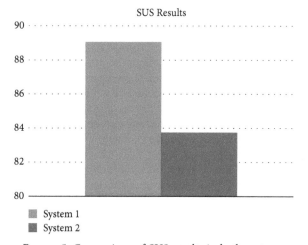

FIGURE 5: Comparison of SUS results in both systems.

attempts (attempts needed to finish each level), *time* (time (seconds) needed to finish each level), *mistakes* (number of mistakes), *keystrokes* (number of keystrokes), and *stress* (number of times a key was pressed too fast in a short time) as the dependent variables.

We performed a standard t-test [41] to compare the means of each dataset and test the null hypothesis that there was no significant difference in the students' performance when using emotion recognition to adapt the system behaviour. We used $\alpha = 0.05$ as our limit for statistical significance, with significant results reported below.

Regarding *keystrokes* ($t = 0.97$; $p = 0.666$), *mistakes* ($t = -1.51$; $p = 0.26$), and *stress* ($t = 1.13$; $p = 0.51$), t-test results confirmed the null hypothesis was false and, thus, that the two datasets are significantly different.

Although the dependent variables *time* ($t = 0.44$; $t = 1.31$) and *attempts* ($t = -0.42$; $t = 1.33$) were similar in both datasets, the efficiency (considered as the lowest number of actions a user needs to finish each level) is greater in users of System 1, even though both users of System 1 and System 2 finished within a similar time frame, what helped the first ones to make less mistakes. The outcomes of the evaluation shown in Figure 4 indicate a clear improvement when using System 1 as the number of mistakes increases in users of System 2 at higher difficulty levels.

Finally, Table 4 and Table 5 show the results of the SUS scores per system and participant. The final value is between 0 and 100, 100 being the highest degree of user's satisfaction. As we can see, System 1 users rated the application with a higher level of satisfaction compared to the level obtained by users of System 2, as shown in Figure 5.

5. Conclusions and Final Remarks

Emotion detection, together with Affective Computing, is a thriving research field. Few years ago, this discipline did not even exist, and now there are hundreds of companies working exclusively on it, and researchers are investing time and resources on building affective applications. However, emotion detection has still many aspects to improve in the coming years.

Applications which obtain information from the voice need to be able to work in noisy environments, to detect subtle changes, maybe even to recognize words and more complex aspects of human speech, like sarcasm.

The same applies for applications that detect information from the face. Most people use glasses nowadays, which can greatly complicate accurate detection of facial expressions.

Applications able to read body gestures do not even exist now, even though it is a source of affective information as valid as the face. There are already applications for body detection (Kinect), but there is no technology like *Affectiva* or *Beyond Verbal* for the body yet.

Physiological signals are even less developed, because of the imposition of sensors that this kind of detection requires.

However, some researchers are working on this issue so physiological signals can be used as the face or the voice. In a not too distant future, reading the heartbeat of a person with just a mobile with Bluetooth may not be as crazy as it may sound.

Previous technologies analyze the impact of an emotion in our bodies, but what about our behaviour? A stressed person usually tends to make more mistakes. In the case of a person interacting with a system, this will be translated in faster movements through the user interface, or more mistakes when selecting elements or typing, and so on. This can be logged and used as another indicator of the affective state of a person.

All these technologies are not perfect. Humans can see each other and estimate how other people are feeling within milliseconds, and with a small threshold error, but these technologies can only try to figure out how a person is feeling according to some input data. To get more accurate results, more than one input is required, so multimodal systems are the best way to guarantee results with the highest levels of accuracy.

In this paper, we present an educational software application that incorporates affective computing by detecting the users' emotional states to adapt its behaviour to the emotions detected. Assessing this application in comparison with another version without emotion detection, we can conclude that the user experience and performance is higher when including a multimodal emotion detection system. Since the system is continuously adapting itself to the user according to the emotions detected, the level of difficulty adjusts much better to their real needs.

On the basis of the outcomes of this research, new challenges and possibilities in other kind of applications will be explored; for example, we could "stress" a user in a game if the emotions detected show that the user is bored. The application could even introduce dynamically other elements to engage the user in the game. What is too simple bores a user, whereas what is too complex causes anxiety. Changing the behaviour of an application dynamically according to the user's emotions, and also according to the nature of the application, increases the satisfaction of the user and helps them decrease the number of mistakes.

As future work, among other things, we aim to improve the mobile aspects of the system and explore further the challenges that the sensors offered by mobile devices bring about regarding emotion recognition, especially in educational settings.

Acknowledgments

This research work has been partially funded by the regional project of JCCM with reference SBPLY/17/180501/000495, by the Scholarship Program granted by the Spanish Ministry of Education, Culture and Sport, and the predoctoral fellowship with reference 2017-BCL-6528, granted by the University of Castilla-La Mancha. We would also like to thank the teachers and pupils from the primary school "Escolapios" who collaborated in the assessment of the system.

References

[1] R. W. Picard, *Affective Computing*, MIT Press, Cambridge, UK, 1997.

[2] E. Johnson, R. Hervás, C. Gutiérrez, T. Mondéjar, and J. Bravo, "Analyzing and predicting empathy in neurotypical and nonneurotypical users with an affective avatar," *Mobile Information Systems*, vol. 2017, Article ID 7932529, 11 pages, 2017.

[3] S. Koelstra, C. Muhl, M. Soleymani et al., "DEAP: a database for emotion analysis using physiological signals," *IEEE Transactions on Affective Computing*, vol. 3, no. 1, pp. 18–31, 2012.

[4] R. Kaliouby, "We need computers with empathy," *Technology Review*, vol. 120, no. 6, p. 8, 2017.

[5] S. L. Marie-Sainte, M. S. Alrazgan, F. Bousbahi, S. Ghouzali, and A. W. Abdul, "From mobile to wearable system: a wearable RFID system to enhance teaching and learning conditions," *Mobile Information Systems*, vol. 2016, Article ID 8364909, 10 pages, 2016.

[6] M. Li, Y. Xiang, B. Zhang, and Z. Huang, "A sentiment delivering estimate scheme based on trust chain in mobile social network," *Mobile Information Systems*, vol. 2015, Article ID 745095, 20 pages, 2015.

[7] B. Ovcjak, M. Hericko, and G. Polancic, "How do emotions impact mobile services acceptance? A systematic literature review," *Mobile Information Systems*, vol. 2016, Article ID 8253036, 18 pages, 2016.

[8] P. Williams, "Emotions and consumer behavior," *Journal of Consumer Research*, vol. 40, no. 5, pp. viii–xi, 2014.

[9] E. Andrade and D. Ariely, "The enduring impact of transient emotions on decision making," *Organizational Behavior and Human Decision Processes*, vol. 109, no. 1, pp. 1–8, 2009.

[10] R. W. Picard, "Affective computing: challenges," *International Journal of Human-Computer Studies*, vol. 59, no. 1-2, pp. 55–64, 2003.

[11] R. W. Picard, "Affective computing," Tech. Rep. 321, M.I.T Media Laboratory Perceptual, Computing Section, Cambridge, UK, 1995.

[12] I. Morgun, *Types of Machine Learning Algorithms*, 2015.

[13] J. García-García, V. Penichet, and M. Lozano, "Emotion detection: a technology review," in *Proceedings of XVIII International Conference on Human Computer Interaction*, Cancún, México, September 2017.

[14] S. Casale, A. Russo, G. Scebba, and S. Serrano, "Speech emotion classification using machine learning algorithms," in *Proceedings of IEEE International Conference on Semantic Computing 2008*, pp. 158–165, Santa Monica, CA, USA, August 2008.

[15] Beyond Verbal, "Beyond verbal–the emotions analytics," May 2017, http://www.beyondverbal.com/.

[16] Vokaturi, May 2017, https://vokaturi.com/.

[17] T. Vogt, E. André, and N. Bee, "EmoVoice—a framework for online recognition of emotions from voice," in *Perception in Multimodal Dialogue Systems*, E. André, L. Dybkjær, W. Minker, H. Neumann, R. Pieraccini, and M. Weber, Eds., Springer, Berlin, Heidelberg, Germany, 2008.

[18] Good Vibrations, "Good vibrations company B.V.–recognize emotions directly from the voice," May 2017, http://good-vibrations.nl.

[19] Microsoft, "Microsoft cognitive services–emotion API," May 2017, https://www.microsoft.com/cognitive-services/en-us/emotion-api.

[20] Affectiva, "Affectiva," May 2017, http://www.affectiva.com.

[21] nViso, "Artificial intelligence emotion recognition software-nViso," May 2017, http://nviso.ch/.

[22] Kairos, "Face recognition, emotion analysis & demographics," May 2017, https://www.kairos.com/.

[23] H. Binali and V. Potdar, "Emotion detection state of the art," in *Proceedings of the CUBE International Information Technology Conference on-CUBE '12*, pp. 501–507, New York, NY, USA, September 2012.

[24] IBM, "Tone analyzer," May 2017, https://tone-analyzer-demo.mybluemix.net/.

[25] Receptiviti, May 2017, http://www.receptiviti.ai/.

[26] Bitext, "Bitext API," May 2017, https://api.bitext.com.

[27] U. Krčadinac, "Synesketch: free open-source textual emotion recognition and visualization," May 2017, http://krcadinac.com/synesketch/.

[28] A. Kleinsmith and N. Bianchi-Berthouze, "Affective body expression perception and recognition: a survey," *IEEE Transactions on Affective Computing*, vol. 4, no. 1, pp. 15–33, 2013.

[29] R. W. Picard, "Future affective technology for autism and emotion communication," *Philosophical Transactions of the Royal Society B: Biological Sciences*, vol. 364, no. 1535, pp. 3575–3584, 2009.

[30] Universität Augsburg University, "OpenSSI," May 2017, https://hcm-lab.de/projects/ssi/.

[31] E. Politou, E. Alepis, and C. Patsakis, "A survey on affective computing," *Computer Science Review*, vol. 25, pp. 79–100, 2017.

[32] N. Hardeniya, *NLTK Essentials*, Packt Publishing Limited, Birmingham, UK, 2015.

[33] G. Hackeling, *Mastering Machine Learning with Scikit-Learn*, Packt Publishing Limited, Birmingham, UK, 2014.

[34] J. Howse, P. Joshi, and M. Beyeler, *OpenCV: Computer Vision Projects with Python*, Packt Publiser Limited, Birmingham, UK, 2016.

[35] P. Ekman and E. Rosenberg, *What the Face Reveals: Basic and Applied Studies of Spontaneous Expression Using the Facial Action Coding System (FACS)*, Oxford University Press, Oxford, UK, 2005.

[36] J. Tao and T. Tan, "Affective computing: a review," *Lecture Notes in Computer Science*, vol. 3784, pp. 981–995, 2005.

[37] J. M. Garcia-Garcia, "emoCook," December 2017, https://emocook.herokuapp.com/.

[38] A. Bar, "What web can do today. An overview of the device integration HTML5 APIs," June 2018, https://whatwebcando.today/device-motion.html.

[39] Usability.gov, "Usability.gov-improving the user experience," January 2018, https://www.usability.gov/get-involved/blog/2015/02/working-with-kids-and-teens.html.

[40] A. All, E. P. Nuñez Castellar, and J. Van Looy, "Assessing the effectiveness of digital game-based learning: best practices," *Computers & Education*, vol. 92-93, pp. 90–103, 2016.

[41] D. Garson, *Significance Testing: Parametric and Nonparametric*, Statistical Associates Publishing, Blue Book Series, Asheboro, NC, USA, 2012.

10

Artificial Intelligence to Prevent Mobile Heart Failure Patients Decompensation in Real Time: Monitoring-Based Predictive Model

Nekane Larburu [1,2] **Arkaitz Artetxe,**[1,2] **Vanessa Escolar,**[3] **Ainara Lozano,**[3] **and Jon Kerexeta**[1]

[1]*Vicomtech, Paseo Mikeletegi 57, 20009 Donostia/San Sebastian, Spain*
[2]*Biodonostia Health Research Institute, P. Doctor Begiristain s/n, 20014 San Sebastian, Spain*
[3]*Hospital Universitario de Basurto (Osakidetza Health Care System), Avda. Montevideo 18, 48013 Bilbao, Spain*

Correspondence should be addressed to Nekane Larburu; nlarburu@vicomtech.org

Guest Editor: Giovanna Sannino

Rapid advances in ICT and collection of large amount of mobile health data are giving room to new ways of treating patients. Studies suggest that telemonitoring systems and predictive models for clinical support and patient empowerment may improve several pathologies, such as heart failure, which admissions rate is high. In the current medical practice, clinicians make use of simple rules that generate large number of false alerts. In order to reduce the false alerts, in this study, the predictive models to prevent decompensations that may lead into admissions are presented. They are based on mobile clinical data of 242 heart failure (HF) patients collected for a period of 44 months in the public health service of Basque Country (Osakidetza). The best predictive model obtained is a combination of alerts based on monitoring data and a questionnaire with a Naive Bayes classifier using Bernoulli distribution. This predictive model performs with an AUC = 67% and reduces the false alerts per patient per year from 28.64 to 7.8. This way, the system predicts the risk of admission of ambulatory patients with higher reliability than current alerts.

1. Introduction

Since these early days, the advances on ICT have given a huge opportunity to telemedicine applications and new e-Health services [1]. Along with this phenomenon are the large quantities of mobile data that are being collected and processed these days. The growth in these two areas are leading in advanced health-care systems that not only provide continuous support to clinicians or informal care givers (e.g., family members), but also to patients. In this context, telemedicine systems that monitor ambulatory patients and guide them in their daily routine are emerging. Nevertheless, often all the potential of the mobile-health data used to support clinical professionals and patients is not sufficiently exploited. Other times, the exploited clinical data, in the form of, for example, predictive models to identify patients at high risk, are not applied in a real setting to support clinicians and patients.

Studies suggest that artificial intelligence by means of predictive models and telemonitoring systems for clinical support and patient empowerment may improve several pathologies [2], such as heart failure.

Heart failure (HF) is a clinical syndrome caused by a structural and/or functional cardiac abnormality. HF patients suffer decompensations, which is defined by Mangini et al. [3] as a clinical syndrome in which a structural or functional change in the heart leads to its inability to eject and/or accommodate blood within physiological pressure levels, thus causing a functional limitation and requiring immediate therapeutic intervention [3]. Hence, decompensations may lead in hospital admissions, which in this study are defined as emergency admissions and hospital admissions, and home interventions. As Ponikowski et al. presented in [4], the prevalence of HF depends on the definition applied, but it is approximately 1-2% of the adults in developed countries, rising to more than 10% among people >70 years of age. Hence, due to the aging population, an increase in the number of HF patients is expected in the future. Therefore, predicting the risk of a patient to suffer a decompensation may prevent admissions and readmissions, improving both patient care and hospital management, which has a high impact on costs and clinical professionals time. The first step to predict the risk of decompensation is to telemonitor ambulatory patients. Next, we need reliable systems to assess the risk. Most telemedicine systems apply alerts or rule-based systems to detect potential complications of ambulatory patients [5–8]. But these usually contain large number of false alerts, and hence, these systems are not trustworthy (Table 1).

Our hypothesis is that with the usage of artificial intelligence (AI) by means of, for instance, predictive models, it is possible to detect decompensations of ambulatory patients and reduce false alerts. In this context, this research extends the study for readmissions detection [9] and presents predictive models of a telemedicine system for heart failure patients, called INCAR. INCAR has been developed to (i) be generally applicable in HF patients, (ii) improve the clinical practice by developing an accurate system that detects the risk of decompensation and suggest actions to prevent them on time, (iii) allow professionals to maintain an efficient and personalized support and follow-up of patient, (iv) give patients support when required and guide them in risk situations, informing clinicians accordingly, and (v) reduce HF patients admission and readmission rate, which have a high economic impact.

This paper focuses on the development of predictive models to detect decompensations, and it is structured as follows: First, the *Related Work* section summarizes the state of the art on telemedicine systems for heart failure and the role of predictive models on telemedicine systems. *Materials* section presents the database used in this study and the characteristics of the dataset. *Methods* presents the applied methods to assess the risk of an ambulatory HF patient to suffer a decompensation that may lead into admission. In *Results*, the outcomes obtained for each of the developed predictive models is presented. Finally, *Discussion* presents the results and limitations of the study, and *Conclusion* gives a summary of the contributions and future work.

2. Related Work

2.1. Telemedicine Systems for HF Patients. Being HF a disease with high prevalence and high readmission rate, the usage of telemedicine systems in this area is common [7]. Chaudhry et al. [2] telemonitored patients by means of telephone-based interactive voice-response system and concluded that the simple phone-based telemonitorization does not improve the outcomes (i.e., readmission, death). Nevertheless, most of current telemonitoring systems do not simply implement telephone-based monitorization, but also the transmission of mobile health data, such as bodyweight, heart rate, and blood pressure [7]. Besides, more advanced noninvasive systems transfer electrocardiograph (ECG) tracings, oxygen saturation, and physical activity (e.g., pedometer)

data. Apart from noninvasive telemedicine systems, invasive systems enable the transfer of variables measured invasively, such as transthoracic impedance and pulmonary and left atrial pressures. But literature studies do not present significantly better results when implementing invasive measurements into their telemedicine systems in terms of HF decompensation prevention. Nonetheless, some benefits have been presented when applying impedance instead of weight for detecting HF patients early decompensation, as presented by Abraham et al. [5] and Gyllensten et al. [6].

2.2. Alerts for HF Patients. Most studies implement "simple" alerts to prevent decompensations based on these data. One of the implemented techniques is *Rule of Thumb (RoT)* based on simple rules (i.e., when a measurement goes beyond or below a given threshold or when they are based on simple difference between the current value of an attribute and a previous measurement that occurs a predefined number of days in the past) [5–8]. Other studies, such as Zhang et al. [7], Gyllensten et al. [6], and Ledwidge et al. [8], make also use of more sophisticated techniques, such as the *Moving Average (MA)* or similar techniques that calculate the variations applied to usually weight. The *Cumulative Sum (CUMSUM)*, applied by Adamson et al. [10], is typically used for detecting changes and implies that when a continuous variation of a measurement is produced over time, that tendency will result in an alert. Additionally, Gilliam et al. [11], apply the *multivariate method*, which consists on the usage of several data elements that are incorporated into a multivariate logistic regression model to form the probability of an event occurring. From the studied papers, we could conclude that each type of alert may work best depending on the applied attribute. For instance, techniques related with MA work best when applied to weight. On the other hand, CUMSUM is one of the best methods when applied to transthoracic or intrathoracic impedance.

Table 1 presents the results of different studies that determine whether a monitored HF patient will have a decompensation, usually implementing alerts. Due to the large number of days that do not end in an admission, even when the computed specificity values are high, the number of false positives could remain too high for the clinical practice, so it is not an optimal testing value in this scope. Taken into account this limitation, based on the literature studies, we could consider the number of false alerts per patient per year (FA/pt-y) as de facto standard to determine the number of false positives. However, as shown in Table 1, some of the studies present the specificity value for determining how well the no admissions are detected using own techniques to compute it.

2.3. Predictive Models on Telemedicine Systems. As shown above, most telemedicine systems apply alerts or rule-based systems to detect potential complications of ambulatory patients. This is not only present in the context of HF, but also in diabetes, atrial fibrillation, and other clinical domains [12, 13]. Hence, there is a lack of the usage of collected data that could lead in more accurate solutions by means of, for instance, predictive models.

TABLE 1: Summary of decompensation detection studies.

Study	Data type	Dataset	Method	Results
Zhang et al. [7]	Weight	135 patients; 1964 days monitoring	RoT	Se = 58.3%, Sp = 54.1%
			MACD	Se = 20.4%, Sp = 89.4% (AUC = 0.55%)
Gyllensten et al. [6]	Weight	91 patients; 10 months	RoT	Se = 20%, Sp = 90%
			MACD	Se = 33%, Sp = 91%
			CUMSUM	Se = 13%, Sp = 91%
	Noninvasive transthoracic bioimpedance	91 patients; 10 months	RoT	Se = 13%, Sp = 91%
			MACD	Se = 13%, Sp = 91%
			CUMSUM	Se = 13%, Sp = 91%
Adamson et al. [10]	Blood pressure	274 patients	CUMSUM	Se = 83.1%, FA = 4.1/pt-y
Abraham et al. [5]	Intrathoracic impedance	156 patients; 537 ± 312 days	RoT	Se = 76.4%; FA = 1.9/pt-y
	Weight	156 patients; 537 ± 312 days	RoT	Se = 21%; FA = 4.3/pt-y
Ledwidge et al. [8]	Weight	87 patients; 23.9 ± 12 weeks	RoT	Se = 21%; Sp = 86%
			HeartPhone algorithm (based on MA)	Se = 82%; Sp = 68%
Gilliam et al. [11]	Multivariate	201 patients		Se = 41%; FA = 2/pt-y

Several studies in the context of HF develop predictive models to determine whether a patient will be readmitted within 30 days after discharge [14–20]. These predicting models make use of baseline information of patients, such as age, sex, or left ventricular injection fraction, but not daily (or weekly) tele-monitored patient mobile data, such as weight, heart rate, or blood pressure, which could be crucial for detecting and preventing an ambulatory patient admission. In several telemedicine studies applied in diverse pathologies, such as chronic obstructive pulmonary disease [21, 22] and preeclampsia [23], predictive models have been successfully applied. However, in the context of HF, limited studies apply predictive models. Lafta et al. [24] is one of these studies that using several telemonitored attributes (i.e., heart rate, systolic blood pressure, diastolic blood pressure, mean arterial pressure, and oxygen saturation) applied basic time series prediction algorithm, regression-based time series prediction algorithm, and hybrid time series prediction algorithm. The obtained results showed that up to 75% and 98% of accuracy values could be obtained across different patients under three algorithms, but still the accuracy value is not objective enough to determine how well the system performs.

The presented study goes beyond the state of the art and applies classifiers based on alerts applied in current medical practice and state-of-the-art studies. Additionally, this study makes use of baseline information and ambulatory tele-monitored information to build an integral telemedicine system that applies predictive models with double goal: assess ambulatory patients' admission risk to provide both patients and clinicians the appropriate guidance to prevent potential decompensations that may lead to hospital admissions.

3. Materials

3.1. Database. The public hospital OSI Bilbao-Basurto (Osakidetza), located in Basque Country (Spain), has been gathering HF patients' information from June 2014 until February 2018 (44 months) to closely monitor HF patients. For the present study, the dataset contained a cohort of 242 HF patients. Clinicians have collected baseline data (i.e., information collected by a clinician when the patient is diagnosed, Table 2), ambulatory patient monitored data (i.e., information collect from three to seven times per week, Tables 3 and 4), and patients admissions information (i.e., emergency admissions, hospital admissions, and home care interventions that are associated to HF associated with a patient decompensation).

Besides vital signs, a questionnaire is also included into the telemonitoring system to ask patients about their condition, with potential impact on decompensation prediction (Table 4).

3.2. Characteristics of Ambulatory Patients Dataset. In the whole study, 242 patients have been enrolled from June 2014 until February 2018. Of these 242 patients, one patient has been excluded as it is a cirrhotic man who often has interventions of evacuational paracentesis due to a liver pathology not related to HF. There is an average follow-up of 13.5 ± 9.11 months. In this time period, there have been 254 decompensations of which 202 are considered as predictable, since 52 decompensations do not have previous telemonitoring information (i.e., less than 3 times in the last week before the decompensation).

4. Methods

Following the methodology applied for the generation of the predictive models is presented: (i) training and testing dataset construction, (ii) application of alerts implemented in current clinical setting, (iii) selection of the alerts for the study, (iv) generation of the dataset to apply the machine learning classifiers, and (v) the application and comparison of different classifiers.

TABLE 2: Baseline characteristics of the study population.

Characteristics	Description	Median ± SD (percentage)
Age	The age of the patient (years)	78 ± 10.9
Height	The height of the patient (mm)	162.37 ± 10.34
Sex	The sex of the patient (men/women)	57% men
Smoker	If the patient smokes, did smoke, and now do not or never has smoked	15.35% do smoke, 22% did smoke (not now)
LVEF	Left ventricular ejection fraction (%)	42.4 ± 15.21
First diag	Years since first diagnosis	5.8 ± 7.04
Implanted device	If implanted device (peacemaker, implanted cardioverter defibrillator, and cardiac resynchronisation therapy)	22.7%
Need oxygen	If the patient needs oxygen	4.7%
Barthel	Barthel scale	82.98 ± 15.23
Gijón [25]	Sociofamily assessment scale in the elderly that allows the detection of risk situations or social problems.	7.47 ± 2.29
Laboratory		
Urea	Urea (mg/dl)	75.12 ± 37.8
Creatinine	Creatinine (mg/dl)	1.3 ± 0.54
Sodium	Sodium (mEq/L)	140.12 ± 4.14
Potassium	Potassium (g/dl)	4.28 ± 0.74
Haemoglobin	Haemoglobin (g/dl)	13 ± 9.6
Comorbidities		
Rhythm	If sinus rhythm, AF or atrial fluter	Sinus: 37.1%
Atrial fibrillation	If the patient has atrial fibrillation (AF)	57.4%
Pacemaker	If the patient has a pacemaker	14.5%

TABLE 3: Ambulatory patients monitored characteristics of the study population.

Characteristics	Description
SBP	Systolic blood pressure (mmHg)
DBP	Diastolic blood pressure (mmHg)
O2Sat	Oxygen saturation (%)
HR	Heart rate (bpm)
Weight	Body weight (kg)

4.1. Splitting Training and Testing Datasets. To build and test a predictive model, the clinical data are divided in training and testing datasets. The training dataset is used to develop the model, and once it is finished, the resulting model is tested with the testing dataset. This way, the overfitting is prevented, and it is possible to check whether the created model will generalize well. The whole dataset is from telemonitored patients starting from June of 2014 until February 2018. The training dataset contains 132 predictable decompensations (i.e., with at least 3 monitorizations in the last week before a decompensations) out of 174 patients, with an average follow-up per patient of 13.47 ± 7.47 months. The testing set contains 70 predictable decompensations out of 162 patients, with an average follow-up per patient of 5.41 ± 3.48 months.

4.2. Applied Alerts for Ambulatory Patients Admission. The alerts implemented in current medical practice are used as a filtering method to obtain the instances for training and building the classifiers. This way, we discard the days when

there is no sign of destabilization of any attribute, leading into a more balanced dataset. Therefore, this section presents the different types of alerts that are implemented in medical practice and their performance to select the optimal ones to be applied in our study.

4.2.1. Generic Alerts. The following tables describe the alerts that are being implemented in OSI Bilbao-Basurto Hospital and their sensitivity (Se) and false alerts per patient per year (FA/pt-y) when applied to the training dataset. They are differentiated into "yellow" and "red" alerts, being these last ones more restrictive and, therefore, more critical.

Simple Rules. Table 5 presents the rules based on each parameter individually. The alerts' thresholds presented in Table 5 are the generic ones. But based on personalized clinical cases, clinicians modified some patients' alerts thresholds. For example, if a patient's O2Sat values are always lower than 90, but the patient is stable, the O2Sat alerts are adapted. This study uses the adapted alerts.

Weight Tendency. Besides simple rules, OSI Bilbao-Basurto Hospital also checks the tendency of weight values in order to trigger an alarm (Table 6). This weight change "red" ("yellow") alert performs with a Se value of 0.52 (0.64) and a FA/pt-y of 9.55 (16.38).

Questionnaire. Additionally, OSI Bilbao-Basurto Hospital clinicians make use of the questionnaire (Table 4) and apply the following alert based on the answers from the questionnaire: if three or more answers are the wrong ones, the questionnaire alert would trigger. This alert achieves

TABLE 4: Ambulatory patients questionnaire.

n	Tag	Question	Possible answer
1	Well-being	Comparing with the previous 3 days, I feel:	B/W/S*
2	Medication	Is the medication affecting me well?	Yes/No
3	New medication	During the previous 3 days, did I take any medication without my clinicians' prescription?	Yes/No
4	Diet and exercise	Am I following the diet and exercise recommendations provided by my clinician and nurse?	Yes/No
5	Ankle	In the last 3 days, my ankles are:	B/W/S*
6	Walks	Can I go walking like previous days?	Yes/No
7	Shortness of breath	Do I have fatigue or shortness of breath when I lay down in the bed?	Yes/No
8	Mucus	Do I notice that I started coughing of with phlegm?	Yes/No

*B/W/S = better/worse/same.

TABLE 5: Simple rules implemented by Osakidetza.

Parameter to study	Threshold number	Type of alert	Se	FA/pt-y
SBP	<95 >150	Yellow	0.28	11.4
	<85 >180	Red	0.08	1.4
DBP	<60 >100	Yellow	0.23	9.1
	<50 >110	Red	0.04	0.9
HR	<55 >90	Yellow	0.30	11.2
	<50 >110	Red	0.08	1.4
O2Sat	<94	Yellow	0.15	3
	<90	Red	0.39	13.5

TABLE 6: Weight alerts implemented by Osakidetza.

Parameter to study	Time period	Minimum (kg)	Maximum (kg)	Type of alert
Weight change	5 days	1	2	Yellow
	3 days	1	25	Red
	5 days	2	25	Red

a Se of 0.31 and FA/pt-y of 9.55. To determine which are the questions that perform best, Table 7 presents the Se and FA/pt-y for each of them based on each possible answer.

The answers of "Worse" in the questions of n1 and n5 (Table 4) result in very good predictors of the decompensations considering Se and FA/pt-y values. Questions n3, n4, n6, and n7 also have predictive power, but not as good as n1 and n5. The other questions cannot be considered as alerts, because of their low/null prediction capacity (Table 7).

4.2.2. Implemented Alerts Based on Moving Average. As presented in the *Related Work* section, weight-associated alerts have been improved, and hence, tendency rules for weight have been substituted for a more advanced method,

based on moving average. Moving Average Convergence Divergence (MACD) algorithm calculates the difference between the average value taken from two windows and generates an alert when this difference exceeds a prespecified threshold. Following the same moving average (MA) concept, a similar method is implemented which consists on the following (Figure 1):

(i) a: immediate previous days (starting from the checking day and continuing backwards) over which the average is calculated

(ii) b: previous days (starting from at least the latest day from a and continuing backwards) over which the average should be calculated

(iii) d: distance between the last day of a and first day from b

(iv) Difference threshold (THR): size of difference between a and b average that should generate an alert

In Figure 2, different scores for each possible variable's value for the MA alert are presented. The tested and illustrated results are from all possible combinations of the following variable's values: $a = (2, 3, 7)$, $b = (3, 4, 7, 14)$, $d = (0, 1, 3, 7)$, and THR = $(0.2, 0.5, 0.75, 1, 1.5, 1.8, 2, 3)$.

After representing all the results of the MA algorithm and applying the Youden index [26], the optimal value of these combinations is the one obtained with $a = 2$, $b = 3$, $d = 0$, and THR = 0.75 (green dot in Figure 2). This alert achieves Se value of 0.56 and FA/pt-y of 11.06 in the training set, similar to the results of the already alert-implemented weight alert. But based on the literature [6–8], this latest one is best.

4.3. Selection of Alerts for Instances Generation. To obtain the right dataset of instances, the best combination of alerts is sought. Once the alerts are selected, when at least one of these alerts is triggered, the patient data of that day are used to build the dataset for machine learning model building (see *Built Dataset for Machine Learning Classifiers*). In Table 8, the results of the combinations of different alerts are presented.

R1 refers to the sum of MA weight alert and the two best alerts from the questionnaire related to *ankle* (n5) and

TABLE 7: Questionnaire questions' performance.

n	Tag	Answer	Se	FA/pt-y
1	Well-being	Same	0.88	120.7
		Better	0.42	90.8
		Worse	0.37	2.7
2	Medication	Yes	1	210
		No	0.05	3.8
3	New medication	Yes	0.15	5.3
		No	1	209
4	Diet and exercise	Yes	1	203
		No	0.22	11.18
5	Ankle	Same	0.86	114
		Better	0.44	96
		Worse	0.35	2.9
6	Walks	Yes	0.99	196
		No	0.37	18
7	Shortness of breath	Yes	0.41	19.93
		No	0.96	194
8	Mucus	Yes	0.44	60.5
		No	0.84	153.5

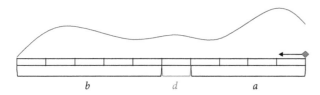

FIGURE 1: Representation of the applied MA algorithm.

well-being (n1) (Tables 4 and 7). If some of these alerts are triggered, R1 is also triggered. R2 refers to the R1 plus the yellow alerts of SBP, DBP, O2Sat, and HR (Table 5). Finally, R3 refers to R2 plus the questions n3, n4, n6, and n7 from Table 4. Since R2 (Table 8) detects almost all decompensations (95%), though FA/pt-y is quite high (FA/pt-y = 51.12), this is the one used to generate the instances for the machine-learning classifiers.

4.4. Built Dataset for Machine-Learning Classifiers. Next, the attributes that are considered for each of the instances and that are applied in the classifiers are presented. Note that the applied attributes come from (i) the telemonitoring dataset, (ii) the baseline dataset, and (iii) the readmission dataset.

(i) *Telemonitoring dataset:*

  (a) The value of SBP, DBP, HR, O2Sat attributes, and, in the case of the weight, the values of the MA algorithm

  (b) The number of consecutive alerts for each type of alert:

    (1) Yellow alerts: the number of yellow alerts that have been triggered in the previous consecutive days related to SBP, DBP, HR, and O2Sat (4 attributes)

    (2) Red alerts: the number of red alerts that have been triggered in the previous consecutive days related to SBP, DBP, HR, and O2Sat (4 attributes)

    (3) MA: the number of alerts that have been triggered in the previous consecutive days for the MA algorithm (1 attribute)

  (c) Questionnaires: the answers of the 8 questions of the questionnaire, shown in Table 7 (8 attributes)

(ii) *Baseline dataset:* the baseline information of the patient shown in Table 2 (24 attributes)

(iii) *Readmissions dataset:* whether in the moment of the instance is about a readmission, i.e., if the last 30 days, the patient has discharged because HF (1 attribute)

4.5. Applied Machine-Learning Classifiers. In this section, we briefly describe the main classification algorithms that were used during the experiments carried out in this work. Since classifier definitions are well known in the literature, we will provide just a summary overview about them.

4.5.1. Naïve Bayes. Naive Bayes methods follow the "naive" assumption that the components of the feature vectors are statistically independent, so that the posterior probability of the class can be approximated as

$$p(y \mid x) = \frac{p(y)\prod_i^n (p(x_i \mid y))}{p(x)}, \tag{1}$$

where $p(x_i)$ is the likelihood of the i-th feature, and $p(y)$ the a priori probability of the class. The Gaussian Naive Bayes assumes that the likelihood follows a Gaussian distribution, where the mean and standard deviation of each feature are estimated from the sample. On the other hand, the Bernoulli Naive Bayes assumes Bernoulli's distribution in the parameters, and hence, it estimates the probability of $p(x_i \mid y)$ following this last distribution.

4.5.2. Decision Tree. Decision Trees (DTs) [27, 28] are built by recursive partitioning of the data space using a quantitative criterion (e.g., mutual information, gain-ratio, gini index), maybe followed by a pruning process to reduce overfitting. Tree leaves correspond to the probabilistic assignment of data samples to classes. One of the most popular implementations of the algorithm is C4.5 [27], which is an extension of the previous ID3 [29] algorithm. At each node, the algorithm selects the feature that best splits the samples according to the normalized information gain.

4.5.3. Random Forest. Random forest [30] is an ensemble classifier consisting of multiple decision trees trained using randomly selected feature subspaces. This method builds multiple decision trees at the training phase. Often, a pruning process is applied to reduce both tree complexity and training data overfitting. In order to predict the class of a new instance, it is put down to each of these trees. Each tree

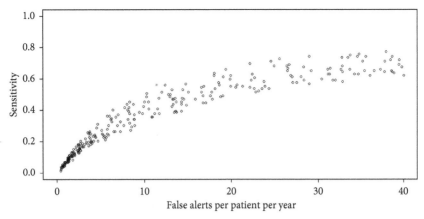

FIGURE 2: Representation of several MA to determine the Younden index.

TABLE 8: Inclusion Criterion performance.

Rules	Description of the rules	Se	FApy
R1	Weight + ankle + well-being	0.79	15.33
R2	R1 + yellow	0.95	51.12
R3	R2 + questionnaire alerts	1	84.5

gives a prediction (votes) and the class having most votes over all the trees of the forest will be selected (majority voting). The algorithm uses the bagging method [31], where each tree is trained using a random subset (with replacement) of the original dataset. In addition, each split uses a random subset of features.

4.5.4. Support Vector Machine. Support vector machines (SVMs) [32, 33] look for the set of support vectors that allow to build the optimal discriminating surface in the sense of providing the greatest margin between classes. In this way, the decision function can be expressed in terms of the support vectors only:

$$f(x) = \text{sign}\left(\sum \alpha_i y_i K(s_i, x) + w_0\right), \qquad (2)$$

where $K(x_i, x_j) \equiv \emptyset(x_i)T\emptyset(x_j)$ is a kernel function, α_i is a weight constant derived from the SVM process, and the s_i is the support vector [33]. Nonlinear kernel functions filling some conditions allow to map a nonlinearly separable discrimination problem into a linearly separable equivalent problem in higher dimensional space.

4.5.5. Neural Network. Multilayer Perceptron (MLP) is a neural network that consists of at least three layers of nodes, namely: (i) an input layer, (ii) one or more hidden layers, and (iii) an output layer. The input layer consists of a set of neurons that represents input features. The hidden layer transforms the outputs of the input layer by means of nonlinear activation functions. The output layer collects the values of the hidden layer and builds the output value. The model is trained using backpropagation, and it can classify data that is not linearly separable.

4.5.6. Class Balancing. In this work, like in many other supervised classification problems, imbalanced class distribution leads to important performance evaluation issues and problems to achieve desired results. The underlying problem with imbalanced datasets is that classification algorithms are often biased towards the majority class and hence, there is a higher misclassification rate of the minority class instances. Although there are several methods that can be used to tackle the class imbalance problem, we have followed an oversampling approach. Random oversampling is the simplest oversampling method, which consists of randomly replicating minority class samples. Despite its simplicity, this method leads easily to overfitting, since it generates exact copies of existing instances [34]. In order to deal with such problems, we have used a more sophisticated technique, namely, synthetic minority oversampling technique (SMOTE). This method over samples the minority class by creating synthetic instances based on its nearest neighbours [35].

Depending on the percentage of synthetic samples that want to be generated (in respect to the original minority class instances), some, or all, minority samples are selected. Having specified beforehand the number of nearest neighbours k, for each sample, the k nearest neighbours are found using the Euclidean distance. Once the nearest samples are selected, a random value between 0 and 1 is generated and multiplied to the distance of each feature between the actual instance and the neighbour. In other words, the vector of coefficients of a random convex linear combination is generated and applied to the k nearest neighbours to create a new sample.

5. Results

This section presents the results obtained after the development of the machine learning classifiers presented in *Applied Machine-Learning Classifiers* and the final results of the selected classifier in the testing dataset.

5.1. Validation Method. Although there are many ways to assess the generalization ability of a ML model, such as cross-validation, time series can be problematic for such validation techniques, as there is an ordered timeline factor to be considered. Henceforth, we use cross-validation on a rolling basis [36], as it is explained in Figure 3.

The training set is separated in the five sets shown in Figure 3. The number written inside the blocks is the number of decompensations corresponding to that period, which is the reason why the dates (on top) are chosen. The splits are not exactly equitable, since all the predecessors of a decompensation must fit within the same block. In Step 1, the classifier is trained in the first block (55 decompensations) and tested in the next block (17 decompensations) getting the score for Step 1. Following, in Step 2, the classifier is trained in Step 1 and tested in the new one (19 decompensations), getting the score for Step 2. Repeating the same with Step 3 and Step 4, we get four scores. It is supposed that the first step is the more unstable, as there are less data to train the classifiers, but, while the training set increases, it is believed that the results will become stable, and the score will converge to its real testing value.

The score value used to test the classifiers is the area under the ROC curve (AUC) [37], a measure that evaluates the balance between sensitivity and specificity and that gets an accurate estimation even in moderately imbalanced datasets, which is our case. The AUC value is used to check how well the classifiers perform and consequently select the best one. To test the global predictive model, we use Se and FA/pt-y which are the ones used in the literature.

5.2. Classifiers Comparison.

In this section, the results of the classifiers explained in *Applied Machine-Learning Classifiers* are presented applied for the training dataset. Additionally, the rolling cross-validation method, presented above, is applied to avoid the overfitting. This way, the classifier(s) with best outcomes and generalizable (and therefore, stable) can be selected for the predictive model.

In Figure 4, the AUC values of each classifier are illustrated for each of the steps defined in the rolling cross-validation method. The points are the mean of the AUCs achieved in each case, with its standard deviation drawn with whiskers. High standard deviation value indicates that the classifier is less generalizable, while low standard deviation hints a stable classifier.

It is expected that the AUCs values converge as the number of steps grow, although with the available dataset, there are a trend of significative improvement in the second step and a worsening trend in the third one. However, Figure 4 clearly shows that the best classifiers are Naïve Bayes (NB) with Bernoulli method and the random forest (RF). NB classifier has lower AUC value than RF, but the standard derivation is almost negligible, and the trend through the steps is more stable. Hence, it is expected that its performance will not vary significantly over time with new data. RF gets the best scores, but is unstable, and it has high standard derivations. Henceforth, NB with Bernoulli method and RF classifiers are selected to validate the models.

Decision tree and SGD classifiers give the lowest results. The other three classifiers (NB with Gaussian distribution, SVM, and MLP classifiers) perform better, but not as good as the selected two.

5.3. Final Results

5.3.1. Alerts Performance.

Since the alerts are used to generate the instances for the machine-learning classifiers (see *Selection of Alerts for Instances Generation*), first, the performance of these in the testing dataset is presented (Table 9).

Comparing these results (Table 9) with the obtained in the training set (Table 5), the weight-associated alerts get worse result. In the case of the questionnaire alerts (Table 10), there is a general worsening comparing with the training set (Table 7). Hence, it is possible to get worse results than the expected when testing the predictive model in the testing set.

5.3.2. Validation Results.

In the current medical practice, their alerts all together obtain the following results: Se = 0.76 and FA/pt-y = 28.64 with the red alerts plus the questionnaire alert, and Se = 1 and FA/pt-y = 88.41 with the yellow alerts plus the questionnaire alert.

After applying the R2 alerts in the testing dataset (see *Selection of Alerts for Instances Generation* section), the selected machine learning classifiers achieved the following AUC values: NB with Bernoulli, AUC = 0.67 and RF, AUC = 0.62. As it was expected, NB with Bernoulli maintains the AUC value in accordance with the results obtained in the training dataset, and in the case of RF, due to the classifier instability, the score deteriorates (Figure 4).

Once the classifier is selected and trained, the results are given depending on the probability of the patient to suffer a decompensation. For that, the probability given by the classifier (0 if none of the alerts of the inclusion criterion is triggered) is split in terciles. Each tercile is associated with a colour: if the probability is less than 0.33, "green" group; if the probability is between 0.33 and 0.66, "yellow" group; and if it is upper than 0.66, "red" group. This way, the clinicians can base their decisions on the risk group. Setting the probabilities of the classifiers to the risk groups, the results achieved are the next (Table 11).

As presented in Table 11, the RF classifier results in a poor predictive model. However, NB reaches acceptable scores comparing with literature studies that have similar attributes (see section *Alerts for HF Patients*). Comparing the obtained results in the "red" group to the current medical practice, though NB (in the red group) gets 38% less of Se (0.76 ⟶ 0.47) value, it achieves 72% less of FA/pt-y (28.64 ⟶ 7.8) value. Henceforth, this predictive model improves the results of the actual alerts method and it is more reliable.

6. Discussion

Current medical practice may use sensitive alerts, that although they detect most of the decompensations due to their high sensitivity, they also have too many false alerts. Therefore, the main goal of this study is the reduction of these false alerts. This study has shown an improvement from current alerts system implemented in the hospital. The system reduces the number of false alerts notably, from 28.64 FA/pt-y of the current medical practice to even 7.8 FA/pt-y

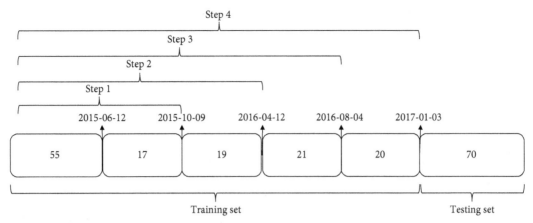

Figure 3: Cross-validation on a rolling basis applied in the study.

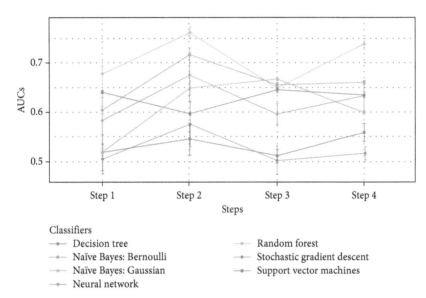

Figure 4: AUC values of the classifiers (colours) depending on the steps (axis *x*).

Table 9: Alerts' performance in the testing set.

Alert tag	Colour alert	Se	FA/pt-y
Weight	—	0.4	13.36
SBP	Yellow	0.49	18.9
	Red	0.1	2.44
DBP	Yellow	0.34	15.4
	Red	0.07	1.3
HR	Yellow	0.37	19.7
	Red	0.06	3.1
O2Sat	Yellow	0.5	27.27
	Red	0.2	4.2

Table 10: Questionnaire alerts' performance in the testing set.

Question tag	Answer	Se	FA/pt-y
Well-being	Worse	0.25	2.9
New Medication	Yes	0.13	5.2
Diet and exercise	No	0.16	8.75
Ankle	Worse	0.13	2.86
Walks	No	0.4	24.7
Shortness of breath	Yes	0.43	21.9

Table 11: Results of the predictive models.

Group	Random forest		Naïve Bayes	
	Se	FA/pt-y	Se	FA/pt-y
Green	1	79	0.75	59.4
Yellow	0.08	1.29	0.41	13.2
Red	0	0.11	0.47	7.8

for the "red" group, which is denoted as the most restrictive group. This last result is achieved with the predictive model built by applying NB with Bernoulli to the combination of telemonitoring alerts and questionnaire alerts (R2). However, as expected, the application of machine learning techniques entails a decrement on sensitivity values. The result obtained in this study for the "red" group is Se = 0.47, while the alerts used in the current medical practice applied to the same testing dataset achieve Se = 0.76. Despite this Se worsening, it is notorious that the FA/pt-y has much higher decrement, with which we conclude that this new predictive model improves the current medical practice. Moreover, when comparing the obtained results with the state of the art, the Se values are similar or better to these studies that do not consider transthoracic impedance (Table 1). Especially considering that in the SoA, most of the studies reduce the real FA/pt-y concatenating the neighbour alerts, since they assume that once an alert has been triggered, the clinician will take action, and hence, next consecutive alerts will not be triggered.

The current study also presents some limitations. Firstly, as presented in *Characteristics of Ambulatory Patients Dataset*, there are patients that did not monitor regularly. As a consequence, from 254 decompensations during this telemonitored period, only 202 of them had 3 or more measurements during the last week previous to the admission, and hence, could be used in our study. The rest did not have even 3 measurements, and hence, they were not predictable.

Secondly, as the clinical data used in the study are from Caucasian patients, the model may perform differently in different settings, such as in non-Caucasian population. Finally, we must stress that heart failure is a very complex disease with multiple factors, and its predictiveness is complex. Nevertheless, larger amount of data and the registration of all type of decompensations is key to improve the current model.

7. Conclusion and Future Work

This article presents the methodology to develop predictive models for HF decompensations prediction based on ambulatory patients' telemonitored data, extending the study for readmissions detection [9].

The results on these studies have been successfully implemented in a telemedicine system, called INCAR. This way INCAR provides the patient with the confidence of being monitored and guided with an advanced technology and clinical professionals' supervision.

Currently, new devices that monitor physical activity and sleeping quality are incorporated in the telemonitoring program in order to determine whether these features could have an impact in the results and improve the outcome. To finish, we will study the possibility of including in the telemonitoring plan a new device that monitors transthoracic impedance, and explore raising deep learning techniques, which have demonstrated their good performance and may improve the presented results.

Acknowledgments

This work has been funded by Basque Government by means of the RIS3 Program, under reference No 2017222015.

References

[1] O. Hamdi, M. A. Chalouf, D. Ouattara, and F. Krief, "eHealth: survey on research projects, comparative study of tele-monitoring architectures and main issues," *Journal of Network and Computer Applications*, vol. 46, pp. 100–112, 2014.

[2] S. I. Chaudhry, J. A. Mattera, J. P. Curtis et al., "Telemonitoring in patients with heart failure," *New England Journal of Medicine*, vol. 363, no. 24, pp. 2301–2309, 2010.

[3] S. Mangini, P. V. Pires, F. G. M. Braga, and F. Bacal, "Decompensated heart failure," *Einstein*, vol. 11, no. 3, pp. 383–391, 2013.

[4] P. Ponikowski, A. A. Voors, S. D. Anker et al., "2016 ESC Guidelines for the diagnosis and treatment of acute and chronic heart failure: the task force for the diagnosis and treatment of acute and chronic heart failure of the European Society of Cardiology (ESC) developed with the special contribution of the Heart Failure Association (HFA) of the ESC," *European Heart Journal*, vol. 37, no. 27, pp. 2129–2200, 2016.

[5] W. T. Abraham, S. Compton, G. Haas et al., "Intrathoracic impedance vs daily weight monitoring for predicting worsening heart failure events: results of the Fluid Accumulation Status Trial (FAST)," *Congestive Heart Failure*, vol. 17, no. 2, pp. 51–55, 2011.

[6] I. C. Gyllensten, A. G. Bonomi, K. M. Goode et al., "Early indication of decompensated heart failure in patients on home-telemonitoring: a comparison of prediction algorithms based on daily weight and noninvasive transthoracic bio-impedance," *JMIR Medical Informatics*, vol. 4, no. 1, p. e3, 2016.

[7] J. Zhang, K. M. Goode, P. E. Cuddihy, J. G. F. Cleland, and TEN-HMS Investigators, "Predicting hospitalization due to worsening heart failure using daily weight measurement: analysis of the Trans-European Network-Home-Care Management System (TEN-HMS) study," *European Journal of Heart Failure*, vol. 11, no. 4, pp. 420–427, 2009.

[8] M. T. Ledwidge, R. O'Hanlon, L. Lalor et al., "Can individualized weight monitoring using the HeartPhone algorithm improve sensitivity for clinical deterioration of heart failure?," *European Journal of Heart Failure*, vol. 15, no. 4, pp. 447–455, 2013.

[9] J. Kerexeta, A. Artetxe, V. Escolar, A. Lozano, and N. Larburu, "Predicting 30-day readmission in heart failure using machine learning techniques," in *Proceedings of the 11th International Joint Conference on Biomedical Engineering Systems and Technologies*, Funchal, Portugal, January 2018.

[10] P. B. Adamson, M. R. Zile, Y. K. Cho et al., "Hemodynamic factors associated with acute decompensated heart failure: part 2—use in automated detection," *Journal of Cardiac Failure*, vol. 17, no. 5, pp. 366–373, 2011.

[11] F. R. Gilliam III, G. A. Ewald, and R. J. Sweeney, "Feasibility of automated heart failure decompensation detection using remote patient monitoring: results from the decompensation detection study," *Innovations in Cardiac Rhythm Management*, vol. 3, pp. 735–745, 2012.

[12] G. García-Sáez, M. Rigla, I. Martínez-Sarriegui et al., "Patient-oriented computerized clinical guidelines for mobile decision support in gestational diabetes," *Journal of Diabetes Science and Technology*, vol. 8, no. 2, pp. 238–246, 2014.

[13] M. Peleg, Y. Shahar, S. Quaglini et al., "Assessment of a personalized and distributed patient guidance system," *International Journal of Medical Informatics*, vol. 101, pp. 108–130, 2017.

[14] B. J. Mortazavi, N. S. Downing, E. M. Bucholz et al., "Analysis of machine learning techniques for heart failure

readmissions," *Circulation: Cardiovascular Quality and Outcomes*, vol. 9, no. 6, pp. 629–640, 2016.

[15] K. Zolfaghar, "Predicting risk-of-readmission for congestive heart failure patients: a multi-layer approach," 2013, http://arxiv.org/abs/1306.2094.

[16] B. Zheng, J. Zhang, S. W. Yoon, S. S. Lam, M. Khasawneh, and S. Poranki, "Predictive modeling of hospital readmissions using metaheuristics and data mining," *Expert Systems with Applications*, vol. 42, no. 20, pp. 7110–7120, 2015.

[17] N. Meadem, N. Verbiest, K. Zolfaghar et al., "Exploring preprocessing techniques for prediction of risk of readmission for congestive heart failure patients," in *Proceedings of International Conference on Knowledge Discovery and Data Mining (KDD), Data Mining and Healthcare (DMH)*, vol. 150, Chicago, IL, USA, August 2013.

[18] H. M. Krumholz, Y.-T. Chen, Y. Wang, V. Vaccarino, M. J. Radford, and R. I. Horwitz, "Predictors of readmission among elderly survivors of admission with heart failure," *American Heart Journal*, vol. 139, no. 1, pp. 72–77, 2000.

[19] R. Amarasingham, B. J. Moore, Y. P. Tabak et al., "An automated model to identify heart failure patients at risk for 30-day readmission or death using electronic medical record data," *Medical Care*, vol. 48, no. 11, pp. 981–988, 2010.

[20] S. Sudhakar, W. Zhang, Y.-F. Kuo, M. Alghrouz, A. Barbajelata, and G. Sharma, "Validation of the readmission risk score in heart failure patients at a tertiary hospital," *Journal of Cardiac Failure*, vol. 21, no. 11, pp. 885–891, 2015.

[21] M. van der Heijden, B. Lijnse, P. J. F. Lucas, Y. F. Heijdra, and T. R. J. Schermer, "Managing COPD exacerbations with telemedicine," in *Artificial Intelligence in Medicine*, pp. 169–178, Westview Press, Boulder, CO, USA, 2011.

[22] M. S. Mohktar, *A Decision Support System for the Home Management of Patients with Chronic Obstructive Pulmonary Disease (COPD) Using Telehealth*, Graduate School of Biomedical Engineering, University of New South Wales, Kensington, NSW, Australia, 2012.

[23] M. Velikova, P. J. F. Lucas, and M. Spaanderman, "A predictive bayesian network model for home management of preeclampsia," in *Artificial Intelligence in Medicine*, pp. 179–183, Westview Press, Boulder, CO, USA, 2011.

[24] R. Lafta, J. Zhang, X. Tao et al., "An intelligent recommender system based on predictive analysis in telehealthcare environment," *Web Intelligence*, vol. 14, no. 4, pp. 325–336, 2017.

[25] M. T. Alarcón and J. I. González-Montalvo, "La escala sociofamiliar de Gijón, instrumento útil en el hospital general," *Revista Española de Geriatría y Gerontología*, vol. 33, no. 1, pp. 178-179, 1998.

[26] W. J. Youden, "Index for rating diagnostic tests," *Cancer*, vol. 3, no. 1, pp. 32–35, 1950.

[27] J. R. Quinlan, *C4.5: Programs for Machine Learning*, Morgan Kaufmann Publishers, Burlington, MA, USA, 1993.

[28] L. Breiman, R. Friedman, R. Olshen, and C. Stone, *Classification and Regression Trees 1984*, Wadsworth and Brooks, Monterey, CA, USA, 1984.

[29] J. R. Quinlan, "Induction of decision trees," *Machine Learning*, vol. 1, no. 1, pp. 81–106, 1986.

[30] L. Breiman, "Random forests," *Machine Learning*, vol. 45, no. 1, pp. 5–32, 2001.

[31] L. Breiman, "Bagging predictors," *Machine Learning*, vol. 24, no. 2, pp. 123–140, 1996.

[32] C. J. C. Burges, "A tutorial on support vector machines for pattern recognition," *Data Mining and Knowledge Discovery*, vol. 2, pp. 121–167, 1998.

[33] V. N. Vapnik, "An overview of statistical learning theory," *IEEE Transactions on Neural Networks*, vol. 10, no. 5, pp. 988–999, 1999.

[34] V. López, A. Fernández, S. García, V. Palade, and F. Herrera, "An insight into classification with imbalanced data: empirical results and current trends on using data intrinsic characteristics," *Information Sciences*, vol. 250, pp. 113–141, 2013.

[35] N. V. Chawla, K. W. Bowyer, L. O. Hall, and W. P. Kegelmeyer, "SMOTE: synthetic minority over-sampling technique," *Journal of Artificial Intelligence Research*, vol. 16, pp. 321–357, 2002.

[36] C. Bergmeir and J. M. Benítez, "On the use of cross-validation for time series predictor evaluation," *Information Sciences*, vol. 191, pp. 192–213, 2012.

[37] A. P. Bradley, "The use of the area under the ROC curve in the evaluation of machine learning algorithms," *Pattern Recognition*, vol. 30, no. 7, pp. 1145–1159, 1997.

11

Developing a Mobile APP-Supported Learning System for Evaluating Health-Related Physical Fitness Achievements of Students

Ching-Hsue Cheng and **Chung-Hsi Chen**

Department of Information Management, National Yunlin University of Science and Technology, 123, University Road, Section 3, Douliou, Yunlin 64002, Taiwan

Correspondence should be addressed to Chung-Hsi Chen; d10123004@yuntech.edu.tw

Academic Editor: Habib M. Fardoun

This study developed a mobile APP support learning system to compare the effects of two different learning approaches based on students' health-related physical fitness (HRPF) achievements, self-efficacy, and system usability. There were 90 participants from four physical education classes in an elementary school of Taiwan who were assigned to the experimental and control groups. An 8-week experiment was conducted to evaluate the two different learning approaches. The experimental results showed that a mobile APP support learning approach could improve the students' HRPF achievements. Furthermore, this study found that the self-efficacy and system operations affect the students' HRPF achievements. To sum up, the combination of traditional and a mobile APP support learning system is an effective approach that would help students to improve their HRPF achievements. The findings can provide the key factors of assisted learning design and students' HRPF achievements for the teachers and the related educators as references.

1. Introduction

Improving health-related physical fitness (HRPF) is a desired learning performance for school-aged children to achieve, and children who are physically fit are more likely to be active people [1]. According to the previous study, there is strong evidence base of physical fitness having a "protective effect" to health [2]. Moreover, World Health Organization (WHO) recommended that 5–17-year-old children should engage in 60 minutes of physical fitness training every day [3]. Schools are a valuable learning environment for promoting and attracting children's physical fitness. In particular, physical education course offers children the opportunity to engage in HRPF training and develop the fundamental movement skills [4]. The United Kingdom's Association of Physical Education has recommended that elementary school children should engage in physical fitness training for at least 50% of physical education course time [5].

In the field of HRPF, many psychological theories are utilized to improve students' motivation for HRPF achievements, and the self-efficacy theory is one among them. Self-efficacy was a key concept in Bandura's social cognitive theory and was proposed by Bandura [6], and it is considered one of the most important determinants in the context of students' learning outcomes [7]. Self-efficacy varies with individuals, and students with higher self-efficacy are more likely to overcome obstacles in the learning process. How to improve the lower self-efficacy of learning is an important issue.

In recent years, many scholars have employed mobile and cloud-based technologies as tools to assist learning. Mobile technology refers to characteristics of a device to handle information access, communication, and business transactions while in a state of motion [8]. Mobile and cloud-based technologies can assist students' learning and provide opportunities to the teachers develop learning activities

[9, 10]. Moreover, the use of mobile and cloud-based technologies can reach a greater level of learning motivation and achieve higher learning outcomes [11].

In addition, some scholars had proposed evidence that the use of assisted technology can achieve higher learning outcomes, such as use of web-based assisted learning system in education [12] and use of mobile-based assisted learning system in education [13, 14]. Further, previous research suggested that researchers could attempt to adopt with mobile devices combined with cloud database to make the peer feedback process more efficient [15]. Also Hwang et al. [16] and Sung et al. [17] reported that game-based learning approach effectively enhanced the students' learning effects, so increasing the entertainment part of APP could be considered, such as enhancing the interface affinity to make the user interface more interesting, so as to enhance students' system satisfaction.

However, there is no research to investigate whether using a mobile APP support learning influences the physical fitness of elementary school students; it indicates a knowledge gap. In order to fill this gap, this paper develops a mobile APP support learning system named My-Fitness which assists students improve their HRPF achievements. My-Fitness is a mobile phone-based system which has a simple and convenient feature, and students can upload data to the cloud and teachers can also use the Firebase on the web to assess students' learning achievement and quickly give advice and encouragement to students.

2. Literature Review

In order to set a cornerstone for understanding this study, we briefly introduce some theoretical views and information based on literature analysis.

2.1. Health-Related Physical Fitness (HRPF).
Many kinds of literature attempted to define physical fitness. Pate [18] reviewed the previous literature and reported that they all focused on defining physical fitness in terms of movement capacities. In addition, according to previous studies, Caspersen et al. & Vancampfort et al. [19, 20] indicated that physical fitness was a set of attributes that people had or achieved related to the ability to perform physical activity, and it could reduce the risk of developing diseases associated with physical inactivity. Furthermore, physical fitness referred to the individual's health and energy status, enabling participation in various types of physical activity in daily life.

A study by Pate [21] reported that in the mid 1970s, a trend toward HRPF began, and in 1980, a new test was developed. Pate & Caspersen et al. [19, 21] further indicated that the health-related components of physical fitness were cardiorespiratory endurance, body composition, muscular strength, muscular endurance, and flexibility. In another study, Chen et al. [1], also pointed out that people with good physical fitness would have the characteristics of better cardiorespiratory endurance, muscle strength, muscular endurance, body composition, and flexibility.

2.2. Mobile Technology.
Mobile devices were the ideal terminal to deliver the message that enabled students to access learning materials anywhere and at anytime, and it could also provide opportunities for the teachers to develop learning activities [9]. Moreover, some researchers had indicated that mobile technologies were creating a wide range of education applications, such as providing personalized learning objectives and schedules [13], understanding the unique characteristics of each communication medium [22], and providing learning support by a wireless device [23].

Other researchers reviewed 110 papers published between 1993 and 2013 that investigated the effects of integrating mobile devices into students' learning showed that the overall effect of using mobile devices appears to be effective [24]. Therefore, in this study, a mobile APP support learning approach was proposed to investigate whether it could improve students' learning achievement.

2.3. Cloud-Based Learning.
Cloud-based learning has become increasingly popular in every field. Cloud-based paradigm provided a virtual resource pool (hardware, development platforms, or services) that was available through the network. Cloud services would make users more efficient and allowed users to access their information anytime and anywhere seamlessly [25]. In addition, cloud-based learning was very practical for teachers and students, and it allowed teachers to administer the entire learning process easily so that students could learn effectively [26].

2.4. Self-Efficacy Theory.
Self-efficacy was a key concept in Bandura's social cognitive theory and proposed by Bandura, and later he further mentioned that self-efficacy was one's belief in his or her ability to execute a particular task or behavior [6]. Self-efficacy referred to individuals' beliefs about their ability to fulfill tasks and achieve goals, while motivating individuals to represent their internal state or condition, activating their behavior, and directing their goals [27]. A study by Hsia et al. [15] indicated that self-efficacy may be a determinant of learning achievement, and they had further pointed out that observing learning progress of peers may be an important factor in influencing students' self-efficacy.

Dinthervan et al. [28] reviewed more than 30 previous studies and attempted to investigate the possible situational and instructional factors within educational contexts affecting students' self-efficacy; they found that factors such as reward, goal setting, modeling, task strategy, self-monitoring, and evaluation could improve student self-efficacy within elementary and secondary educational levels. Accordingly, we attempted to investigate the influence of students' self-efficacy on HRPF achievements.

2.5. System Usability.
Holden and Rada [29] suggested that usability was especially important in the field of education, and the learning support system should be in compliance with flexibility and individual needs. In addition, Harrati et al. [30] indicated that usability was the extension by which

a product could be readily used by a particular user to achieve specified goals with effectiveness, efficiency, and satisfaction in a specified context of use. A system must be accepted by users; otherwise, it was easy to fail, and the accepted critical factor was system usability. Wagner et al. [31] reviewed four studies which further reported that usability was related to personal error reduction and positive attitude, and could increase the user's intention to use the computer.

Brooke [32] proposed a system usability scale (SUS) to measure usability. The SUS included 10 items which were rated on a five-point scale ranging from strongly disagree to strongly agree, five of which are positive statements and the rest are negative. Additionally, the SUS score was in the range of 0 to 100, and it was the sum of all scores for the 10 items then multiplied by 2.5, as shown in the equation as follows [30]; these scores for individual item among the 10 items did not make sense. When the SUS score was higher than 70, the SUS was acceptable by most users. The SUS has been proved of a very powerful measure of system usability [33]. The formula of SUS score was defined as follows:

$$SUS = 2.5 \times \left[\sum_{n=1}^{5} \left(U_{2n-1} - 1 \right) + \left(5 - U_{2n} \right) \right]. \quad (1)$$

3. Research Method and Questions

In this section, we describe the research design, research questions, operational definitions, participants, experimental procedure and learning activities, measuring tools, and system development.

3.1. Research Design, Research Questions, and Operational Definitions

3.1.1. Research Design. This experiment was done mainly to explore whether the use of a mobile APP support system could help elementary school students to enhance HRPF achievements and self-efficacy. A research framework proposed in this study is depicted in Figure 1. In order to compare the achievement of different learning approaches, the students were randomly assigned to experimental and control group. The experimental group students randomly came from two classes including 48 students, and the control group students were from the other two classes including 41 students. In addition, the students in the experimental group were provided with a mixed learning strategy that combined with traditional and mobile APP-supported learning, and those in the control group were provided with traditional learning. In the meantime, the two groups were taught by the same teacher, and they had the same learning content and schedule. Moreover, several measuring tools were used to evaluate the students' self-efficacy, system usability, and HRPF achievements, and quantitative methods were used to analyze the collected data.

3.1.2. Research Questions. The purpose of this study focuses on the effect of different learning approaches on students'

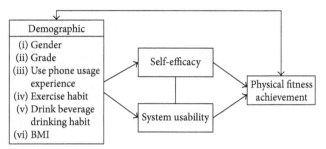

FIGURE 1: Research framework.

HRPF achievements. In the meantime, the demographic variables how to affect students' self-efficacy and system usability and the impact of self-efficacy and system usability on students' HRPF achievements are also explored. The research questions include:

(1) Is there any difference between the different demographic variables and the students' HRPF achievements?

(2) Is there any difference between the different demographic variables and the students' self-efficacy?

(3) Is there any difference between the different demographic variables and the students' system usability?

(4) Does the students' self-efficacy affect the HRPF achievements?

(5) Does the different perception of system usability result in different HRPF achievements?

(6) Is there any correlation between the students' self-efficacy and system usability?

3.1.3. Operational Definitions. The research variables include a dependent variable of HRPF achievements and independent demographic variables, self-efficacy, and system usability. The definitions and references of variables are listed in Table 1.

3.2. Participants. The students were from four classes of the 3th–6th graders at an elementary school in southern Taiwan, and 90 students took part in a HRPF lesson. A total of 89 students fully completed the course, while one was unable to put through the activities, including 40 males and 49 females with an average age of 10.1 years.

3.3. Experimental Procedure and Learning Activities. The experimental procedure is shown in Figure 2; in this experiment, the students took learning activities for 8 weeks, which consisted of two 40-minute courses per week. At the beginning of the first week, the teacher carried out HRPF introduction including video viewing and PPT teaching. Following that, the teacher conducted HRPF achievements pretest which involved 800-meters sprinting/walking, standing long jump, one-minute bent-knee sit-ups, and sit and reach; then each student completed the self-efficacy and system usability questionnaire.

TABLE 1: Operational definitions.

HRPF achievements	The HRPF achievements are evaluated by the Sports Administration Ministry of Education (SAMOE) normative physical fitness scores [34]
Demographic gender	There are gender differences in information technology use and implementation [35]. Gender differences will produce different results in rehabilitation [36]
Demographic grade (experimental and control group)	Nursing students who were older had better physical fitness [37]. The participants who had assistant system support would have a better learning achievement [14]
Demographic phone usage experience	Participants' experiences in using similar gaming systems. Experience will affect a user's self-efficacy [6]
Demographic exercise habit	Exercise habit will affect a person's confidence [38]. Exercise training should be beneficial to physical fitness [39]
Demographic beverage drinking habit	A caffeine-containing energy drink might be an effective ergogenic aid to improve physical performance and accuracy in male volleyball players [40]
Demographic BMI	With lower BMI had better physical fitness [37]. Fitness capacity therefore decreased progressively as the BMI increased [41]
Self-efficacy	Holden & Rada (2011) said that both self-efficacy and perceived usability had a positive correlation in education [29]
System usability	Usability is not a quality that exists in any real or absolute sense [32]

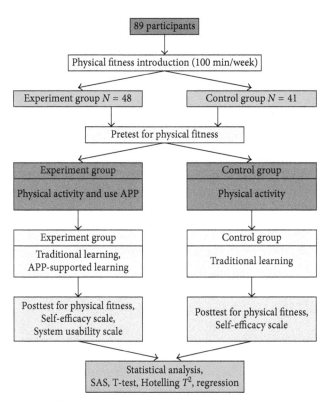

FIGURE 2: Flow of the two learning activities.

TABLE 2: Demographic variables.

Variable	Statement
Gender	Denote the student gender: male or female
Grade	Denote the student grade: 3–6
Experience in using phone	How many times does the student use phone per week?
Exercise habit	How many minutes does the student exercise per week?
Beverage drinking habit	How many cans of beverages does the student drink per week? Denote the student BMI: low
BMI	BMI or high BMI

HRPF achievements over the past week and then provide comments and encourage to them.

On the other hand, the control group students were guided by the same teacher to practice and complete their learning tasks. In the meantime, their HRPF achievements over the past week were recorded in the manuscript and commented by the teacher. Finally, in the last week, the posttest for HRPF achievements, self-efficacy, and system usability was conducted after the learning activities were completed.

3.4. Measuring Tools. These measurement tools include demographic scale, SAMOE normative physical fitness scores, self-efficacy scale, and SUS as shown in each section.

3.5. Demographic Scale. In this study, the demographic variables included gender, grade, phone usage experience, exercise habit, and beverage drinking habit that are employed to explore the demographic relationship with self-efficacy and system usability. Table 2 describes the demographic variables.

3.6. Evaluation of HRPF. In order to evaluate the HRPF achievements, this study uses the SAMOE normative physical fitness scores as shown in Table 3 to measure the

In the second to seventh week, the teacher in the experimental group randomly assigned 3 or 4 students into a team, in which they were equipped with a mobile phone to interact with the APP. Furthermore, each team was asked to elect a team leader who was responsible for handling mobile phone and arranging the use of mobile phone. In the meantime, the students were asked to present their achievement weekly and upload them to the cloud by the APP. Then they could browse and compare their achievement with those of the other classmates on the mobile phone. Moreover, the teacher used the mobile phone or cloud to assess their

TABLE 3: HRPF-based test.

HRPF items	Activities	Descriptions
Body composition	Body mass index	The weight in kilogram/the square of height in meters
Body flexibility	Sit and reach	The students sat on a wooden board with their feet apart (about 30 centimeters) and kept knees straight. Place the measuring tape between the students' legs and their heels were aligned with the 25 cm mark of the measuring tape. The students naturally stretch forward as far as possible and then measure the score. A higher score indicates greater flexibility
Muscular strength	Standing long jump	The students stand behind the take-off line on the ground then attempt to jump as far as possible. Measure the distance from the take-off line to the nearest contact landing point (heel). Three attempts are allowed and the longest distance in the three attempts is then recorded
Muscular endurance	One-minute bent-knee sit-ups	The students were asked to lie on the floor with the knees bent to form a 45 degree angle and two hands across the chest and palms on the shoulders. During one minute of testing, hold the students' ankles to keep the heel in contact with the floor and make them to try their best to complete the sit-ups test
Cardiorespiratory endurance	800-meter sprinting/walking ability	To measure 800-meter sprinting/walking ability. The teacher encouraged the students to do their best and complete the test. When the students were unable to run they could replace to walk. The shorter the time, the better the cardiopulmonary endurance

TABLE 4: Self-efficacy scale.

1. I believe I will receive an excellent grade in this physical fitness test
2. I'm certain I can master the skills being taught in this physical fitness program
3. I'm certain I can understand the most difficult part presented in the physical fitness program
4. I'm confident I can understand the most complex part presented by the instructor in this physical fitness program
5. I'm confident I can catch the basic movements taught in this physical fitness program
6. I'm confident I can do an excellent job on the movements in this physical fitness achievement
7. I expect to do well in this physical fitness test
8. Considering the difficulty of this physical fitness program, the teacher, and my skills, I think I will do well in this achievement

TABLE 5: System usability scale [32].

Questionnaire item	Strongly disagree	Strongly disagree
(1) I think that I would like to use this system frequently	1	5
(2) I found the system unnecessarily complex	1	5
(3) I thought the system was easy to use.	1	5
(4) I think that I would need the support of a technician to be able to use this system	1	5
(5) I found that various functions of this system were well integrated	1	5
(6) I thought there was too much inconsistency in this system	1	5
(7) I would imagine that most people would learn to use this system very quickly	1	5
(8) I found the system very cumbersome to use	1	5
(9) I felt very confident using the system	1	5
(10) I needed to learn a lot of things before I could use this system	1	5

HRPF level by students. In addition, we use the physical fitness test items published by Taiwan Ministry of Education, and the content was listed on the website [34]. The students' HRPF results will be converted into scores based on the SAMOE normative physical fitness scores.

3.7. Self-Efficacy Scale. This self-efficacy questionnaire was obtained from Wang and Hwang [12]; then this study was modified as self-efficacy questionnaire of HRPF achievements, as shown in Table 4. This self-efficacy scale has 8 items and uses a five-point Likert-type scale where 1 represents "strongly disagree" and 5 "strongly agree." After calculating the score for each student's response, the higher the participants' rating score is obtained, the higher the self-efficacy is achieved. Self-efficacy is concerned with the achievement capability of individuals.

3.8. System Usability Scale. The SUS is based on the mobile APP system and developed by John Brooke at Digital Equipment Corporation in the UK [32], including 5 items of positive response in odd-numbered items and 5 of negative

response in even-numbered items, as given in Table 5. The SUS is a simple questionnaire rated by 5-point Likert-type scale, where 5 represents "strongly agree" 1 represents "strongly disagree," is employed to evaluate the usability of a mobile APP system. The higher the rating score the mobile APP system gets, the higher the acceptance it has.

3.9. System Development. Mobile devices have been widely used in various fields, and in this study, we use Android Studio and Eclipse as development tools to develop the mobile APP named My-Fitness on the mobile platform. Table 6 describes the system development process. The interface of the mobile APP is simple, clear, in which the functions consist of HRPF input, HRPF historical record, HRPF ranking, and HRPF related knowledge, and the students can easily learn how to use correctly.

TABLE 6: APP development lifecycle.

(1) Analyze
  (i) Develop tool (Eclipse and Android Studio)
  (ii) Program language (Java)
  (iii) SDK (Android SDK)
  (iv) Database selection (Firebase)
  (v) Functional planning
(2) Design
  (i) Interface of the HRPF APP (on screen)
  (ii) Design the system process of the HRPF APP
(3) Develop/implement
  (i) Use the selected tools, program language, and SDK to develop the HRPF APP
(4) Testing
  (i) Recruit volunteers for testing the HRPF APP usability
  (ii) Bug fixes

Furthermore, the mobile APP provided hyperlinks of physical fitness-related knowledge and interesting games. We arranged suitable hyperlinks for the students to inquire and also provided many hyperlinks of interesting physical fitness games that could inspire students' curiosity and learning motivation. The physical fitness games consist of introduce activity games, development activity games, and recovery activity games, in which the students could get a lot of physical fitness knowledge and right learning methods by using the mobile APP, so as to enhance HRPF achievements. The interface and content of the mobile APP are shown in Figures 3–5. During the learning activities, we allow the students to use the afore-mentioned functions freely.

Additionally, the students could enter their fitness scores and view their present and historical record achievement by the mobile phone. During the learning activities that the system will rank students' HRPF ordering, they could compare the achievement with those of other classmates and know their own ranking in class timely. Meanwhile, their own achievement could be compared with the SAMOE normative physical fitness scores to know their own relative position of physical fitness. Moreover, the teacher could view and manage students' login status and evaluate their achievement by the Firebase on the web. The system architecture is given in Figure 6.

4. Results and Findings

In this study, the students' self-efficacy and system usability are measured, and the effects of different groups on HRPF achievements are analyzed. After analysis, the internal consistency Cronbach's alpha of the self-efficacy scale reaches 0.9 in the collected questionnaire. The descriptive statistics include frequencies, means, and standard deviations, to present the trends of demographic variables; furthermore, the Hotelling T^2 method, t-test method, and regression analyses are used to analyze the collected data. In addition, all of the research questions are statistically tested by using SAS software.

4.1. Analysis of HRPF Achievements in the Different Demographic Variables. For investigating the effects of different demographic variables on students' HRPF achievements, the analysis of Hotelling T^2 method is adapted. The analysis results show that all demographic variables (gender, grade, phone usage experience, exercise habit, beverage drinking habit, BMI) have different effects on students' HRPF achievements as shown in Table 7. Different groups on grade, exercise habit, BMI have significant differences on HRPF achievements ($P < 0.05$). The students who accepted a mobile APP-supported learning approach (mean = 66.56) have better HRPF achievements than those who learn in traditional learning approach (mean = 50.73). In addition, the mean values for exercise habit group and low BMI group for the questionnaire rating are 66.02 and 63.51, respectively, and 55.09 and 45.53 for no exercise habit group and high BMI group, respectively. The results reveal that the students with exercise habit or low BMI will achieve better HRPF achievements.

4.2. Analysis of Self-Efficacy in the Different Demographic Variables. For investigating the effects of different demographic variables on students' self-efficacy, analysis of the Hotelling T^2 method is adapted. The results show that all demographic variables (gender, grade, phone usage experience, exercise habit, beverage drinking habit, and BMI) have different effects on students' self-efficacy as shown in Table 8. Exercise habit group has a significant difference on students' self-efficacy ($P < 0.05$). The mean value for exercise habit group is 4.42, and 3.74 for no exercise habit group. The result reveals that the students with exercise habit would have better self-efficacy.

4.3. Analysis of System Usability in the Different Demographic Variables. For testing the effects of different demographic variables on students' system usability, the statistical t-test is performed. The analysis results show that all demographic variables (gender, grade, phone usage experience, exercise habit, beverage drinking habit, and BMI) have different effects on students' usability as shown in Table 9. Exercise habit group and BMI group have significant differences in students' perception of system usability ($P < 0.05$). The mean values of exercise habit group and low BMI group were 78.5 and 77.1 for the questionnaire rating, respectively, and 69.7 and 64.7 for no exercise habit group and high BMI group, respectively. The results reveal that the students with exercise habit or low BMI would have more confidence in learning an unfamiliar system. In addition, most of the student's system usability scores are greater than 74, indicating that My-Fitness is a successful and accepted system.

4.4. Regression Analysis of Self-Efficacy on HRPF Achievements. A linear regression analysis is employed to predict the results of students' self-efficacy and HRPF achievements in the pretest and posttest, as shown in Table 10. The results reveal that students' self-efficacy significantly predicted HRPF achievements in the pretest ($R^2 = 0.131$, $P < 0.01$) and posttest ($R^2 = 0.083$, $P < 0.01$). This implies that the higher the self-efficacy they obtain, the better HRPF achievements they present.

FIGURE 3: The interface of APP.

4.5. Regression and Hotelling T^2 Snalyses of System Usability on HRPF Achievements. A linear regression analysis is utilized to predict the result between students' system usability and HRPF achievements in the posttest, as shown in Table 11. The result reveals that the students' system usability significantly predicted HRPF achievements in the posttest ($P < 0.01$).

FIGURE 4: The content of APP. Note: The APP provided hyperlinks of physical fitness related knowledge and interesting games [42]. The students could get a lot of physical fitness knowledge and right learning methods by using the mobile APP [43].

In addition, the Hotelling T^2 method is performed to test the statistically significant difference in different system usability on students' HRPF achievements, as shown in Table 12. The low perception of system usability and high perception of system usability have a significant difference on students' HRPF achievements ($P < 0.01$). The mean value

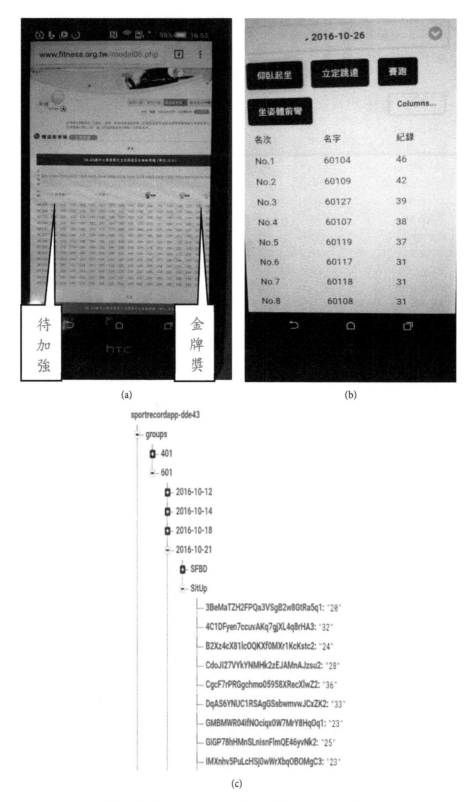

(a) (b)

(c)

FIGURE 5: Ranking scores. Note: The students check the normative physical fitness scores table based on gender, age to know their own relative position of physical fitness. HRPF ranking offers the students a place to view and compare the achievements with the other team members and classmates and could timely know their own ranking in the class. The teacher could assess their achievements by the Firebase on the web. Firebase is a mobile and web application platform that includes tools and infrastructure and is designed to help people build high-quality APP.

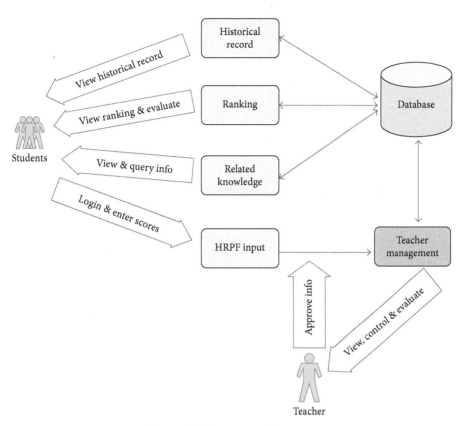

FIGURE 6: The system architecture.

TABLE 7: The results of Hotelling T^2 of students' HRPF achievements for the demographic variables (all participants).

	Variables	Descriptive statistics			Hotelling T^2	
		N	Mean	SD	F value	P
Gender	Male	40	55.37	21.30	1.42	0.23
	Female	49	62.44	18.50		
Grade	Grade 3, 5	41	50.73	16.61	10.02	0.01*
	Grade 4, 6	48	66.56	19.92		
Phone usage experience	Weekly phone using times <5	54	56.85	20.73	0.59	0.66
	Weekly phone using times ≥5	35	63.00	18.51		
Exercise habit	Weekly exercise minutes ≤80	55	55.09	18.59	2.68	0.03*
	Weekly exercise minutes >80	34	66.02	20.63		
Beverage drinking habit	Beverage consumption per week (cans) <3	55	58.45	18.49	0.22	0.92
	Beverage consumption per week (cans) ≥3	34	60.58	22.47		
BMI	BMI ≤ 2	68	63.51	17.72	11.65	0.01*
	BMI > 2	21	45.53	21.22		

Grade 3, 5: control group, grade 4, 6: experimental group; exercise habit: weekly exercise minutes >80, no exercise habit: weekly exercise minutes ≤80; low BMI: BMI ≤ 2, high BMI: BMI > 2, divide by the median, *$P < 0.05$, **$P < 0.01$.

of low perception of system usability group is 51.02 for the questionnaire rating, and 79.71 for the high perception of system usability group. This implies that the higher the perception of system usability, the better HRPF achievements they present. In addition, usability average score is more than 74 points, which indicates that most students are satisfied with My-Fitness. And this study employs the usability average score to assign students to two groups.

4.6. Correlation Analysis between Self-Efficacy and System Usability. A Pearson correlation coefficient is used to test

the relationship between self-efficacy and system usability, as shown in Table 13. There is a significant and positive correlation between the self-efficacy and system usability ($r = 0.41$, $P < 0.01$). This implies that the students had the higher self-efficacy would handle the mobile APP system more easily.

4.7. Effects of Different Learning Approaches on Students' Self-Efficacy and HRPF Achievements. In terms of different learning approaches on students' self-efficacy and HRPF achievements, they are utilized to prove that students will

TABLE 8: The results of Hotelling T^2 for the demographic variables on students' self-efficacy (experimental group).

	Variables	Descriptive statistics			Hotelling T^2	
		N	Mean	SD	F value	P
Gender	Male	22	4.22	0.67	0.75	0.64
	Female	26	4.02	0.64		
Grade	Grade 4	21	4.00	0.73	1.25	0.29
	Grade 6	27	4.20	0.78		
Phone usage experience	Weekly phone using times <5	25	4.10	0.62	1.06	0.41
	Weekly phone using times ≥5	23	4.12	0.70		
Exercise habit	Weekly exercise minutes ≤80	22	3.74	0.62	5.10	0.01**
	Weekly exercise minutes >80	26	4.42	0.50		
Beverage drinking habit	Beverage consumption per week (cans) <3	24	4.12	0.69	1.04	0.42
	Beverage consumption per week (cans) ≥3	24	4.10	0.63		
BMI	BMI ≤ 2	38	4.13	0.64	1.20	0.32
	BMI > 2	10	4.01	0.73		

Exercise habit: weekly exercise minutes >80, no exercise habit: weekly exercise minutes ≤80, divide by the median, $^{*}P < 0.05$, $^{**}P < 0.01$.

TABLE 9: The results of t-test for the demographic variables on students' system usability (experimental group).

	Group	Descriptive statistics			t-test	
		N	Mean	SD	T value	P
Gender	Male	22	75.0	15.21	0.21	0.83
	Female	26	74.1	12.84		
Grade	Grade 4	21	75.2	14.7	0.31	0.75
	Grade 6	27	73.9	13.3		
Phone usage experience	Weekly phone using times <5	25	74.3	13.3	−0.12	0.90
	Weekly phone using times ≥5	23	74.7	14.6		
Exercise habit	Weekly exercise minutes ≤80	22	69.7	11.7	−2.29	0.02*
	Weekly exercise minutes >80	26	78.5	14.4		
Beverage drinking habit	Beverage per week (cans) <3	24	75.2	11.7	0.34	0.73
	Beverage per week (cans) ≥3	24	73.8	15.8		
BMI	BMI ≤ 2	38	77.1	13.7	2.67	0.01**
	BMI > 2	10	64.7	9.1		

System usability is only rating in posttest, and it's score is a number, exercise habit: weekly exercise minutes >80, no exercise habit: weekly exercise minutes ≤80; low BMI: BMI ≤ 2, high BMI: BMI > 2, $^{*}P < 0.05$, $^{**}P < 0.01$.

TABLE 10: Regression analysis between students' self-efficacy and HRPF achievements (all participants).

Dependent variable	Independent variables	R^2	P.E.V.	Std. error	T	Sig.
Pretest-fitness	Self-efficacy	0.131	9.808	2.703	3.630	0.000
Posttest-fitness	Self-efficacy	0.083	7.684	2.728	2.820	0.006

P.E.V.: parameter estimated value, $^{**}P < 0.01$.

TABLE 11: Regression analysis between students' usability and HRPF achievements (experimental group).

Dependent variable	Independent variables	R^2	P.E.V.	Std. error	T	Sig.
Fitness	Usability	0.511	1.030	0.148	6.950	0.000

P.E.V.: parameter estimated value, system usability is only rating in posttest, it's score is a number, $^{**}P < 0.01$.

TABLE 12: The result of Hotelling T^2 between students' system usability and HRPF achievements (experimental group).

Dimension	Group	Descriptive statistics			Hotelling T^2	
		N	Mean	SD	F value	P
Fitness	System usability ≤ 74	22	51.02	16.1	13.44	0.01**
	System usability > 74	26	79.71	11.6		

Low perception of system usability ≤ 74, high perception of system usability > 74, divide by the median, $^{**}P < 0.01$.

TABLE 13: Correlation between students' self-efficacy and system usability.

	Self-efficacy	
	r	P value
Usability	0.41	0.01**

**P < 0.01.

have more confidence to complete the learning process and achieve better HRPF achievements. A Hotelling T^2 is employed to examine the statistically significant differences on students' self-efficacy and HRPF achievements, as shown in Table 14. The results show that the different learning approaches have significant differences on students' self-efficacy ($P < 0.05$) and HRPF achievements ($P < 0.01$).

4.7. The Findings. This study attempts to investigate how self-efficacy levels affect HRPF achievements and reinforce self-efficacy by using a mobile APP system to make the process more interesting in order to enhance the HRPF achievements. The students in the experimental group were provided with a mixed mode that combined with traditional learning and a mobile APP-supported learning to assist their learning, and those in the control group were provided with traditional learning that attempts to ascertain whether the use of a mobile APP system is significantly effective to enhance the HRPF achievements. From data analysis and results, there are three main findings as follows.

4.7.1. Demographic Variables. The demographic variables related to these discussions are summarized as follows:

(1) There is no significant difference in different genders on the students' HRPF achievements, self-efficacy, and system usability.

The experimental results are quite different from that of scholars [35], who reported that men were more confident and had better achievement and be able to learn information technology than women. The possible reason may be due to the different ideas between elementary school students and other age groups.

(2) A mobile APP-supported learning approach (grade 4, 6) help the students enhance HRPF achievements

The students who accepted a mobile APP-supported learning approach would have better HRPF achievements than those who learned with traditional learning approach. The result conformed to the study of [14], the participants who had assistant system support would have a better learning achievement.

(3) There was no significant difference in different phone using experiences on the students' HRPF achievements, self-efficacy, and system usability.

The experimental results were different from that of scholars [35], who indicated that experience in using

information system would affect user's self-efficacy. The reason for the difference might be due to the attitude toward using. Elementary school students might only use the phone for fun or entertainment.

(4) Exercise habit help the students enhance HRPF achievements, self-efficacy, and have more confidence in learning an unfamiliar system.

This finding is consistent with the viewpoint of other scholars [38, 39] who had indicated that the participants with regular exercise habit would have more confidence in learning an unfamiliar system and get better results.

(5) There was no significant difference in different beverage drinking habit on the students' HRPF achievements, self-efficacy, and system usability.

This result was different from the viewpoint of other scholars [40].

(6) Low BMI help the students enhance HRPF achievements and had more confidence to learn an unfamiliar system

The students who have low BMI would have the confidence to complete the instructional process and learn a new system. This result is consistent with the viewpoint of other scholars [37, 41] who had indicated that the participants with lower BMI had better physical fitness.

4.7.2. Effects of Different Learning Approaches on Students' Self-Efficacy and HRPF Achievements. In order to test whether My-fitness is helpful for students' self-efficacy and HRPF achievements, a control group was utilized to compare with the experimental group. The results indicate that the experimental group will have more confident to complete the learning process and achieve higher HRPF achievements, as shown in Figure 7. This implies that a mobile APP-supported learning approach is effective and successful in the physical education course of an elementary school.

4.7.3. Self-Efficacy and System Usability. The result shows that there is a positive correlation between self-efficacy and system usability, as shown in Table 13. When these students think that My-Fitness is acceptable, accepting a new learning approach make them feel that it is not difficult, which will give students a higher confidence to complete the learning process. This is consistent with previous studies which arrived to similar conclusions that self-efficacy and perception of system usability had a positive correlation [29].

4.7.4. The Results of Different Learning Approaches on Students' Pretest and Posttest HRPF Achievements. We found that two different learning approaches can improve students' HRPF achievements, but a mobile APP-supported learning approach (experimental group) has a much better effect than the traditional learning approach (control group), as shown in Table 14. The increased mean value of experimental group

TABLE 14: The results of different learning approaches on students' self-efficacy and HRPF achievements (all participants).

Dimension	Group	Descriptive statistics			Hotelling T^2	
		N	Mean	SD	F value	P
Fitness	Control group	41	43.62	18.20	3.38	0.01**
	Experimental group (pretest)	48	54.76	21.96		
	Control group	41	50.73	16.61	10.02	0.01**
	Experimental group (posttest)	48	66.56	19.92		
Self-efficacy	Control group	39	3.99	0.84	2.14	0.04*
	Experimental group	47	4.09	0.65		

We removed 3 outliers during the Self-efficacy analysis process, *$P < 0.05$, **$P < 0.01$.

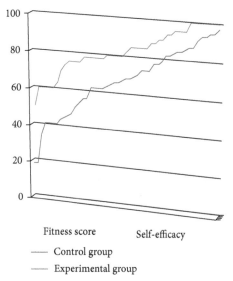

FIGURE 7: The difference of HRPF score and self-efficacy between the control group and the experimental group.

was 11.8, and 7.11 for the control group. The result conformed to the study of [14], the participants who had assistant system support would have a better learning achievement.

5. Conclusions and Suggestions

This study has developed a mobile APP-support learning system named My-Fitness which assists students to improve their HRPF achievements. It allows students to strengthen their self-efficacy based on the mobile phone-based environment and give them a more efficient and complete learning process. In addition, My-fitness is a mobile phone-based system which has a simple and convenient feature, and students can upload data to the cloud; teachers can also use the Firebase on the web to evaluate students' learning achievement and quickly give advice and encouragement to students. The results showed that most students were satisfied with the system and confirmed that My-Fitness did enhance their learning achievement.

This study explores several factors of HRPF achievements, such as self-efficacy, a user-friendly system, and various demographic variables. The results confirmed that these factors actually affect students' learning achievement. My-Fitness is indeed able to enhance students' self-efficacy

and improve their learning achievement. In addition, this study found that regular exercise can achieve better HRPF achievements.

These findings can provide a good reference for assisted tool design of relevant studies and help teachers implement efficient instruction and improve students' HRPF achievements. This paper aims to investigate how My-Fitness improves students' HRPF achievements and effectiveness. The contributions of this paper are listed as follows:

(1) A mobile APP-supported learning system was proposed to improve students' learning achievement and make students to easily use it.

(2) The HRPF achievements by proposed system were investigated.

(3) The self-efficacy and system usability influence effectiveness of HRPF achievements and the students' acceptance of a mobile APP support learning approach.

To sum up, the major contribution of this study is to propose a new application of a mobile APP-supported learning system in the field of physical fitness with elementary school children.

In the future, a mobile APP-supported learning approach can be further improved by providing a database with categorized candidate feedback through which students are able to receive an immediate feedback from the system, and the system can be a good support to improve students' HRPF achievements. Furthermore, a new perspective can enhance the work. Fu et al. [44] found that efforts to improve user satisfaction will increase the therapeutic impact, and Kaipio et al. [45] also indicated that physicians' low ratings for their electronic health record (EHR) systems considerably hinder the efficiency of EHR use and physician's routine work. Therefore, it is worth trying to explore students' perceptions using other questionnaire tools, such as Questionnaire for User Interaction Satisfaction (QUIS) suggested by Johnson et al. [46] to make the APP more reliable.

References

[1] W. Chen, S. Mason, A. Hammond-Bennett, and S. Zalmout, "Manipulative skill competency and health-related physical

fitness in elementary school students," *Journal of Sport and Health Science*, vol. 5, no. 4, pp. 491–499, 2016.

[2] C. E. Barlow, H. W. Kohl, L. W. Gibbons, and S. N. Blair, "Physical Fitness, Mortality and Obesity," *International Journal of Obesity*, vol. 19, no. 4, pp. S41–S44, 1995.

[3] World Health Organization, "Physical activity and young people," 2017, http://www.who.int/dietphysicalactivity/factsheet_young_people/en/.

[4] A. P. Hills, D. R. Dengel, and D. R. Lubans, "Supporting public health priorities: recommendations for physical education and physical activity promotion in schools," *Progress in Cardiovascular Diseases*, vol. 57, no. 4, pp. 368–374, 2015.

[5] J. L. Hollis, A. J. Williams, R. Sutherland et al., "A systematic reviewand meta-analysis of moderate-to-vigorous physical activity levels in elementary school physical education lessons," *Preventive Medicine*, vol. 86, pp. 34–54, 2016.

[6] A. Bandura, "Self-efficacy: toward a unifying theory of behavioral change," *Psychological Review*, vol. 84, no. 2, pp. 191–215, 1977.

[7] L. De Grez and M. Valcke, "Student response system and how to make engineering students learn oral presentation skills," *International Journal of Engineering Education*, vol. 29, no. 4, pp. 940–947, 2013.

[8] Y. Yuan, N. Archer, C. E. Connelly, and W. Zheng, "Identifying the ideal fit between mobile work and mobile work support," *Information & Management*, vol. 47, no. 3, pp. 125–137, 2010.

[9] C. K. Looi, P. Seow, B. Zhang, H. J. So, W. Chen, and L. H. Wong, "Leveraging mobile technology for sustainable seamless learning: a research agenda," *British Journal of Educational Technology*, vol. 41, no. 2, pp. 154–169, 2010.

[10] F. Khaddage, W. Muller, and K. Flintoff, "Advancing mobile learning in formal and informal settings via mobile app technology: where to from here, and how?," *Educational Technology & Society*, vol. 19, no. 3, pp. 16–26, 2016.

[11] C. H. Su and C. H. Cheng, "A mobile gamification learning system for improving the learning motivation and achievements," *Journal of Computer Assisted Learning*, vol. 31, no. 3, pp. 268–286, 2015.

[12] S. L. Wang and G. J. Hwang, "The role of collective efficacy, cognitive quality, and task cohesion in computer-supported collaborative learning," *Computers & Education*, vol. 58, no. 2, pp. 679–687, 2012.

[13] H. Y. Sung, G. J. Hwang, S. Y. Liu, and I. H. Chiu, "A prompt-based annotation approach to conducting mobile learning activities for architecture design courses," *Computers & Education*, vol. 76, pp. 80–90, 2014.

[14] C. L. Lai and G. J. Hwang, "An interactive peer-assessment criteria development approach to improving students' art design performance using handheld devices," *Computers & Education*, vol. 85, pp. 149–159, 2015.

[15] L. H. Hsia, I. Huang, and G. J. Huang, "Effects of different online peer-feedback approaches on students' achievement skills, motivation and self-efficacy in a dance course," *Computers & Education*, vol. 96, pp. 55–71, 2016.

[16] G. J. Hwang, T. C. Hsu, C. L. Lai, and C. J. Hsueh, "Interaction of problem-based gaming and learning anxiety in language students' English listening performance and progressive behavioral patterns," *Computers & Education*, vol. 106, pp. 26–42, 2017.

[17] H. Y. Sung, G. J. Hwang, C. J. Lin, and T. W. Hong, "Experiencing the analects of confucius: an experiential game-based learning approach to promoting students' motivation and conception of learning," *Computers & Education*, vol. 110, pp. 143–153, 2017.

[18] R. R. Pate, "The evolving definition of physical fitness," *QUEST*, vol. 40, no. 3, pp. 174–179, 1988.

[19] C. J. Caspersen, K. E. Powell, and G. M. Christenson, "Physical activity, exercise, and physical fitness: definitions and distinctions for health-related research," *Public Health Reports*, vol. 100, no. 2, pp. 126–131, 1985.

[20] D. Vancampfort, M. Probst, A. Daenen et al., "Impact of antipsychotic medication on physical activity and physical fitness in adolescents: an exploratory study," *Psychiatry Research*, vol. 242, pp. 192–197, 2016.

[21] R. R. Pate, "A new definition of youth fitness," *Physician and Sportsmedicine*, vol. 11, no. 4, pp. 77–83, 1983.

[22] K. Hyewon, L. MiYoung, and K. Minjeong, "Effects of mobile instant messaging on collaborative learning processes and outcomes: the case of South Korea," *Educational Technology & Society*, vol. 17, no. 2, pp. 31–42, 2014.

[23] K. E. Chang, Y. J. Lan, C. M. Chang, and Y. T. Sung, "Mobile-device-supported strategy for Chinese reading comprehension," *Innovations in Education and Teaching International*, vol. 47, no. 1, pp. 69–84, 2010.

[24] Y. T. Sung, K. E. Change, and T. C. Liu, "The effects of integrating mobile devices with teaching and learning on students' learning performance: a meta-analysis and research synthesis," *Computer and Education*, vol. 94, pp. 252–275, 2016.

[25] S. Marston, Z. Li, S. Bandyopadhyay, J. Zhang, and A. Ghalsasi, "Cloud computing-the business perspective," *Decision Support Systems*, vol. 51, no. 1, pp. 176–189, 2011.

[26] K. Paul, K. N. Chen, and L. Gloria, "When cloud computing meets with semantic web: a new design for e-portfolio systems in the social media era," *British Journal of Educational Technology*, vol. 41, no. 6, pp. 1018–1028, 2010.

[27] R. Franken, *Human Motivation*, Wadsworth, Florence, KY, USA, 6th edition, 2006.

[28] M. Dinthervan, F. Dochy, and M. Segers, "Factors affecting students' self-efficacy in higher education," *Educational Research Review*, vol. 6, no. 2, pp. 95–108, 2011.

[29] H. Holden and R. Rada, "Understanding the influence of perceived usability and technology self-efficacy on teachers' technology acceptance," *Journal of Research on Technology in Education*, vol. 43, no. 4, pp. 343–367, 2011.

[30] N. Harrati, I. Bouchrika, A. Tari, and A. Ladjailia, "Exploring user satisfaction for e-learning systems via usage-based metrics and system usability scale analysis," *Computers in Human Behavior*, vol. 61, pp. 463–471, 2016.

[31] N. Wagner, K. Hassanein, and M. Head, "The impact of age on website usability," *Computers in Human Behavior*, vol. 37, pp. 270–282, 2014.

[32] J. Brooke, "SUS: a "quick and dirty" usability scale," in *Usability Evaluation in Industry*, Taylor and Francis, London, UK, 1996.

[33] K. Orfanou, N. Tselios, and C. Katsanos, "Perceived usability evaluation of learning management systems: empirical evaluation of the system usability scale," *International Review of Research in Open and Distributed Learning*, vol. 16, no. 2, pp. 227–246, 2015.

[34] Sports Administration Ministry of Education in Taiwan, *Sports Administration Ministry of Education (SAMOE) Normative Physical Fitness Scores*, Sports Administration Ministry of Education in Taiwan, Taipei, Taiwan, 2016.

[35] I. J. Reinen and T. Plomp, "Information technology and gender equality: a contradiction in terminis?," *Computers & Education*, vol. 28, no. 2, pp. 65–78, 1997.

[36] H. K. Vincent, A. P. Alfano, L. Lee, and K. R. Vincen, "Sex and age effects on outcomes of total hip arthroplasty after

inpatient rehabilitation," *Archives of Physical Medicine and Rehabilitation*, vol. 87, no. 4, pp. 461–468, 2006.

[37] P. Klainin-Yobas, H. G. He, and Y. Lau, "Physical fitness, health behaviour and health among nursing students: a descriptive correlational study," *Nurse Education Today*, vol. 35, no. 12, pp. 1199–1205, 2015.

[38] J. F. Sallis, R. B. Pinski, R. M. Grossman, T. L. Patterson, and P. R. Nader, "The development of self-efficacy scales for health-related diet and exercise behaviors," *Health Education Research*, vol. 3, no. 3, pp. 283–292, 1988.

[39] M. E. Platta, I. Ensari, R. W. Motl, and L. A. Pilutti, "Effect of exercise training on fitness in multiple sclerosis: a meta-analysis," *Achieves of Physical Medicine and Rehabilitation*, vol. 97, no. 9, pp. 1564–1572, 2016.

[40] J. Del Coso, A. Pérez-López, J. Abian-Vicen, J. J. Salinero, B. Lara, and D. Valadés, "Enhancing physical performance in male volleyball players with a caffeine-containing energy drink," *International Journal of Sports Physiology and Performance*, vol. 9, no. 6, pp. 1013–1018, 2014.

[41] A. Ambarish and B. Neha, "Co-relation between physical fitness index (PFI) and body mass index in asymptomatic college girls," *Journal of Exercise Science & Physiotherapy*, vol. 11, no. 2, pp. 129–133, 2015.

[42] Physical Fitness Teaching Resource," 2018, http://home.ied. edu.hk/~teachpe/Physical_Fitness/4_lesson_act.html.

[43] Textbook Resource Center," 2018, https://market.cloud.edu. tw/content/primary/gym/yl_bc/teach/run.htm.

[44] H. Fu, S. K. McMahon, C. R. Gross, T. J. Adam, and J. F. Wyman, "Usability and clinical efficacy of diabetes mobile applications for adults with type 2 diabetes: a systematic review," *Diabetes Research and Clinical Practice*, vol. 131, pp. 70–81, 2017.

[45] J. Kaipio, T. Lääveri, H. Hyppönen, S. Vainiomäki, J. Reponen, and A. Ku, "Usability problems do not heal by themselves: national survey on physicians' experiences with EHRs in Finland," *International Journal of Medical*, vol. 97, pp. 266–281, 2017.

[46] T. R. Johnson, J. Zhang, Z. Tang, C. Johnson, and J. P. Turley, "Assessing informatics students' satisfaction with a web-based courseware system," *International Journal of Medical Informatics*, vol. 73, no. 2, pp. 181–187, 2004.

12

Wearable Sensor-Based Location-Specific Occupancy Detection in Smart Environments

Md Abdullah Al Hafiz Khan ⓘ, Nirmalya Roy ⓘ, and H. M. Sajjad Hossain ⓘ

Department of Information Systems, University of Maryland Baltimore County, Baltimore, MD, USA

Correspondence should be addressed to Md Abdullah Al Hafiz Khan; mdkhan1@umbc.edu

Academic Editor: Yuh-Shyan Chen

Occupancy detection helps enable various emerging smart environment applications ranging from opportunistic HVAC (heating, ventilation, and air-conditioning) control, effective meeting management, healthy social gathering, and public event planning and organization. Ubiquitous availability of smartphones and wearable sensors with the users for almost 24 hours helps revitalize a multitude of novel applications. The inbuilt microphone sensor in smartphones plays as an inevitable enabler to help detect the number of people conversing with each other in an event or gathering. A large number of other sensors such as accelerometer and gyroscope help count the number of people based on other signals such as locomotive motion. In this work, we propose multimodal data fusion and deep learning approach relying on the smartphone's microphone and accelerometer sensors to estimate occupancy. We first demonstrate a novel speaker estimation algorithm for people counting and extend the proposed model using deep nets for handling large-scale fluid scenarios with unlabeled acoustic signals. We augment our occupancy detection model with a magnetometer-dependent fingerprinting-based localization scheme to assimilate the volume of location-specific gathering. We also propose crowdsourcing techniques to annotate the semantic location of the occupant. We evaluate our approach in different contexts: conversational, silence, and mixed scenarios in the presence of 10 people. Our experimental results on real-life data traces in natural settings show that our cross-modal approach can achieve approximately 0.53 error count distance for occupancy detection accuracy on average.

1. Introduction

Localized commercial (university, office, mall, cineplex, restaurant, etc.) and residential (apartment, home, etc.) building occupancy detection and estimation at room/zone-level granularity in real time can provide meaningful insights into many smart environment applications, such as green building, social gathering, and event management. Smartphone-based participatory and citizen sensing applications have adhered to the promise of building such applications by utilizing various context-sensing sensors on board. Different sensors can be exploited individually or in tandem to build a variety of such novel applications to satisfy the myriad requirements of differing smart environment applications. For example, potential benefit from microphone sensor-based application is the assessment of social interaction and active engagement among a group of people by leveraging their conversational contents

[1] and speaker identification and characterization of social settings [2–4]. To enumerate the number of people in a conversational episode, such as during a social gathering, interactive lecture session, or in a restaurant or shopping mall environment, various speaker-counting paradigms have been explored [5–8]. Most of the recent studies which focus on conversational data features to extract high-level occupancy information assume that all of the users need to take turns at some point. While this specific scenario is feasible, it is not ideal. To tackle this ideal situation, researchers have proposed using arrays of microphone sensors, video cameras, or motion sensors for identifying microscopic occupancy information in real time [9–12] which is obtrusive in nature. We envision to move one step further by considering a more natural environment where people may spontaneously participate or abstain from any conversational phenomenon. We posit to augment the smartphone-based locomotive sensing model in

the absence of any conversational episode along with acoustic sensing-based audio inference model to precisely capture the characteristic of a natural environment and accurately estimate the occupancy count. To further pinpoint the occupancy, we integrate the smartphone's magnetometer sensor-based location-sensing model. In pursuit of these goals, we design a model which opportunistically exploits both the audio and motion data, respectively, from the smartphone's microphone and accelerometer sensor to infer the number of people present in a gathering and their semantic location information as supplemented by the magnetometer sensor on the smartphone. We also introduce a crowdsourcing model to reduce the effort for obtaining semantic location information at scale.

In particular, we propose a zero-hassle ambient and infrastructure-less mobile sensing (a.k.a. smartphone) based approach by exploiting only the smartphone's sensors to provide significantly greater visibility on real-time occupancy and its semantic location [13, 14]. The key challenge in this case is to effectively estimate the number of people in a crowded and noncrowded environment either in the presence of any conversational data or not. Such a hybrid sensing approach could potentially furnish more fine-grained occupancy profiling to better serve many participatory sensing applications while saving smartphones' battery power by advocating a distributed sensing strategy. Main contributions of this paper are summarized as follows:

(i) We propose an online acoustic sensing-based linear time adaptive people-counting algorithm based on real-life conversational data which promotes a unified strategy of considering both overlapped and nonoverlapped conversational data in a natural environment. We propose to select opportunistically the minimal number of microphone sensors which can substantially reduce the energy consumption of smartphones. Our proposed people-counting algorithm can dynamically select the length of the audio segment compared to the other existing work [6].

(ii) We also propose an offline data-driven people-counting algorithm which uses the deep neural network-based clustering approach. We optimize the deep network by learning the feature space and cluster membership jointly. We allocate the cluster dynamically to determine the number of people present in a conversation. Our proposed model dynamically provides beneficial frames to the occupancy-counting module. We perform extensive evaluation in the presence of 10 domestic users to validate our model performance.

(iii) Although the acoustic sensing-based approach holds great promises in inferring the number of occupants, it fails in the absence of any conversational data. Therefore, we propose the augment motion sensing-based counting strategy with our acoustic sensing-based people-counting algorithm which works on extreme modality of either of the data sources, be it acoustic or locomotive.

(iv) We design a magnetometer sensor-based localization technique at zone/room-level granularity to infer the location of a conversing group. We propose a novel crowdsourcing model to map the magnetic signature of different locations and collect a large number of annotated location information to tag the occupancy with its semantic location information [13].

2. Related Work

We review the most relevant literatures on the occupancy inference problem in the context of conversational sensing, localization, and speaker estimation which are smartphone-based.

2.1. Speaker Sensing. Occupancy estimation is an important enabler of various applications such as HVAC (heating, ventilation, and air-conditioning) controlling [12, 15–17] and social interaction [18]. For example, Nikdel et al. [15] quantified the energy consumption using building occupancy information. Aftab et al. [12] predicted occupancy from video sensing using object-tracking techniques from a scene and controlling the HVAC system in real time. In addition, various speaker-sensing algorithms have been proposed in the recent past using acoustic sensing [19, 20]. Valle [19] proposed a hybrid occupancy estimation model by combining the Gaussian mixture model (GMM) and hidden Markov model (HMM). A large number of previous works have used the smartphones' microphone to opportunistically analyze audio for context characterization. For example, SpeakerSense [4] performs speaker identification and SoundSense [21] classifies sounds from macro- to microcontexts. They have often in common employing the supervised speaker-learning techniques. In contrast, our model's occupancy-counting process is entirely unsupervised. Our proposed model anonymously estimates the number of people from the smartphones' acoustic cum locomotive sensing model where we have employed unsupervised learning techniques to cluster different forms of acoustic signatures. For example, Ofoegbu et al. [22] have built a model from mean and covariance matrices of the linear predictive cepstral coefficient (LPCC) of voice segments in conversations and used the Mahalanobis distance to determine whether two models belong to the same or different speakers. Iyer et al. [23] have performed speaker clustering using distance of the feature vectors extracted from different speakers and finally applied the modified k-means algorithm with distance metric data. However, their experiments for occupant estimation were on telephonic conversational data, where multiple participants were present, and voices were frequently overlapped and intertwined with the noisy environment. Sell et al. [20] predicted the number of occupants present using acoustic signals by employing the agglomerative hierarchical clustering (AHC) algorithm. Our proposed model performs speaker counting without any predefined environmental setup and collects data from natural conversation. Our proposed speaker-counting algorithm is close to [24] and [6] where smartphone-based speaker counting has been proposed in a controlled scenario where all the participants spoke actively. Xu et al. [6] used a fixed-length audio segment (3 sec) where each segment corresponds to an individual, but we performed this audio segmentation dynamically to increase the accuracy of occupancy inference. Xu et al. [6] also classified a few segments as undetermined, but our system never discards segments as undetermined which is achieved only through employing dynamic segmentation. Therefore, our proposed audio-based occupancy inference model tackles a richer problem, where none of the speakers are discarded for handling the computational challenges. Crowd++ [6] was proposed to combine pitch with MFCC to compute the number of people with an average error distance of 1.5 speakers. On the other hand, our MFCC-based proposed model improved the average error distance by a factor of two (0.76 speakers) [13, 14].

However, the major disadvantage of the MFCC-based acoustic approach is that the MFCC discards a lot of information present in the speech sound. Therefore, we need to develop a robust system which could potentially capture discriminative speaker information and establish correlation between features. Recently, deep learning methods become the state of the art in many acoustic applications such as object recognition [25] and speaker recognition [26]. In particular, deep learning algorithm helps design robust audio-related acoustic signal modeling like speech recognition [27] and phone recognition [28] with better accuracy. The major reason behind the popularity of the deep model is the capability of learning features from a large data set automatically and easing the rely on handcrafted features. Deep models are capable of learning robust features from both labeled and unlabeled data. Among the different deep learning methods, the deep neural network (DNN) has been recently shown to be effective in speech recognition applications [29, 30]. Milner and Hain [31] concatenated features from all the audio channels and helped train the DNN (deep neural network) model with this mixed feature to predict the number of speakers using audio signals. Mohamed [32] used the DNN to recognize phone from the acoustic data and showed that DNN performance is superior than that of the Gaussian mixture model (GMM). However, the features were learned from the labeled information provided during the training phase. Unsupervised feature learning has also been investigated in [33] where the convolutional deep belief network (CDBN) has been applied to audio data for gender detection, phone recognition, and speech identification applications. Deep learning-based clustering techniques have also been employed in [34, 35], where convolutional neural network- (CNN-) generated features were fed into the hierarchical clustering methods to cluster nonoverlapped utterances. Xie [35] proposed offline DNN-based clustering techniques and used the k-means clustering algorithm to initialize the cluster centroids. However, our DNN-based method automatically helps cluster the speakers and train the models in an unsupervised fashion. Our model dynamically determines the total number of clusters present in the given input audio stream and employs the offline DNN-based clustering strategy to help achieve an average error count distance of ≈ 0.53 which is 30% higher than that of our MFCC-based iterative speaker-counting approach.

2.2. Indoor Localization. Wang et al. [36] proposed an unsupervised indoor localization approach exploiting environmental identifiable artifacts and specific signatures on single or multiple sensing dimensions using smartphones' different sensor readings (mainly from accelerometer, compass, gyroscope, and WiFi APs). Track [37] deployed reusable beacons around the place of an event and utilized the location of the beacons in conjunction with the smartphone contact list and applied crowdsourcing techniques to infer users' location. Chung et al. [38] measured geomagnetic field in a way which is spatially varying but temporally stable, using an array of e-compasses to infer location. However, they used a bunch of sensors or sensor arrays for location detection, whereas our model only used the smartphones' magnetometer sensor to infer semantic location information of a gathering at zone/room-level granularity. Subbu et al. [39] used magnetic fingerprints with dynamic time-warping algorithm to predict location information with 92% accuracy. Our model used the standard random forest algorithm and achieved 98% accuracy to detect high-level semantic location information of any gathering. IndoorAtlas location technology [40] utilized anomalies of ambient magnetic fields for indoor positioning. This platform provides the functionality for participatory sensing where the crowd can contribute by war-driving magnetic signatures of an unexplored location.

3. Overall System Architecture

We envision developing a minimally invasive cost-free robust mobile system for counting the number of people present at any time in any environment and enlighten their semantic location information. Our model boosts these capabilities by employing smartphones' magnetometer, microphone, and accelerometer sensors. Our system, as shown in Figure 1, comprises two subsystems, one deployed on the smartphone and the other in the server. Using only acoustic sensing, it is not always possible to predict the correct number of the occupants present in a specific location as some people get involved in a conversation, while others remain silent. For example, in a classroom scenario, while professor lectures, some of the students participate, but the majority of the students remain silent. Sensed data are stored in a *data sink* (sink) for posterior analysis in the mobile part of our proposed architecture consisting of an accelerometer and a magnetometer. In our model, we propose to utilize microphone sensor-based acoustic sensing in conjunction with accelerometer sensor-based locomotive sensing for occupancy detection. For this joint collaborative sensing, acoustic sensed data are being fed to the filter to collect acoustic fingerprint (AFP), consisting of content-based audio. The AFPs being collected from all smartphones are sent to the *"estimate proximity"* module residing on the server which helps distinguish the audio signals in vicinity and approximate the inclusion of a group of smartphones to form a single clique. Finally, the *"optimum node"* module elects the clique leader (the most informative smartphone) to record the audio data and notifies the condition of deactivation to the other smartphones from capturing the duplicate audio signal. It also helps in sorting the smartphone list based on their audio signal strength which is

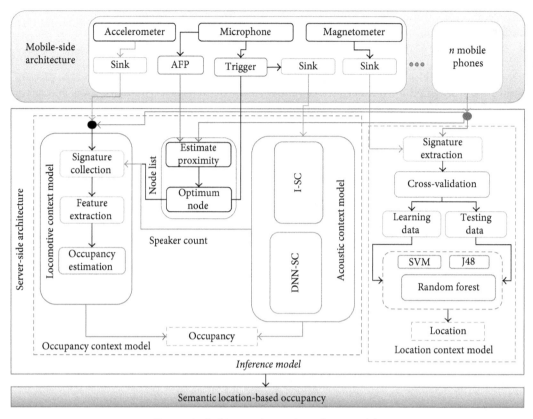

FIGURE 1: Architectural overview of our model.

eventually utilized by the locomotive "signature collection" module to opportunistically check on and trigger the accelerometer sensor on the smartphones [41]. The server-side architecture consists of two main logical subcomponents: (i) occupancy context model and (ii) location context model. These models together form the inference engine of our proposed semantic location-sensitive occupancy detection system.

3.1. Occupancy Context Model. It has two submodules: acoustic context model and locomotive context model.

3.1.1. Acoustic Context Model (ACM). Our acoustic context model has two independent inference modules: (i) iterative speaker count (I-SC) and (ii) deep neural network- (DNN-) enabled speaker count (DNN-SC). We employ these modules for inferring occupancy.

(i) Iterative speaker count (I-SC): this module serves as the core processor for occupancy counting. It takes the raw audio signal as input and generates the MFCC as features and then measures the similarities between the audio frames and segments. Based on these similarity measures, it decides whether those speech segments are generated from distinct or the same speaker. It keeps track of all the segments and their identities with respect to a specific person and finally helps count the total number of existing speakers during a conversational episode.

(ii) Deep neural network- (DNN-) enabled speaker count (DNN-SC): it accepts raw audio signals and produces features such as MFCC, ZCR, and so on and deploys the deep neural network (DNN) to infer occupancy.

3.1.2. Locomotive Context Model (LCM). It comprises (i) signature collection, (ii) feature extraction, and (iii) occupancy estimation modules. The signature collection module receives the total number of people count from the *ACM module* and the sorted smartphone list from the *optimum module* to opportunistically select a single smartphone's microphone sensor. Based on these two inputs, the *LCM module* makes decision on which smartphones' sensors are needed for further occupancy estimation. The feature extraction module calculates the accelerometer sensor magnitude and feeds that into the occupancy estimation module, which infers binary occupancy for each smartphone and finally helps counting the total number of people present in a conversational cum silent environment.

3.2. Location Context Model. Our location context model consists of two submodules: (i) signature extraction and (ii) location estimation. In the signature extraction phase, we compute the feature vectors from smartphone's magnetometer sensor data. In the location estimation phase, we use those feature sets for cross-validation to construct training and testing sets. After producing training and testing sets, we apply machine learning techniques to infer location.

4. Design Methodology

In this section, we describe the details of our model design framework. We present an acoustic augmented locomotive sensing model for counting the number of people present in a conversing, nonconversing natural environment. We posit a magnetometer sensor-based fingerprinting methodology to semantically localize the gathering.

4.1. Occupancy Estimation Using Acoustic Signature. We compute the total number of speakers present in a conversation using two methods: (i) iterative speaker counting (I-SC) and (ii) deep neural network- (DNN-) enabled speaker counting (DNN-SC). We discuss the details of our approach below.

4.1.1. Iterative Speaker Count (I-SC). This module has three submodules: (i) preprocessing, (ii) feature extraction, and (iii) occupancy estimation. Figure 2 shows the stacked pipeline of our iterative speaker count module.

(1) Preprocessing: this module is the most trivial phase for acoustic signal processing. This module helps to perform the filtering and select the audio segment length dynamically. It finally helps remove all the noises and silences and produce smooth conversational data which are later passed to the feature extraction module.

(2) Feature extraction: this is the main basis for extracting all types of features which is utilized in the speaker estimation module. This module takes conversational samples and processes them through a series of data cleaning and feature extraction steps. It helps making frames from samples to calculate various features like MFCC, pitch, and so on. These features are later used by the speaker estimation module.

(3) Speaker estimation: in this section, we describe our iterative occupancy estimation using our proposed acoustic sensing model. We look into the specific cases where all the occupants have been conversing. We first attempted to calculate the number of speakers engaged and consider three different phases to compute the number of personnel present. First, we propose to create dynamic segments from the raw audio data and assume that each segment belongs to an individual person. We attempt to detect every speaker change point in the entire audio signal spectrum and assign one segment to one person to increase the counting performance of our occupancy detection algorithm. A speaker change point depicts the stopping point of one speaker and the starting point of another speaker. Speaker change point detection algorithms have been investigated extensively [42–44]; however, it is a complex process to detect the speaker change point in conversational speech because utterance lengths can be extremely short, speaker changes may occur frequently, overlaps between the speakers may happen, and surrounding environment can be noisy. We create

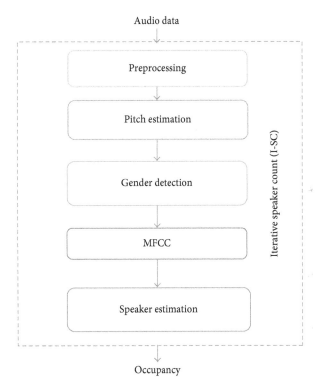

FIGURE 2: Processing pipeline of iterative speaker count (I-SC).

segments from the raw audio dynamically. Details of our preprocessing are discussed in Section 5.4.1.

Assume that the entire audio signal has N segments $\{S_1, S_2, \ldots, S_N\}$ and consider a segment which contains m frames, and each segment consists of frames $\{F_1, F_2, \ldots, F_m\}$. We calculated the MFCC for each frame where each segment has corresponding MFCC feature vectors as $\{M_1, M_2, \ldots, M_m\}$. We also computed the pitch for each segment to apprehend gender in the conversational data. Segment pitches are represented as $\{P_1, P_2, \ldots, P_m\}$, where the average pitch for male falls between 100 and 146 Hz, whereas the female pitch is within 188 to 221 Hz, as demonstrated in [45]. Segments which fall within the male frequency are marked as male and similarly for female. These two sets are then passed to our proposed people-counting heuristic algorithm. Before passing these male and female segments for checking similarity measures, we calculated intracosine angle of each segment to sort out both male and female segments. Next, we have checked the similarity among intersegments whether it falls within our predefined threshold, θ_{th}, or not. If these segments have been similar, then we have merged them to make a new segment and continued to check for the next segment with this newly created segment. If these segments have been dissimilar, then we have moved forward and picked another segment to check the similarity with the next one. The pseudocode of our proposed people-counting heuristic has been shown in Algorithm 1.

4.2. Deep Neural Network- (DNN-) Enabled Speaker Count (DNN-SC). Our deep neural network-based speaker count (DNN-SC) model comprises four submodules: (i) preprocessing,

Input: set of segments, $S = \{S_1, S_2, \ldots, S_N\}$, total number of segments $= N$
Output: Number of distinct speakers, N_s
$F_s = \{\}$ ▷ Selected empty frame set
for i from $1 : N$ **do**
$m_i = $ Compute_MFCC(S_i) ▷ Compute MFCC vectors
Insert(M, m_i) ▷ Insert m_i into MFCC set M
Sort(M) ▷ Sort MFCC set and keep sorted MFCC set into the same set M
PS $= \{\}$ ▷ Initialize Person Set which contains similar person in sets PS$_j$
for i from $1 : N$ **do**
for j from $(i + 1) : N$ **do**
$\theta = $ Cosine_Similarity(M_i, M_j)
if $(\theta \leq \theta_{\text{th}})$ **then**
Insert(PS_i, M_j)
else
$i = j$
Insert(PS, PS_i) ▷ PS denotes Person Set
$N_S = $ Count_Elements(PS)
return N_S

ALGORITHM 1: People count (S, N).

(ii) feature extraction, (iii) gender detection, and (iv) DNN model. In the preprocessing module, raw audio signals are segmented dynamically based on the confidence score of segments where each segment contains one speaker's voice information. Detailed segmentation is discussed in Section 5.4.1. Frames are generated from these raw audio segments, and selected frames are admitted into the feature extraction module which helps extract mel-frequency cepstral coefficients (MFCCs), zero crossing rate (ZCR), and spectral flux (SF). The DNN establishes nonlinear correlation among these features. In the gender detection module, male and female segments are differentiated using pitch information calculated with the help of the YIN [46] algorithm. Gender-specific audio features are passed into the deep neural network to count the total number of speakers present in the audio conversation data. Figure 3 shows our DNN-SC architecture. Next, we discuss the details of our frame selection algorithm and DNN-based speaker-counting algorithm.

4.2.1. Frame Selection Algorithm. Frames are created from the raw audio segments which may have important human voice information, silence, white noise, and so on. Since we are interested only in the voice information, therefore, we need to discard unwanted frames to improve the performance of our people-counting algorithm. These unwanted (unvoiced, silence, etc.) frames can occur at any time due to the different positions of the phone or contexts of the environment. Silence or unvoiced frames have low energy levels. Energy is obtained by calculating root mean square (RMS) values of the frames. Spectral entropy is also a good indicator of unwanted audio frames. White noise or silence has a flat spectrum and has high entropy, whereas low entropy represents human voice information. Entropy of the frame is obtained by calculating the normalized fast Fourier transform (FFT) spectrum of the frame. We represent the spectral entropy mathematically as follows:

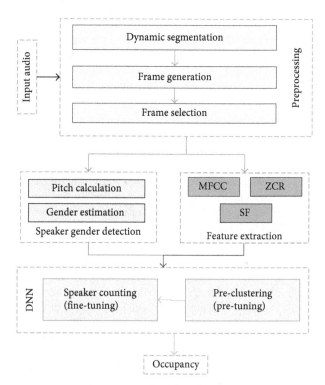

FIGURE 3: DNN-SC architecture.

$$h_f = -\sum_{j=1}^{m} p_j \log p_j. \tag{1}$$

Our frame selection algorithm selects frames based on the RMS and entropy values calculated above. Voiced frames have high RMS values when the recorded audio sample sound is high. However, the sound of the audio samples may be low due to the microphone or phone's position. In this case, we use entropy to admit or discard a frame. Since we are only interested in voiced frames, we focus on increasing

```
Input: Frames, F = {F_1, F_2, ..., F_m}
Output: Selected frames, F_s
F_s = {}          ▷ Selected empty frame set
for i ← 1; i < |F|; i ← i + 1 do
r_i ← compute_rms(F_i)          ▷ Calculate RMS value of a given frame
h_i ← compute_entropy(F_i)      ▷ Calculate entropy using (1)
if (r_i ≤ rth) or (h_i ≥ hth) then
Insert(F_s, F_i)          ▷ Add the selected frame F_i to the set F_s
return F_s
```

ALGORITHM 2: Frame selection (F).

the true positive. Therefore, we use two thresholds, RMS threshold (rth) and entropy threshold (hth) to admit or discard a frame. These thresholds are determined based on the empirical analysis of the given acoustic signals. We admit a frame when either the RMS value crosses the threshold (rth) or the entropy h_f value is lower than the threshold (hth). These threshold values depend on the phone, microphone, and context. We empirically determine this threshold from the collected audio data. The complete procedure is summarized in Algorithm 2.

4.2.2. DNN-Based Speaker Counting. We construct four layers of the deep neural network (DNN) [29] to cluster the entire audio signal. Figure 4 shows the building block of our DNN network. DNN is a feed-forward artificial neural network that has one or more hidden layers between the input and the output layers. Restricted Boltzmann machine (RBM) is the basic building block of a DNN, and each RBM is stacked one after another to form the network. An RBM is a type of Markov random field (MRF) and has one visible layer and one output layer. Each layer is composed of binary stochastic units. All units from the visible layer are connected to the hidden layer units, but there are no visible-visible or hidden-hidden unit connections. Each hidden unit's output depends on all of the visible units and the corresponding connection weights and a bias factor. The probability distribution function is defined using these weights and biases of the units and the joint distribution of the visible (**v**) and hidden (**h**) state vectors. This is defined as an energy function that is represented as follows:

$$E(\mathbf{v}, \mathbf{h}; \theta) = -\sum_{i=1}^{N_v} \sum_{j=1}^{N_h} w_{ij} v_i h_j - \sum_{i=1}^{N_v} b_i v_i - \sum_{j=1}^{N_h} a_j h_j, \quad (2)$$

where N_v and N_h represent the number of visible and hidden units, respectively, $\theta = (w, b, a)$ is the model parameter, w is the weight, and a and b are the biases of the visible and hidden units, respectively. Each RBM helps construct hidden units from the given visible units and reconstruct the visible units from the constructed hidden units. The visible vector probability is defined as follows:

$$p(\mathbf{v}|\theta) = \frac{\sum_{\mathbf{h}} e^{-E(\mathbf{v}, \mathbf{h})}}{\sum_{\mathbf{u}} \sum_{\mathbf{h}} e^{-E(\mathbf{v}, \mathbf{h})}}, \quad (3)$$

The conditional distributions of the visible and hidden layers are defined as follows:

$$p(v_i = 1|\mathbf{h}; \theta) = \sigma\left(\sum_{j=1}^{N_h} w_{ij} h_j + b_i\right),$$

$$p(h_j = 1|\mathbf{v}; \theta) = \sigma\left(\sum_{i=1}^{N_v} w_{ij} v_i + a_j\right). \quad (4)$$

Our first layer, RBMs' visible units, is constructed using Gaussian visible units [47] that use real-valued features which are extracted from the audio signal. The remaining RBM layers employ rectified linear unit (ReLU) activation functions to produce the binary output. This DNN model helps determine the total number of speakers present in a conversation in two phases: (i) preclustering and (ii) speaker counting, where the former is responsible for preclustering the audio segments and the latter designates the appropriate number of speakers or clusters present in the provided audio data.

(1) Preclustering. This module combines consecutive segments that are from the same speaker. We train the DNN network in a greedy layerwise basis before uniting the smaller segments into a larger one. The unlabeled audio data are leveraged to train the model using the contrastive divergence (CD) algorithm [48] which calculates the gradient and updates the model weights as follows:

$$\Delta w_{ij} = \langle v_i h_j \rangle_{\text{data}} - \langle v_i h_j \rangle_1, \quad (5)$$

where $v_i h_{j\text{data}}$ is the expectation of the training data and $v_i h_{j1}$ is the expectation calculated from the distribution of samples using the Gibbs sampling method [47].

Raw audio features—MFCC, zero crossing rate (ZCR), and spectral flux—are placed side by side to form a feature vector from a raw audio segment. These feature vectors are used to train each RBM one after another in a greedy layerwise fashion to find the correlation between features and distinct vocal tract characteristics of the speaker. Once pretraining is completed, each raw audio segment produces a binary feature vector which is then used to form clusters using the forward clustering method. Assuming that the binary feature vector set for the audio segment is $\{f_{S_1}, f_{S_2}, \ldots, f_{S_N}\}$, where N is the number of segments present in the audio signal, the raw audio segment set is

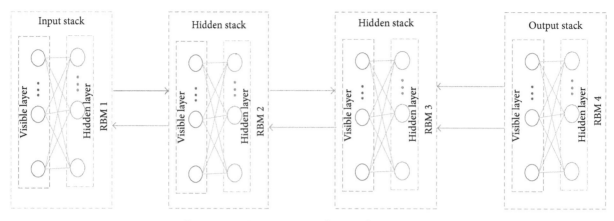

FIGURE 4: RBM structure used in speaker counting.

represented as $\{S_1, S_2, \ldots, S_N\}$. In the forward clustering method, we pick the first segment's feature vector f_{S_1} and calculate the cosine distance against f_{S_2}. If the distance is smaller than the similarity threshold δ_s, we merge these two raw segments into a new raw audio segment S_1 and compute the cluster centroid C_1 by taking an average of these two segments' feature vectors. We then calculate raw features from the merged segment again and formulate the binary feature vector with the help of the pretrained network. Next, we compare this newly computed feature vector f_{S_1} with f_{S_3}. If these two are similar, we then repeat merging and updating the centroid. Otherwise, we begin comparing f_{S_3} with f_{S_4} and form a new centroid. In this forward pass, we merge consecutive similar speaker audio segments to form a bigger segment since the smaller voice segments have a high likelihood of dissipating from the same speaker. The longer segments help increase the likelihood of distinguishing the different speakers and thus help increase the clustering performance. After this forward clustering, we merge the segment set $\{S_{C_1}, S_{C_2}, \ldots, S_{C_k}\}$ and their corresponding cluster centroid set $\{C_1, C_2, \ldots, C_k\}$ of the inferred speakers, where k represents the number of newly inferred speakers. In this preclustering technique, we have both longer (combined) and shorter segments as $\{S_1, S_2, \ldots, S_{k+m}\}$, where k is the number of smaller nonmerged segments. We compute the pitch for each of these longer segments, where each centroid is associated with pitch information. We assume that the pitch set is $\{P_{S_1}, P_{S_2}, \ldots, P_{S_k}\}$ which helps determine the gender of the speaker. Figure 5 shows the schematic diagram of our preclustering method.

(2) Speaker Counting. We count the number of people from the preclustered segments in this step. We employ the DNN for this purpose, but it seeks ground truth label information to compute the gradient of the network parameters. Since we have no label information, we postulate the previously computed centroid set as initial labels for this network. We start with segment S_i and pass this through the network to generate the feature vector $Z_i = z_1, z_2, \ldots, z_{li}$, where l is the total number of output units. The output of each unit, l, is calculated as follows:

$$z_l = \frac{1}{1 + e^{\left(b_l + \sum_j x_j w_{jl}\right)}}, \tag{6}$$

where x_j is the output of the ith unit from the previous layer.

We then compute the cosine distance against all the centroids which have similar gender information as with the current segment. If the cosine similarity distance (δ) is less than the empirically calculated threshold, δ_s, we then compute the new mean centroid C_i across all the similar centroids. This process is repeated for each of the segments. If any segment has the cosine similarity distance $D(C_i, S_j)$, greater than the empirically determined intraspeaker distance threshold, δ_s, and less than the interspeaker distance threshold, δ_d, we discard that segment. While a segment's cosine similarity distance is higher than the threshold, δ_s, we assign that feature vector as a new centroid in the network. Since these intra- and interspeaker cosine distance thresholds depend on the microphone sensitivity, we determine it from our collected samples such that it reduces the total number of false positives. We validate our model by setting $\delta_s = 16$ and $\delta_d = 31$. To optimize the network parameters, $\theta = (w, b)$, we define our network objective function, $J(\theta)$, based on the cosine similarity measure as follows:

$$J(\theta) = \frac{1}{(m+k)} \sum_{j=1}^{k} 1 \cdot \{\delta \le \delta_s\} \cos^{-1}\left(\frac{Z_i \cdot C_j}{|Z_i||C_j|}\right), \tag{7}$$

where $1 \cdot \{\delta \le \delta_s\}$ is the indicator function where it infers one when the cosine distance $\delta < \delta_s$ or $\delta > \delta_s$, otherwise infers zero. We jointly optimize the network parameter θ and cluster centroids using the stochastic gradient descent (SGD) algorithm. The gradient of the objective function, J, with respect to each unit (z_l) of Z_i is calculated as follows:

$$\frac{\delta J(\theta)}{\delta(z_l)} = \frac{-1}{\sqrt{1 - A^2}} \times B, \tag{8}$$

where $A = Z_i \cdot C_j / |Z_i| \cdot |C_j|$ and $B = z_l / |Z_i| \cdot |C_j| - A \cdot c_l / |C_j|^2$.

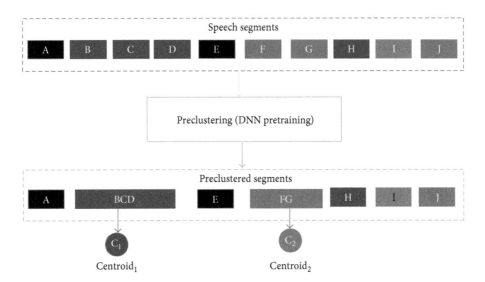

FIGURE 5: Schematic diagram of the preclustering method that shows how individual contingent speech segments combined to form clusters. Different colors represent different speaker's audio. Contingent audios from the same speakers are combined into a new bigger segment and formed a new cluster. A newly formed cluster centroid is shown in circle.

Similarly, we calculate the gradient of the objective function, J, with respect to each component (c_l) of the centroid, C_j. The derivative is represented as follows:

$$\frac{\delta J(\theta)}{\delta(c_l)} = \frac{-1}{\sqrt{1-A^2}} \times D, \qquad (9)$$

where $D = c_l/|Z_i| \cdot |C_j| - A \cdot z_l/|Z_i|^2$. These gradients are passed down to our DNN network and used the standard backpropagation algorithm to optimize networks weights, w, and bias, b. Once the training of our DNN is complete, the total number of clusters represent the total number of speakers present in a conversation.

4.3. Occupancy Estimation Using Accelerometer Signature. In this section, we discuss our locomotive sensing model in the absence of any conversational data or in a mixed environment where a group of people may talk and others listen silently. If a smartphone is stationary for a significant amount of time, the on-board accelerometer sensor produces a steady-state signature which has no variation or spikes in terms of signal amplitude, whereas if there is a movement, it generates a spike or corresponds to a steady-state signal alteration. To detect these abrupt changes in locomotive signal amplitude, we propose to use the change point detection-based technique [49].

Change point detection helps to find the abrupt variation in the movement data stream. Our motivation in this work is to use the change point to find the stray movements by finding abrupt changes in the accelerometer signals. These changes help inferring binary people counting (whether people are present or not). We investigated the offline Bayesian change point [49] detection-based algorithm for inferring the occupant's presence in $\mathcal{O}(n^2)$. Let the observed accelerometer data sequence be $x_{1:N} = \{x_1, x_2, x_3, \ldots, x_N\}$, where N denotes the number of data points over time T. We partition this data sequence into nonoverlapping regions based on *run length* [50]. The length of each partition or time since the last change point occurred is defined as *"run length"*. If there are m partitions, then the partition data set is denoted as $\{\rho_1, \rho_2, \rho_3, \ldots, \rho_m\}$. We also denote $x_{t_i:t_j}$ as the contiguous set of observations between times t_i and t_j inclusively. If the length of the current run at time m is denoted by r_m, then it can be defined as follows:

$$r_m = \begin{cases} 0 & \text{if change point occurs at } (m-1). \\ r_{m-1} + 1 & \text{otherwise} \end{cases} \qquad (10)$$

Change points occur at discrete time points. The conditional probability that a change point occurs at time t_k after the last change point at time t_{k-1} is

$$\pi(t_m|t_{m-1}) = g(t_m - t_{m-1}), \quad \text{where } 0 < m-1 < n,$$
$$\pi(t_m) = \sum_{j=0}^{m-1} g(t_m - t_j)\pi(t_{m-1}), \qquad (11)$$

where $\pi(t_m)$ is the prior probability of a change point at time t_m and depends on the probability distribution of the observed data sequence and the preceding change point.

Change point detection algorithm computes predictive distribution $\pi(x_{n+1}|x_n)$ on a given run length r_m taking the integration over the posterior distribution $\pi(r_n|x_{1:n})$ which is computed using the following equation:

$$\pi(r_n|x_{1:n}) = \frac{\pi(r_n, x1:n)}{\pi(x_{1:n})}. \qquad (12)$$

It also finds out the joint distribution over the run length and the observed data as follows:

$$\pi(r_n, x_{1:n}) = \sum_{r_{n-1}} \pi(r_n, r_{n-1}, x_{1:n})$$
$$= \sum_{r_{n-1}} \pi(r_n|r_{n-1})\pi(x_n|r_{n-1}, x_{1:n})\pi(r_{n-1}, x_{1:n-1}), \qquad (13)$$

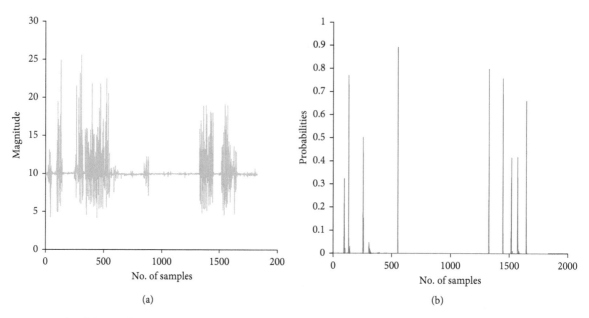

FIGURE 6: Magnitude of the accelerometer signal (a) and change points with probabilities of that signal (b) due to random movement patterns of a person.

where $\pi(x_n | r_{n-1}, x_{1:n})$ is the segment log-likelihood which depends on the data $x_n^{(r)}$ and $\pi(r_n | r_{n-1})$ is the change point probability which can be calculated as follows:

$$\pi(r_n | r_{n-1}) = \begin{cases} H_f(r_{n-1} + 1) & \text{if } r_n = 0, \\ 1 - H_f(r_{n-1} + 1) & \text{if } r_n = r_{n-1} + 1, \quad (14) \\ 0 & \text{otherwise,} \end{cases}$$

where hazard function $H_f(\eta)$ is calculated using $H_f(\eta) = g(\eta) / \sum_{j=\eta}^{\inf} g(j)$. We employ this change point technique in our locomotive sensing model for designing the binary occupancy detection algorithm. It has been built on the basis of the following threefold methodology. First, we calculate a priori probability of two successive change points at a distance d (run length). We use the Gaussian-based log-likelihood model [50] to compute log-likelihood of the data in a sequence $[s, d]$, where no change point has been detected. Second, we calculate log-likelihood for the entire signal $S[t, n]$, log-likelihood of the data sequence $S_s[t, s]$ where no change point has been occurred between t and s and $\pi[i, t]$, and the log-likelihood that the ith change point occurs at time step t. Finally, we calculate the probability of a change point at time step t by summing up the log-likelihoods for that sequence. Figure 6 presents the change points and their probabilities which are being detected successfully in our proposed locomotive sensing model using the smartphone's accelerometer sensor. We filter those change points based on empirically determined threshold probability (δth) and infer the presence of the occupants based on the admitted change point sequence. We also count the number of change points in the data sequence which indicates the movement score that represents how frequent a person moves. The overall algorithm has been summarized in Algorithm 3, and we named it as the locomotive speaker-counting (LSC) algorithm.

4.4. *Location Estimation.* In this scenario, our goal was to explore the possibility of inferring the location at the zone/room level in different commercial and residential buildings by only using the smartphones' magnetometer sensor signals. Intuitively, this is possible as different rooms have magnetic patterns that are distinct based on their unique structures and furniture layouts. This opens up the possibility that a sophisticated machine learning technique may learn to discriminate magnetic signatures belonging to different rooms. In our experiment, we collected the magnetic signature of different rooms, office spaces, and lobby areas in an academic building using the smartphones' magnetometer sensor. In a room, all furniture and metallic objects generally remain fixed in positions and rarely are moved from one place to another. This gives us an intuition that each room has its own magnetic fingerprints which can be utilized to detect that specific room or semantic location.

We notice that the magnetic sensor is sensitive to magnetic fluctuations in indoors specially near pillars and metallic objects. Figure 7 represents this behavior where peaks occur near pillars, elevators, and so on because pillars and elevators emit high magnetic fields. Magnetic fields produced by pillars are different for each floor because of their varying intensity levels. These density characteristics guide in localization because each floor is independent of the structure and height with other levels, from which it is also probable to infer floor-level location. From these empirical observations, we conclude that each room has its unique magnetic fingerprint. We analyze different rooms' data at the university's Information Technology and Engineering (ITE) building for three months. Figure 8 represents this analysis which depicts each room-specific magnetic fingerprint helping to create a coarse localization model for pinpointing the semantic location of gatherings at the zone/room level.

Input: Accelerometer Sensor Data, *data*, Total number of data points = n
Output: Binary Speaker Count
for (t from $1 : n$) **do**
$g[i] = \log(1/(n + 1))$
if $i = 0$ **then** $G[i] = g[i]$
else
$G[i] = \log(\exp(G[i - 1]) + \exp(g[i]))$
$P[n - 1, n - 1] = \text{Gaussian\_log\_likelihood}(data, n - 1, n)$
for (t from $1 : n$) **do**
prob_next_changepoint = Cal_Joint_Dist($data, t, n - 1$)
$P[t, n - 1] = \text{Gaussian\_log\_likelihood}(data, t, n)$
$Q[t] = \log(\exp(P\_next\_run), \exp(P[t, n - 1] + 1 - \exp(G[n - 1 - t])))$
for (i from $1 : n - 1$) **do**
changepoint_prob$[0, t] = (P[0, i] + Q[i + 1] + g[i] - Q[0])$
for (i from $1 : n - 1$) **do**
for (t from $1 : n - 1$) **do**
tmp_sum = (changepoint_prob$[i - 1, i - 1 : t] + P[i : t + 1, t] + Q[t + 1] + g[0 : t - i + 1] - Q[i : t + 1])$
changepoint_prob$[i, t] = \log(\text{sum}(\exp(\text{tmp\_sum})))$
if (changepoint_prob$[i, t]\delta$th) **then**
num_effective_cp = num_effective_cp + 1
if (num_effective_cp > 0) **then**
occupancy = 1
return occupancy

ALGORITHM 3: Locomotive speaker-counting (*data, n*).

We also note that this magnetic signal differs not only for different indoor environments but also for the phone's placement. This distraction has been optimized in two different ways: (i) calibrating magnetic signals and (ii) calculating absolute magnitude.

During our experimentation, we observe that magnitude represents different fingerprints for a separate indoor environment. Figure 9 describes how normalized magnitude of different rooms varies upon the total number of samples. Performing this experimentation over several rooms helps establish the fact that each room represents a different magnitude which may form their own fingerprint. We consider magnitude of the magnetometer because for different persons with distinct movement, it does not deviate much other than little variations. Figure 10 represents these characteristics where the magnetic signature has been collected from two different people in the same room, and both signals delineate the same shape and almost the same magnitude.

From this empirical study, we conclude that, by only using the magnetic signature, it is difficult to estimate fine-grained indoor location in different indoor environments; for this reason, we also consider the mean, standard deviation, and variance of different axes. Based on those feature vectors, we generate two sets of data: training and testing using the cross-validation process. We use the training set to learn indoor characteristics by using different machine learning models and later use the testing set to predict location. To estimate fine-grained semantic location, we use SVM, J48, and random forest classifiers.

4.5. Crowdsourcing Magnetic Model. We propose to use collaborative sensing or crowdsourcing to ease our ground truth data collection and location-mapping process. We have divided the area of interest inside the ITE building as a grid of squared cells (details are provided in Section 5.2). We collected data from most frequently visited grids without any major obstruction. While crowdsourcing the unique characteristics of grid location, it was difficult to choose the right representation of data as analogous magnetic signatures of different grids in different locations were prevalent. As a result, it was deemed necessary to display a potential set of locations from which the crowd would finalize the association of a semantic label with a particular observed magnetic signature pattern. Considering this, we provide the floor information for a specific signature pattern, such that our crowdsourcing model will enable the crowd to choose the appropriate semantic location or room from that specific floor. Nevertheless, the search space remains large as the possibilities of multiple rooms with similar magnetic footprints in a floor are quite abundance. We propose a simple grid-mapping crowdsourcing model which reduces the search space by mapping the magnetic signature pattern of point of occupancy with the existing patterns and sorts the rooms according to the similarity measurement. Our model takes the Manhattan distance and the squared deviation of magnetic magnitude as input parameters for the mapped grids and searches the repository of existing signature pattern database.

Consider a set of cell values found from a test pattern $X = x_1, x_2, x_3, \ldots, x_n$. First, we take x_1 from X and try to map this value with the cell values of existing patterns. We do not assume to have any prior idea regarding the organization of the cells in the test pattern. For mapping signature values, we consider the deviation of ± 2 which has been determined empirically according to our experiments.

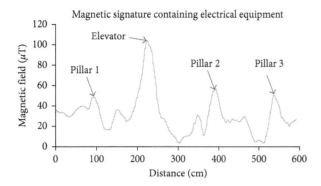

FIGURE 7: Magnetic signature variation for different equipment.

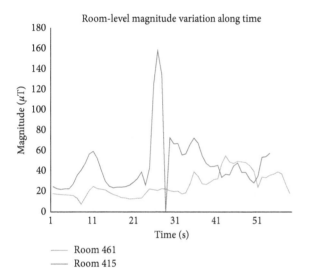

FIGURE 8: Magnetic signature variation along time for two rooms.

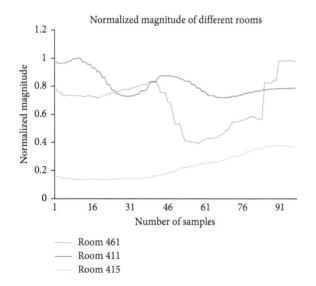

FIGURE 9: Normalized magnitude of the magnetometer for different rooms.

We add patterns which match the similarity value of a cell to our candidate set C and initialize a $n \times n$ distance matrix $\overline{M^{(i)}}$ and a $n \times 1$ deviation matrix $\overline{D^{(i)}}$ for each candidate c_i.

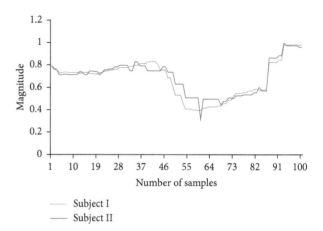

FIGURE 10: Normalized magnitude of the room for different subjects.

$\overline{M^{(i)}}$ records the Manhattan distances between the mapped cells in a candidate pattern C_i, and $\overline{D^{(i)}}$ stores the squared deviation between the mapped cell values. If we find similarities in multiple cell values in a single room signature pattern, we consider them as an individual candidate. We take the next test pattern, x_2, in the next iteration and do the similar operation like x_1, but this time we consider only the candidates in C. In this iteration, if the deviation and distance matrices of a candidate c_j do not get updated, then we discard them from the candidate set and reduce the search space. We recursively perform the same mapping for the remaining grid values and compute the final matching candidate set C_F with their corresponding distance and deviation matrices.

At this stage, it is still possible to have a large number of candidates in C_F. To tighten the search space, next we compute the error measurements for each candidate $E(c_i)$ and sort the candidates with respect to this value assuming that, in an ideal conversational episode, the participants remain in close proximity. We calculate $E(C_i)$ based on the following:

$$E(c_i) = \sum_{p=1}^{m} X_{k,p} \left(\sum_{r=1}^{n} \overline{M^{(i)}}_{a,r} \overline{D^{(i)}}_{r,b} \right)_{p,l}, \qquad (15)$$

where $k = 1$, $l = 1$, $1 \le a \le n$, and $b = 1$.

After calculating the error measurements for each candidate, we sort C_F and choose the first 10 candidates from C_F. We plot the magnetic signature pattern of these candidates and the test pattern. The crowd now have to choose the signature pattern in which they find the test pattern. In our experiments, there were some cases where we observed the empty candidate set. In these cases, we selected the last iteration's candidate set which was not empty. We also asked the crowd that if they found match with multiple candidates, then they have to choose the earliest signature pattern.

5. System Implementation and Evaluation Results

We now discuss the detailed implementation and evaluation of our model framework.

5.1. Tools and Resources. We used Google Nexus-5 with built-in microphone and three-axis accelerometer sensor for our experiments. Our entire system comprises two parts: (i) sensing and (ii) classification and clustering; the first one was implemented on Nexus-5 and the latter on the server. Application software was written in Java which utilizes the Android application programming interface (API) to sense microphone and accelerometer signals. Classification and clustering algorithms and our occupancy-counting algorithm have been implemented on the server side using Python.

We consider the Python-based deep learning platform *Tensorflow* [51] to implement our deep neural network-(DNN-) based clustering algorithm. Features are fed into the DNN in batch with a length of 32. Our DNN comprises 4 layers which represent two hidden layers with 1024 units each, one input layer of 22 units and one output layer of 512 output units. In the pretraining phase, each layer was trained for 100 epochs, and in the fine-tuning phase, each layer was trained for 1000 epochs. The internal architecture of our DNN network is shown in Table 1.

5.2. Data Collection. Magnetic sensor signals are sensed through our Android application and stored temporarily on mobile storage. We first collected magnetic data for the training set and subsequently for the testing set. We divided the room space into small regions, each containing an area of 0.5×0.5 m^2 and was named as the cell. Thus, each room forms the grid containing cells. We collected data from each cell for 5 minutes both clockwise and counterclockwise direction to form the training set. We also maintain fixed height (approximately 4 feet from the floor) when collecting our ferromagnetic fingerprint because it also depends on the height. The partial 3rd floor map along with the sample magnetic data collection path is shown in Figure 11. It shows the sample data collection path of room number 305, where green line shows how the grid forms and red line shows the data collecting path in both directions along the grid. We use a sampling rate of 5 Hz for magnetometer sensor data. We implemented the acoustic sensing and collected conversational data from different places at different times in natural settings. Conversational data have been collected and properly anonymized during the spontaneous lab conversation among the students (without making the occupants aware of it), lab meeting, and general discussions in the lobby/corridor in the presence of a variety of surrounding noise levels. The demographic for our conversational data collection was 1–10 persons (with 5 females and 5 males) in the age group of 18–50 years. The acoustic data were collected at a monosampling rate of 16 kHz at 16 bit pulse-code modulation (PCM).

5.3. Privacy. One of the major concerns of smartphone-based acoustic signal processing is privacy. This concern becomes more serious when the smartphone records the conversation data. Our counting algorithm determines the number of speakers in this environment in an anonymized manner. We used text file as cover in which our recorded audio is embedded. A secret key is induced for the

TABLE 1: DNN internal architecture.

Input units	Output units	Hidden layers	Each hidden layer's unit	Total layers
22	512	2	1024	4

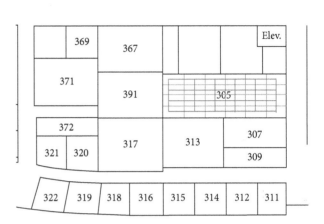

FIGURE 11: Sample magnetic data collection path.

embedding and extraction process which is known by both the sender and the recipient. A steganographic function takes cover file as an argument and then embeds audio file and key to produce *stego* as output which is sent to our server. A reverse steganographic function on our server side takes *stego file* and key as parameters and produces audio file as output. There are different steganographic methods (i.e., LSB coding, parity coding, and phase coding), but we used the simplest method, the least significant bit algorithm, which replaces the least significant bits of some bytes in the cover file to hide a sequence of bytes containing hidden data. To generate the *stego file*, the algorithm first converts each character of the cover file into bit stream followed by converting the audio file into bit streams and finally replacing the LSB bit of the cover file with the bit of the audio in the secret information. We also ensured that the size of the file was not changed during this encoding and it was suitable for any type of audio file formats.

5.4. Preprocessing. In this section, we discuss the details of our preprocessing module.

5.4.1. Acoustic Data Preprocessing. We process the raw audio streams to remove noise and prepare the audio data for the feature extraction module. This module is responsible for segmenting the raw audio signals to extract appropriate frames. These frames contain event information (i.e., voice, noise, and silence) that accounts for further processing.

(1) Dynamic Segmentation. We create segments from the entire audio signal dynamically assuming that each raw audio segment contains single speaker information. We calculate the confidence score for the entire audio segment

which represents the probability of finding the pitch within a segment. We then start finding the confidence score from a small segment (32 ms) and increase the step size in the successive iterations and repeat this up to an audio segment of size 10 seconds. We calculated the variance of this confidence score, and based on a lower variance associated with a specific segment, we selected that segment length as one unit of conversation.

If a segment has over 90% confidence, we considered it. As there are many audio segments with different segment lengths, we have chosen a segment length corresponding to a single person unit associated with a higher confidence score and greater number of audio segments with a lower segment length. Figure 12 shows various confidence scores for different segment lengths. We selected 2.72 sec as the segment length instead of 3.36 sec when both have a confidence score of 1, but the first segment length admitted a greater number of segments than the latter one. We have calculated this confidence score using the YIN [46] algorithm by using nonoverlapping frames and skipped the best local estimate step. This helps to determine on real time the unit audio segment which solely depends on the recorded audio.

As human voice ranges approximately from 300 Hz to 4000 Hz, we filter each of the segments based on that frequency range using the band pass filter. After filtering the raw audio, we have applied the Hamming window to reduce the spectral leakage while creating audio segments.

(2) Framing. We create frames from the filtered audio segments using a fixed-width sliding window. Each frame has a length of 32 ms and 50% overlap. These frames are able to capture the person's subtle vocal characteristics present in the sounds.

5.5. Feature Extraction. We discuss different features relevant to our acoustic, locomotive sensing, and localization technique in this section.

5.5.1. Magnetic Features. For location detection, we used only the magnetometer sensor. The smartphones' magnetic sensor provides three axis values: x-, y-, and z-axis. From these values, we calculated magnitude using $m = \sqrt{x^2 + y^2 + z^2}$. We considered only the resultant magnitude to mitigate variations of the readings resulting from smartphones' different axes based on different positions. We also calculated the mean, variance, and standard deviation of each reading and combined those features to generate the feature vectors.

5.5.2. Acoustic Features. We generated four basic features which are used in the speaker identification—MFCC, pitch, zero crossing rate (ZCR), and spectral flux. Each feature has been described in detail in the following:

    (i) *MFCC* is one of the most significant features which is used for acoustic processing. We followed the following steps to process it: (1) take the Fourier transform of (a windowed excerpt of) a signal, (2)

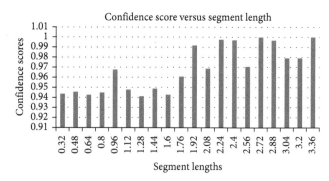

FIGURE 12: Confidence scores for different segment lengths of a sample audio.

map the powers of the spectrum obtained above onto the mel scale using triangular overlapping windows, (3) take the logs of the powers at each of the mel frequencies, and (4) finally, take the discrete cosine transform of the list of mel log powers. We excluded the first coefficient of the MFCC and then chose 20 coefficients as feature vectors. The MFCC feature computation schematic diagram is shown in Figure 13.

    (ii) *Pitch* is defined as the lowest frequency of a periodic waveform. It is the discriminative feature between man and woman. The human voice pitch interval falls within the range of 50 Hz to 450 Hz [45]. We calculated the pitch of different segments using the YIN [46] algorithm.

    (iii) *Zero crossing rate (ZCR)* is defined as the rate at which the signal changes its sign from positive to negative or back [52]. Human voice has both voiced and nonvoiced sounds. Nonvoiced and voiced sounds show lower or higher variations of the ZCR, respectively. Therefore, the ZCR is an important feature to count the number of speakers. The ZCR is calculated as follows:

$$\text{zcr} = \frac{\sum_{k=0}^{n} |\text{sign}(s_i) - \text{sign}(s_{i-1})|}{2}. \tag{16}$$

    (iv) *Spectral flux (SF)* [53] is defined as the l_2 norm of the spectral amplitude difference between the current frame, $F(t)$, and the previous frame, $F(t-1)$, and mathematically represented as follows:

$$\text{SF}_t = \sum_{i=1}^{n} (F(t) - F(t-1))^2. \tag{17}$$

Human speech changes from voice to nonvoice rapidly and thus alters its spectral shape frequently. Spectral flux helps measure these spectral shape changes. Usually, speech has a higher SF value.

5.5.3. Locomotive Features. We considered the magnitude of the accelerometer data as our locomotive feature in order to mitigate calibration.

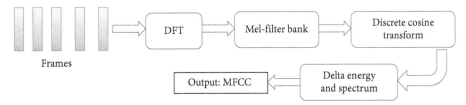

FIGURE 13: MFCC feature extraction.

5.6. Accuracy Metric Definition. To evaluate and compare the performance of our location-sensitive occupancy model, we first define the following metrics:

(i) *Occupancy metric*: we computed the average error count as the normalized predicted occupancy metric represented by $|EC - AC|/N$, where EC, AC, and N denote the estimated people count, actual people count, and number of samples, respectively. We presented only the absolute value in order to avoid any positive or negative contribution.

(ii) *Location metric*: for evaluating location measurement, we consider the following metrics: average precision ($TP/TP + FP$), average recall $TP/TP + FN$, and average F-1 score ($2 \times precision \times recall/precision + recall$), where TP, FP, TN, and FN are the number of instances of true positive, false positive, true negative, and false negative, respectively.

(iii) *Location prediction error*: it is defined as the mean absolute error between predicted and actual values of the estimated variable. This error is expressed as the mean absolute error $= (1/n)\sum_{i=1}^{n} |(f_i - y_i)|$, where f_i is the prediction and y_i is the actual value.

5.7. Occupancy-Counting Results. We evaluated our opportunistic occupancy-counting algorithm in four scenarios: (i) no conversation among occupants, (ii) all occupants are conversing in a single clique, (iii) occupants are conversing in multiple cliques, and (iv) mixed conversing and non-conversing occupants.

(i) *No conversation among occupants*: for the first scenario, when no occupants are involved in a conversation, we used the accelerometer to count the occupancy. Each accelerometer sensor provides binary occupancy indication based on our change point detection algorithm as discussed in Section 4.3, which computes the total number of people present in the environment. Figure 14 shows the total number of people successfully counted using our locomotive speaker-counting (LSC) algorithm. We note that our locomotive sensing model achieves 80% accuracy (8 out of 10 people) in predicting occupancy when most of the users carry their smartphones with them.

(ii) *All occupants are conversing in a single clique*: our opportunistic sensing system plays a critical role when all occupants have been conversing in a single clique. Our system helps to activate a single

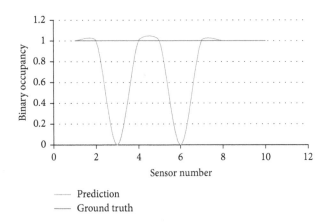

FIGURE 14: Occupancy count for the LSC algorithm.

microphone for occupancy counting and deactivate all other microphones and accelerometer sensors based on the server's feedback. Figure 15 depicts the effect of cosine distant similarity measures on our occupancy-counting algorithm (I-SC) as shown in Figure 1. We noticed that similarity distance angle measures (in degree) play a pivotal role in reducing the error count of occupancy inference. In our experiments, with 3 people conversing, we found that 15-degree similarity measure threshold is an appropriate choice for consideration to reduce the error count for our proposed adaptive people-counting algorithm.

We also have run experiments in an uncontrolled environment (completely in a natural setting) without imposing any restrictions on smartphones' relative positions and distances from each other or from the server. Figure 16 reports the average error count distance as ≈ 0.5 with respect to different positions of the phone. It is noted that when the smartphone is placed on the table and two persons speak, the error count becomes zero, but when three persons start speaking, error count tends to become slightly higher due to the ambient noise and overlapped conversation. Figure 17 shows occupancy-counting results for DNN-SC on different positions of the phone. We notice that the average error count distance for DNN-SC is 0.30 which is 40% less than our I-SC approach as we employ a more selective strategy to select appropriate frames in our frame selection algorithm.

Figure 18 depicts that the error count increases as the single clique leader's distance from other occupants increases. We note that, for a 3-meter distance, the error count becomes close to two which confirms that even for a large

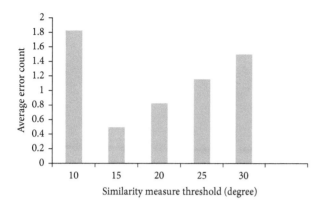

FIGURE 15: Performance with different cosine measures.

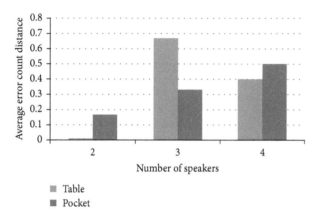

FIGURE 16: Occupancy count over different phone positions for iterative speaker counting (I-SC).

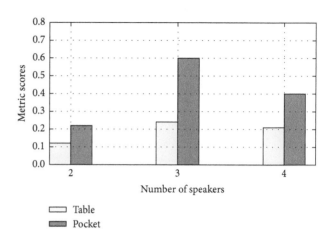

FIGURE 17: Occupancy count over different phone positions for DNN-SC.

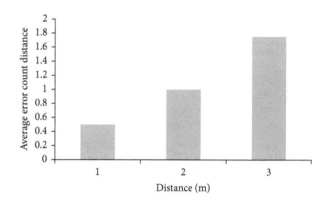

FIGURE 18: People counting versus phone distance for I-SC.

internal distance separation among the conversing occupants, our acoustic sensing model performs quite well. Figure 19 shows the average error count distance with different distances of the phone from the speakers. Note that DNN-SC outperforms I-SC in this case. However, DNN-SC reports similar trends as in I-SC with the increasing distance of the phone from the speakers.

Figure 20 presents the performance of our people-counting algorithm (I-SC) where users speak naturally with overlapped conversations. It is observed that the average error count is 0.1 for 2 people and 1.7 for 10 people when conversing together. Thus, the overall average error count is 0.76 with the number of users present varying from 2 to 10 establishing that our acoustic-based occupancy-counting algorithm performs well even in a crowded environment. Figure 21 presents the performance of our DNN-SC algorithm. We observe that the overall average error count for DNN-SC is 0.5316 with the number of speakers present varying from 2 to 10. Our DNN-SC people-counting algorithm performance improves 30% than our I-SC occupancy-counting algorithm. In Figure 20, we notice that our I-SC algorithm performance decreases with the increase of the number of speakers present in a conversation because of the overlapping segments which span across multiple speakers' voice and limited capabilities of MFCC features to

differentiate these speakers. In Figure 21, we observe the similar trends as in our I-SC method, but DNN-SC helps improve performance with the increasing number of speakers because DNN-SC can capture the hidden correlation between features.

(iii) *Occupants are conversing in multiple cliques*: in our third scenario, where occupants are conversing in multiple cliques (three cliques in our experiment), we deployed three microphones and accelerometer sensors which are chosen based on the proximity measure from the server to infer the occupancy. Figure 22 shows the intragroup count in the presence of conversational data with distinct clique formation. In our experiments, the first group has 5 occupants (2 men and 3 women), the second group has 6 occupants (3 men and 3 women), and the last group has 8 occupants (4 men and 4 women). We observe that the mean error count is ≈ 1 even for our group-based acoustic sensing model which attests the promise of our occupancy detection model in different real-life scenarios.

(iv) *Mixed conversing and nonconversing occupants*: in our last scenario, where some people speak and some people remain silent, we propose to utilize our

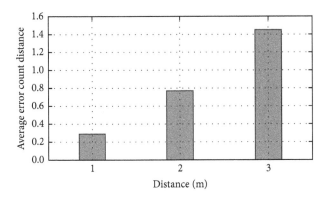

FIGURE 19: People counting versus phone distance for DNN-SC.

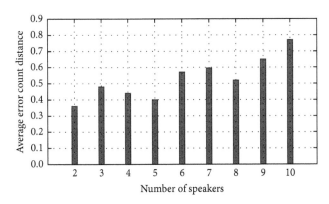

FIGURE 20: Accuracy versus the number of people for I-SC.

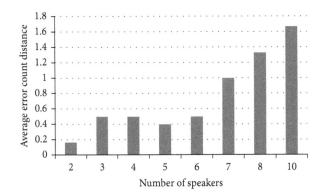

FIGURE 21: Accuracy versus the number of people for DNN-SC.

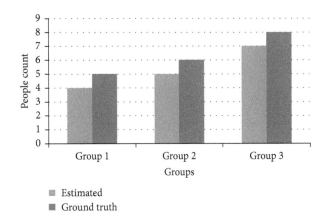

FIGURE 22: People counting versus multiple colocated groups of speakers.

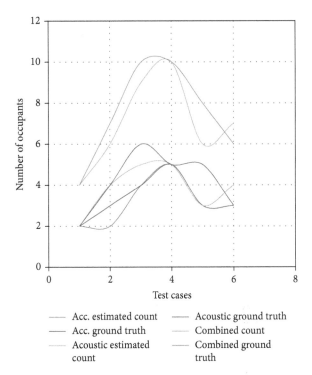

FIGURE 23: Locomotive augmented acoustic occupancy count.

hybrid locomotive cum acoustic sensing model to infer the total number of occupants. For example, consider a scenario where six persons are involved in conversation while four remain silent. For conversing population, we activate either a single microphone sensor if there is a single clique or multiple microphone sensors if there are multiple conversing cliques as determined by our "*estimate proximity*" module implemented on the server. We use mean error count estimation to infer the number of people conversing. To estimate the number of people who are not involved in that conversation, we utilize our locomotive sensing model which postulates binary occupancy using change point detection applied on the accelerometer's signal and finally infers the total number of silent people. Figure 23 plots overall occupancy-counting performance based on our hybrid approach. For example, when there are ten people in which 6 persons converse in a single clique and 4 persons remain silent, our acoustic sensing estimates 5 people out of 6 and locomotive sensing estimates 4 people out of 4, resulting in total of predicting 9 people out of 10. We have compared the performance of our model with Crowd++ framework [6] for counting the number of people. Table 2 shows that the average error count distance for Crowd++ is 1.78, whereas for our model (I-SC

TABLE 2: Comparison of average error count between Crowd++ and our model.

Number of speakers	Crowd++ (error count)	Combined (I-SC + LSC) (error count)	Combined (DNN-SC + LSC) (error count)
2	0.5	0.167	0.22
4	2.33	0.5	0.36
6	2.5	0.83	0.40
Average	1.78	0.5	0.33

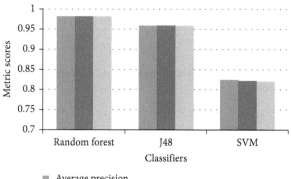

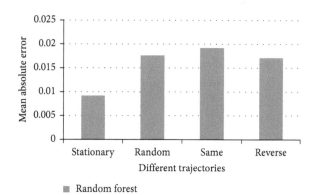

FIGURE 25: Location estimation error versus different trajectories.

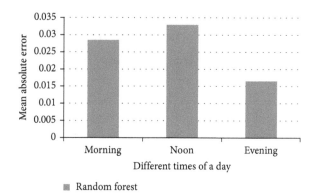

FIGURE 24: Location estimation errors for different classifiers.

FIGURE 26: Location estimation error during different times of a typical day.

+ LSC), it is 0.5, more than a threefold increase in accuracy for inferring the total number of people. From Table 2, we observe that our combined (DNN-SC + LSC) model outperforms the combined (I-SC + LSC) model by approximately 34% in total.

5.8. Location Estimation Results. Figure 24 presents the location estimation error of an occupancy gathering using different classifiers. The random forest classifiers perform best with an average precision, recall, and F1 score of 0.98.

We also validated our location model through different test cases where we consider (i) different trajectories, (ii) different times of a day, and (iii) different rooms with a varying number of occupants.

We conducted our experiments following different trajectories, like keeping mobile phone on the table, following the same or reverse directions when collecting data, and finally, collecting data randomly for a room. We noted that these different movement patterns do not affect much in the performance of our occupancy-gathering location determination model. Figure 25 shows errors for different movement patterns. We find that the stationary pattern shows better accuracy, while moving in the same direction gives higher error rate. Average errors are close to 0.015, which is quite acceptable with a minor number of false positives or true negatives.

Figure 26 depicts the varying nature of the magnetic signature during the different times of a day. We observe that the location estimation of any gatherings is similar during the different times of a typical day. It shows that error ranges approximately from 0.015 to 0.03 due to the global variation

of weather and other magnetic factors making our model as time invariant.

We also ran experiments for location-sensing model with respect to different rooms at different floors in ITE building with a different set and size of the occupants. From Figure 27, we do observe that the mean absolute error approximately varies in the range of 0.015 to 0.04 which has a negligible effect on the performance of our location-sensitive occupancy determination model. We observed some discrepancies between different subjects' data for room 321 and room 461. After investigating, we found that the discrepancies happened due to unusual magnetic inferences of electronic devices present while collecting data for subject II. To evaluate our crowdsourcing model, we ran a simulation of our magnetic crowdsourcing model in the Vowpal Wabbit (VW) toolkit [54]. We implemented our mapping algorithm on the server side and then used the function *active_interactor* of VW to interact with the users. We showed 10 magnetic

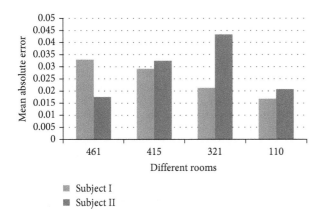

FIGURE 27: Location estimation error in different rooms with different occupancy sizes.

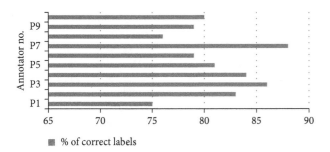

FIGURE 28: Results of our magnetic crowdsourcing model.

signature patterns and 1 test pattern to an user and asked him to choose the magnetic signature pattern in which he/she finds the test pattern. 10 participants participated in the crowdsourcing, and in Figure 28, we show the overall accuracy for each participant when given 15 pattern-matching tasks. Average accuracy of gaining correct annotation for these 15 patterns is ≈81% which is adequately high. Our results indicate that the probability for getting noisy labels is very low, and the crowd-annotated data can be chosen as input to the classifier.

6. Discussion and Future Work

In the current version of our work, we have assumed that people keep their smartphones in the pocket or in the hand which might not be ideal in some cases. In future, our plan is to make our architecture more robust and independent of smartphones' location. The performance of our counting algorithm does not get affected by TV or radio sounds as TV or radio follows different modulation techniques which make it easier for us to remove those external noises from resultant audio signal systems. We have used source separation where significant overlap between human conversation and TV occurs. In the current implementation, location-mapping process is independent of the classification process. In future, we plan to develop and integrate a combined mapping and classification model. We also plan to investigate fine-grained floor-level location using smartphone barometric sensing. We plan to investigate a more

advanced opportunistic sensing model considering microphone, accelerometer, and magnetometer sensor participation not only based on a server-based architecture but also based on an intersmartphone distributed collaborative sensing-based approach.

7. Conclusion

In this paper, we presented an innovative system to infer the number of people present in a specific semantic location which opportunistically exploits the accelerometer and microphone sensor of smartphones for people counting. We proposed an acoustic sensing-based unsupervised clustering algorithm by addressing the underpinning challenges evolving from naturalistic overlapped and sequential conversation to infer the occupancy in an environment. We posit a change point detection-based locomotive sensing model to infer the number of people in the absence of any conversational episode. We implement an opportunistic context-aware client-server-based architecture to leverage smartphones' microphone, accelerometer, and magnetometer sensors and combine our acoustic sensing with locomotive and semantic location-sensing model to better predict the location-augmented occupancy information. We have also demonstrated a novel crowdsourcing model for reducing the effort of collecting location information at the zone/room level at a large scale. Our experimental results hold promises in a variety of natural settings with an average error count distance of 0.76 in the presence of 10 users. We believe that this investigation holds promises and helps to open up many new research directions in this opportunistic multimodal sensing domain.

Acknowledgments

This research was partially supported by the NSF Grant CNS-1544687.

References

[1] L. Rabiner and B.-H. Juang, *Fundamentals of Speech Recognition*, Dorling Kindersley (India) Pvt. Ltd., New Delhi, India, 1993.

[2] T. Choudhury and A. Pentland, "Sensing and modeling human networks using the sociometer," in *Proceedings of the Seventh IEEE International Symposium on Wearable Computers*, Washington, DC, USA, October 2003.

[3] D. Jayagopi, H. Hung, C. Yeo, and D. Gatica-Perez, "Modeling dominance in group conversations using nonverbal activity cues," *IEEE Transactions on Audio, Speech, and Language Processing*, vol. 17, no. 3, pp. 501–513, 2009.

[4] H. Lu, A. J. B. Brush, B. Priyantha, A. K. Karlson, and J. Liu, "SpeakerSense: energy efficient unobtrusive speaker identification on mobile phones," in *Proceedings of the IEEE International Conference on Pervasive Computing*, Seattle, WA, USA, March 2011.

[5] R. Sen, Y. Lee, K. Jayarajah, A. Misra, and R. Krishna Balan, "GruMon: fast and accurate group monitoring for heterogeneous urban spaces," in *Proceedings of the 12th ACM Conference on Embedded Network Sensor Systems (SenSys'14)*, Memphis, TN, USA, November 2014.

[6] C. Xu, S. Li, G. Liu et al., "Crowd++: unsupervised speaker count with smartphones," in *Proceedings of the 2013 ACM International Joint Conference on Pervasive and Ubiquitous Computing (UbiComp'13)*, Zurich, Switzerland, September 2013.

[7] Y. Lee, J. Song, C. Min et al., "SocioPhone: everyday face-to-face interaction monitoring platform using multi-phone sensor fusion," in *Proceedings of the 11th Annual International Conference on Mobile systems, Applications, and Services (MobiSys'13)*, Taipei, Taiwan, June 2013.

[8] A. Alexandridis, N. Stefanakis, and A. Mouchtaris, "Towards wireless acoustic sensor networks for location estimation and counting of multiple speakers in real-life conditions," in *Proceedings of the 2017 IEEE International Conference on Acoustics, Speech and Signal Processing (ICASSP)*, pp. 6140–6144, New Orleans, LA, USA, March 2017.

[9] R. Tomastik, S. Narayanan, A. Banaszuk, and S. Meyn, "Model-based real-time estimation of building occupancy during emergency egress," in *Proceedings of the Pedestrian and Evacuation Dynamics 2008*, Wuppertal, Germany, February 2008.

[10] E. Hailemariam, R. Goldstein, R. Attar, and A. Khan, "Real-time occupancy detection using decision trees with multiple sensor types," in *Proceedings of the Symposium on Simulation for Architecture and Urban Design 2011 (SimAUD 2011)*, Boston, MA, USA, April 2011.

[11] T. H. Pedersen, K. U. Nielsen, and S. Petersen, "Method for room occupancy detection based on trajectory of indoor climate sensor data," *Building and Environment*, vol. 115, pp. 147–156, 2017.

[12] M. Aftab, C. Chen, C.-K. Chau, and T. Rahwan, "Automatic HVAC control with real-time occupancy recognition and simulation-guided model predictive control in low-cost embedded system," *Energy and Buildings*, vol. 154, pp. 141–156, 2017.

[13] M. A. A. H. Khan, H. Hossain, and N. Roy, "Infrastructure-less occupancy detection and semantic localization in smart environments," in *Proceedings of the 12th EAI International Conference on Mobile and Ubiquitous Systems: Computing, Networking and Services*, pp. 51–60, Coimbra, Portugal, July 2015.

[14] M. A. A. H. Khan, H. S. Hossain, and N. Roy, "SensePresence: infrastructure-less occupancy detection for opportunistic sensing applications," in *Proceedings of the 2015 16th IEEE International Conference on Mobile Data Management (MDM)*, vol. 2, pp. 56–61, Pittsburgh, PA, USA, June 2015.

[15] L. Nikdel, K. Janoyan, S. D. Bird, and S. E. Powers, "Multiple perspectives of the value of occupancy-based HVAC control systems," *Building and Environment*, vol. 129, pp. 15–25, 2018.

[16] H. Zou, Y. Zhou, H. Jiang, S.-C. Chien, L. Xie, and C. J. Spanos, "WinLight: a WiFi-based occupancy-driven lighting control system for smart building," *Energy and Buildings*, vol. 158, pp. 924–938, 2018.

[17] Y. Peng, A. Rysanek, Z. Nagy, and A. Schlüter, "Using machine learning techniques for occupancy-prediction-based cooling control in office buildings," *Applied Energy*, vol. 211, pp. 1343–1358, 2018.

[18] A. Lykartsis, S. Weinzierl, and V. Dellwo, "Speaker identification for Swiss German with spectral and rhythm features," in *Proceedings of the 2017 AES International Conference on Semantic Audio*, Erlangen, Germany, June 2017.

[19] R. Valle, "ABROA: audio-based room-occupancy analysis using Gaussian mixtures and hidden Markov models," in *Proceedings of the Future Technologies Conference (FTC)*, pp. 1270–1273, Francisco, CA, USA, December 2016.

[20] G. Sell, A. McCree, and D. Garcia-Romero, "Priors for speaker counting and diarization with AHC," in *Proceedings of the Interspeech 2016*, San Francisco, CA, USA, September 2016.

[21] H. Lu, W. Pan, N. D. Lane, T. Choudhury, and A. T. Campbell, "SoundSense: scalable sound sensing for people-centric applications on mobile phones," in *Proceedings of the 7th International Conference on Mobile Systems, Applications, and Services (Mobisys'09)*, Kraków, Poland, June 2009.

[22] U. O. Ofoegbu, A. N. Iyer, R. E. Yantorno, and B. Y. Smolenski, "A speaker count system for telephone conversations," in *Proceedings of the International Symposium on Intelligent Signal Processing and Communications*, Tottori, Japan, December 2006.

[23] A. N. Iyer, U. O. Ofoegbu, R. E. Yantorno, and B. Y. Smolenski, "Blind speaker clustering," in *Proceedings of the 2006 International Symposium on Intelligent Signal Processing and Communications*, Tottori, Japan, December 2006.

[24] A. Agneessens, I. Bisio, F. Lavagetto, M. Marchese, and A. Sciarrone, "Speaker count application for smartphone platforms," in *Proceedings of the IEEE 5th International Symposium on Wireless Pervasive Computing 2010*, Modena, Italy, May 2010.

[25] C. Szegedy, A. Toshev, and D. Erhan, "Deep neural networks for object detection," in *Advances in Neural Information Processing Systems*, pp. 2553–2561, The MIT Press, Cambridge, MA, USA, 2013.

[26] N. D. Lane, P. Georgiev, and L. Qendro, "DeepEar: robust smartphone audio sensing in unconstrained acoustic environments using deep learning," in *Proceedings of the 2015 ACM International Joint Conference on Pervasive and Ubiquitous Computing*, pp. 283–294, Osaka, Japan, September 2015.

[27] N. D. Lane, S. Bhattacharya, P. Georgiev et al., "DeepX: a software accelerator for low-power deep learning inference on mobile devices," in *Proceedings of the 2016 15th ACM/IEEE International Conference on Information Processing in Sensor Networks (IPSN)*, pp. 1–12, Vienna, Austria, April 2016.

[28] A.-R. Mohamed, G. Dahl, and G. Hinton, "Deep belief networks for phone recognition," in *NIPS Workshop on Deep Learning for Speech Recognition and Related Applications*, vol. 1, p. 39, Whistler, BC, Canada, December 2009.

[29] G. Hinton, L. Deng, D. Yu et al., "Deep neural networks for acoustic modeling in speech recognition: the shared views of four research groups," *IEEE Signal Processing Magazine*, vol. 29, no. 6, pp. 82–97, 2012.

[30] A.-R. Mohamed, D. Yu, and L. Deng, "Investigation of full-sequence training of deep belief networks for speech recognition," in *Proceedings of the Interspeech 2010*, Chiba, Japan, 2010.

[31] R. Milner and T. Hain, "DNN approach to speaker diarisation using speaker channels," in *Proceedings of the 2017 IEEE International Conference on Acoustics, Speech and Signal Processing (ICASSP)*, pp. 4925–4929, New Orleans, LA, USA, March 2017.

[32] A.-R. Mohamed, G. E. Dahl, and G. Hinton, "Acoustic modeling using deep belief networks," *IEEE Transactions on Audio, Speech, and Language Processing*, vol. 20, no. 1, pp. 14–22, 2012.

[33] H. Lee, P. Pham, Y. Largman, and A. Y. Ng, "Unsupervised feature learning for audio classification using convolutional deep belief networks," in *Proceedings of the Advances in Neural Information Processing Systems*, pp. 1096–1104, Vancouver, BC, Canada, December 2009.

[34] Y. Lukic, C. Vogt, O. Dürr, and T. Stadelmann, "Speaker identification and clustering using convolutional neural networks," in *Proceedings of the 2016 IEEE 26th International Workshop on Machine Learning for Signal Processing (MLSP)*, pp. 1–6, Salerno, Italy, September 2016.

[35] J. Xie, R. Girshick, and A. Farhadi, "Unsupervised deep embedding for clustering analysis," in *Proceedings of the International conference on machine learning (ICML'16)*, New York City, NY, USA, 2016.

[36] H. Wang, S. Sen, A. Elgohary, M. Farid, M. Youssef, and R. R. Choudhury, "No need to war-drive: unsupervised indoor localization," in *Proceedings of the Annual International Conference on Mobile systems, Applications, and Services (MobiSys'12)*, Lake District, UK, June 2012.

[37] L. Xiang, T.-Y. Tai, B. Li, and B. Li, "Tack: learning towards contextual and ephemeral indoor localization with crowd-sourcing," *IEEE Journal on Selected Areas in Communications*, vol. 35, no. 4, pp. 863–879, 2017.

[38] J. Chung, M. Donahoe, C. Schmandt, I.-J. Kim, P. Razavai, and M. Wiseman, "Indoor location sensing using geo-magnetism," in *Proceedings of the Annual International Conference on Mobile systems, Applications, and Services (MobiSys'11)*, Washington, DC, USA, 2011.

[39] K. Subbu, B. Gozick, and R. Dantu, "Indoor localization through dynamic time warping," in *Proceedings of the 2011 IEEE International Conference on Systems, Man, and Cybernetics*, Hong Kong, October 2011.

[40] IndoorAtlas, https://www.indooratlas.com/.

[41] L. A. Castro, J. Favela, J. Beltrán et al., "Collaborative opportunistic sensing with mobile phones," in *Proceedings of the 2014 ACM International Joint Conference on Pervasive and Ubiquitous Computing Adjunct Publication (UbiComp'14) Adjunct*, Seattle, WA, USA, September 2014.

[42] J. Ajmera, I. McCowan, and H. Bourlard, "Robust speaker change detection," *IEEE Signal Processing Letters*, vol. 11, no. 8, pp. 649–651, 2004.

[43] D. Liu and F. Kubala, "Fast speaker change detection for broadcast news transcription and indexing," in *Proceedings of the EuroSpeech*, Brighton, UK, 2009.

[44] L. Lu and H.-J. Zhang, "Real-time unsupervised speaker change detection," in *Proceedings of the IEEE Pattern Recognition*, Quebec City, Canada, August 2002.

[45] R. J. Baken and R. F. Orlikoff, *Clinical Measurement of Speech and Voice*, Cengage Learning, Boston, MA, USA, 2000.

[46] A. de Cheveignéb and H. Kawahara, "YIN, a fundamental frequency estimator for speech and music," *Journal of the Acoustical Society of America*, vol. 111, no. 4, pp. 1917–1930, 2002.

[47] A. Fischer and C. Igel, "An introduction to restricted Boltzmann machines," in *Iberoamerican Congress on Pattern Recognition*, pp. 14–36, Springer, Berlin, Germany, 2012.

[48] M. A. Carreira-Perpinan and G. E. Hinton, "On contrastive divergence learning," in *Proceedings of the AISTATS*, vol. 10, pp. 33–40, Barbados, 2005.

[49] P. Fearnhead, "Exact and efficient Bayesian inference for multiple changepoint problems," *Statistics and Computing*, vol. 16, no. 2, pp. 203–213, 2006.

[50] X. Xuan and K. Murphy, "Modeling changing dependency structure in multivariate time series," in *Proceedings of the 24th International Conference on Machine learning (ICML'07)*, New York, NY, USA, April 2007.

[51] Tensorflow, *Deep Learning*, http://www.tensorflow.org.

[52] J. Saunders, "Real-time discrimination of broadcast speech/music," in *Proceedings of the 1996 IEEE International Conference on Acoustics, Speech, and Signal Processing*, vol. 2, pp. 993–996, 1996.

[53] E. Scheirer and M. Slaney, "Construction and evaluation of a robust multifeature speech/music discriminator," in *Proceedings of the 1997 IEEE International Conference on Acoustics, Speech, and Signal Processing*, vol. 2, pp. 1331–1334, 1997.

[54] *Vowpal Wabbit*, http://hunch.net/vw/.

13

SRAF: A Service-Aware Resource Allocation Framework for VM Management in Mobile Data Networks

Kang Liu,[1] **Ruijuan Zheng** ⓘ**,**[1] **Mingchuan Zhang** ⓘ**,**[1] **Chao Han,**[2] **Junlong Zhu,**[1] **and Qingtao Wu** ⓘ[1]

[1]*College of Information Engineering, Henan University of Science and Technology, Luoyang 471000, China*
[2]*Institute of High Energy Physics, Chinese Academy of Sciences, Beijing 100049, China*

Correspondence should be addressed to Ruijuan Zheng; zhengruijuan@haust.edu.cn

Academic Editor: Wenchi Cheng

Service latency and resource utilization are the key factors which limit the development of mobile data networks. To this end, we present a service-aware resource allocation framework, called SRAF, to allocate the basic resources by managing virtual machine (VM). In SRAF, we design two new methods for better virtual machine (VM) management. Firstly, we propose the self-learning classification algorithm (SCA) which executes the service request classification. Then, we use the classification results to schedule different types of VMs. Secondly, we design a sharing mode to jointly execute service requests, which can share the CPU and bandwidth simultaneously. In order to enhance the utilization of resources with the sharing mode, we also design two scaling algorithms, i.e., the horizontal scaling and the vertical scaling, which execute the operation of resource-level scaling and VM-level scaling, respectively. Furthermore, to enhance the stability of SRAF and avoid the frequent operation of scaling, we introduce a Markov decision process (MDP) to control VM migration. The experimental results reveal that SRAF greatly reduces service latency and enhances resource utilization. In addition, SRAF also has a good performance on stability and robustness for different situations of congestion.

1. Introduction

Virtual machine (VM) management based on service awareness is a new method which can greatly reduce the service latency and enhance the resource utilization. In addition, VM management has been widely used in various mobile data networks, such as information center network (ICN), mobile vehicle network (MVN), and mobile cloud network (MCN). However, the resource pools of networks are limited. Moreover, the service latency and resource utilization are interactional. Therefore, how to reduce the service latency and how to enhance the utilization of resources, simultaneously, have been the focus point of research studies, especially for MCN [1, 2].

For this reason, service latency and resource utilization have become main aspects in many research studies. In the perspective of service latency, Reference [3] proposes the Predictable Resource Guarantee Scheduler scheme to realize the proportional sharing of CPU and I/O bandwidth, which

reduces the waiting time in the Xen platform. Reference [2] uses the cloudlet selection strategy to schedule the cloudlets for cutting down the response time. In addition, there are also some studies using the method of cutting down the distance between locations to reduce the latency. For example, Reference [1] aims to find the shortest path between the user and the nearest cloud datacenter for reducing the transmission latency. For reducing the queueing time of requests, Reference [4] uses the method of active communication between controllers to proactively pull the requests when the controller finishes its requests so as to cut down the queue length. On the contrary, in the perspective of resource utilization, Reference [5] presents a smart migration mechanism to implement processor memory optimization based on VM placement. References [6, 7], respectively, use the methods of Bejo and kNN classification schemes to classify the requests for better scheduling. Reference [8] designs a Lyapunov optimization framework to improve the efficiency of the mobile-edge computing. The purpose of the

Lyapunov optimization framework is to minimize the resource overload by VM scheduling.

The works cited above propose many new ideas or methods to realize the optimal request schedule or optimal resource configuration, which can reduce delay or enhance the resource utilization. However, due to the complexity of the cloud network and diversity of mobile devices, the requests are also various and uncontrolled. So, the simple objective studies do not always have a good performance on different factors because many factors are interactional. Therefore, there are some researchers who design efficient methods with an overall framework to optimize these problems. In [9], the authors design a resource sharing framework named "Symbiosis" to realize the sharing of CPU and bandwidth. When one request is working in the CPU, the Symbiosis will make another request to perform the transmission. Moreover, the Symbiosis can efficiently reduce the service latency of the requests. In [6], the authors propose a new classification algorithm named "Bejo" to classify the requests. The classification results are used to perform the VM scheduling. The fitting VM for requests can enhance the utilization of resources. Therefore, we propose a new framework called SRAF to execute requests classification and resources sharing based on the strength of research studies in [6, 9].

For measuring the performance of SRAF, we analyze MCN in detail. The SRAF can be divided into two aspects. Firstly, we propose a self-learning classification algorithm (SCA) to perform the classification operation before the request is sent to the VM. The SCA is designed by two weighting methods, location weighting and feature weighting [10], which can improve the veracity of requests classification. The precise classification results can help the request find a fitting VM so as to reduce the service latency. Secondly, we design a sharing mode (Figure 1) to realize the resource sharing in a VM. Furthermore, in order to improve the utilization of the resources, we also design two VM scaling algorithms, the horizontal scaling and vertical scaling. The former is to realize the resource-level scaling in a VM. When the utilization of CPU or bandwidth is too high, the algorithm will add corresponding resources to the VM for avoiding overload, or otherwise for scaling down. The latter is to perform the VM-level scaling. When all the VMs are busy and the arrival tasks are continuously growing, new VMs are created, or otherwise released.

The contributions of this paper are as follows:

(i) We propose a new framework named "SRAF" to improve the resource utilization and reduce the service latency simultaneously.

(ii) We design the SCA which has the self-learning capacity for updating features so as to classify the service requests. In addition, SCA can improve the accuracy of classification continuously until all the features are learned.

(iii) We introduce a Markov decision process (MDP) to control VM migration so as to reduce the frequent scaling operation and enhance the stability of SRAF.

(iv) We propose a Combination Scheduling Cost Model and a sharing mode for mobile data networks. Combination Scheduling Cost Model can systematically operate VMs scheduling and scaling. Moreover, the resource utilization is improved directly via the sharing mode.

The rest of this paper is organized as follows: Section 2 has a brief introduction of related research works. We particularly introduce the details of each component of SRAF in Section 3. Section 4 shows the overall process of SRAF, including each model and related algorithm. Section 5 presents the feasibility and performance of SRAF by some experiments. Section 6 presents a brief conclusion of this paper.

2. Related Work

To reduce the service delay and enhance the utilization of resources, many studies propose various methods. For example, Reference [3] proposes a new prediction method to reach the objective of resource sharing. Researchers use the prediction method to predict the demand of the next duration time so as to adjust the resources for enhancing the utilization of resources. References [2, 5] use different methods to schedule the cloudlet and VM for reducing the response time and free time of resources, respectively. Due to the diversity of mobile access devices, the requests are also different and uncontrolled. For this reason, some studies design different classification algorithms to classify different requests, which depend on demand or input data for prior disposal [6, 7]. Then, researchers can use some classical and effective methods to reduce the queueing time and improve the utilization of resources, such as shortest job first (SJF) [12] and priority-aware longest job first (PA-LJF) [13]. Moreover, there are also some other new improved methods, such as shortest expected-remaining service time policy (SERSTP) [14] and dynamic-threshold service policy [15], which have a better performance for reducing delay and improving the QoS.

In addition, due to the randomness of service requests, some researchers use the Markov decision process (MDP) to quantify the overall process of cloud service. For example, Reference [16] uses the dynamic Markov decision process to model the process of VM scheduling. Then, the value iteration algorithm is used to find the optimal VM control policy for reducing energy expenditure. Reference [17] also uses the MDP to quantify the overall VM control. It uses the Bellman optimality equation to find a global optimum threshold so as to cut down erratic operation of VMs. To enhance the veracity of task scheduling by MDP, Reference [18] designs a semi-Markov decision process to select some computation-intensive tasks for offloading so as to reduce the computations in mobile devices.

Due to the above analysis, single method or policy can only realize one or two objectives. Therefore, some researchers choose to design an overall framework to handle multiple objectives. For example, References [8, 9] design different frameworks, Lyapunov optimization framework

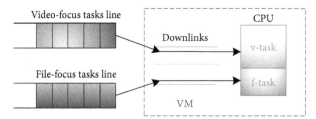

FIGURE 1: The sharing mode of the high-type VMs.

and Symbiosis, to control the scheduling of requests and VMs, respectively. Furthermore, there are also some researchers who add classification methods to enhance the performance of different frameworks for different objectives. For example, Reference [19] uses the reweighting method to label different factors by machine learning. Reference [11] uses the multi-instance learning method (MIL) to quantify different data for precise classification with the probabilistic graphical framework. Reference [10] uses the method of local feature selection to classify the data directly. In addition, many studies also use the sharing method to improve the utilization of resources in different fields. For example, Reference [20] proposes a feasible and truthful incentive mechanism (TIM) to realize the resource sharing with the trade-off between users and service providers. Reference [21] uses the sharing mode to satisfy the resources demand of the remote radio network and central virtual base station so as to maximize downlink of networks.

3. Architecture Design

In this section, we designed a green cloud resource allocation framework called SRAF. The objective is to reduce the delay of scheduling the service request and improve the resource utilization simultaneously. Our framework contains three layers, the User Layer, the Request Manager Layer, and the Resource Provider Layer.

In Figure 2, we show the overall response process of service requests. The User Layer has many users with various mobile terminals which send service requests to the mobile cloud. The Request Manager Layer is the most important layer to receive the service requests from the User Layer. Its main duty is to make optimization management of service requests and VMs. Then, the results are sent to the Resource Provider Layer for VM configuration. The Resource Provider Layer provides basic resources for the service. The Request Manager Layer includes four components:

(1) The History Loads is used to store the requests and their categories which can help the Classification Manager in updating its feature mapping library. The category information comes from the Combination Scheduling Manager when requests are serviced.

(2) The Classification Manager analyzes the information from the requests and classifies the requests into three types, i.e., file-focus tasks, video-focus tasks, and normal tasks, depending on the demand of bandwidth and CPU resources (details are given in Section 4.2).

(3) The Combination Scheduling Manager uses the classification results from the Classification Manager to make combination scheduling of the requests; then, it performs resource allocation and pushes the real information of the requests to the History Loads for updating its features (details are given in Section 4.4) during the operation.

(4) The Monitor Manager monitors the utilizations of the CPU and bandwidth of each active VM to support real-time service information.

The Resource Provider Layer has many resource pools, such as CPU, bandwidth, and memory. This layer provides basic resources for VMs so as to handle these service requests.

4. Model Design and Algorithm Analysis

4.1. System Model. Our goal is to reduce the service delay and enhance the resource utilization by the proposed system architecture, which can choose a suitable VM in the physical machine for the service requests. In other words, we will make a fitting combination of request, VM, and physical host, described as $\langle \text{task}_i \rangle \longrightarrow \langle \text{VM}_j \rangle \longrightarrow \langle \text{host}_k \rangle$, $i \in \{1, 2, \ldots, M\}$, $j \in \{1, 2, \ldots, N\}$, and $k \in \{1, 2, \ldots, Q\}$.

Definition 1. For a set of hosts $\langle \text{host}_k \rangle = \{\text{host}_1, \text{host}_2, \ldots, \text{host}_k, \ldots, \text{host}_Q\}$ and the VM set $\langle \text{VM}_j = \{\text{VM}_1, \text{VM}_2, \ldots, \text{VM}_j, \ldots, \text{VM}_N\}$, the connection of them can be defined as a matrix $U_{Q \times N}$, i.e.,

$$U_{Q \times N} = \left[u_{kj} \right]^{Q \times N}, \tag{1}$$

where if VM_j is not created or released on host_k at the beginning, then we set $u_{kj} = -1$. If VM_j locates on host_k, then $u_{kj} \in (0, 1)$. At the same time, we use the $u_{kj}^{\text{cpu}} \in (0, 1)$ to denote the utilization of CPU on VM_j and use $u_{kj}^{\text{bw}} \in (0, 1)$ to denote the utilization of bandwidth on VM_j.

Definition 2. For a set of tasks $\langle \text{task}_i \rangle = \{\text{task}_1, \text{task}_2, \ldots, \text{task}_i, \ldots, \text{task}_M\}$ and a virtual machine set $\langle \text{VM}_j \rangle = \{\text{VM}_1, \text{VM}_2, \ldots, \text{VM}_j, \ldots, \text{VM}_N\}$, the distribution of the tasks is defined as a matrix $A_{M \times N}$, i.e.,

$$A_{M \times N} = \left[a_{ij} \right]^{M \times N}, \tag{2}$$

where $a_{ij} \in \{0, 1\}$; if $a_{ij} = 1$, then that task_i is distributed on VM_j. If $a_{ij} = 0$, there is no connection between task_i and VM_j.

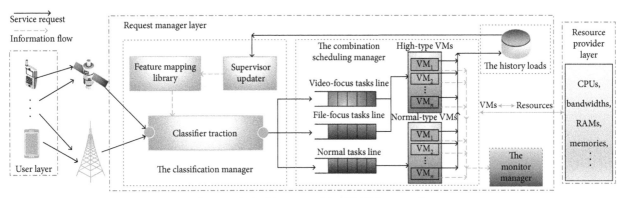

FIGURE 2: Architecture of the SRAF framework.

4.2. Classification Model. Considering the complexity of mobile terminals, we will make the classification from the perspective of service in the mobile cloud network. Therefore, all of the requests can be divided into three kinds of tasks, i.e., the video-focus tasks, file-focus tasks, and normal tasks.

The classification model has three components: (1) *feature mapping library* is used to store the relationship of the features and their corresponding service classes so as to make classification; (2) *classifier traction* is used to classify the tasks according to their input data and the mapping in the feature mapping library; and (3) *supervisor updater* is used to supervise and update the mapping in the feature mapping library based on the information from the History Loads.

In this section, we use self-learning classification algorithm (SCA) to operate classification. SCA is designed by the machine learning technology [22, 23] because we can constantly extend and update the feature mapping library so as to improve the accuracy of the classification by the machine learning. The SCA also improves the traditional classification algorithm by multiple weighting and uses the semisupervised method to update the features in the feature mapping library according to the feedback from the History Loads [24]. SCA uses the method of learning to expand the new relationship of features and service requests so as to enhance the veracity of classification results. Therefore, SCA uses the combination of location weighting, feature weighting, and self-learning methods to determine the final class of each request.

In the process of SCA, we use G_V and G_F to denote the mapping set of videos and files, respectively. We firstly append some typical features into the feature mapping library for the mapping set $G = \{G_V, G_F\}$, such as $G_V = \{\langle tv, video\rangle, \langle dvd, video\rangle, \dots, \langle avi, video\rangle\}$ and $G_F = \{\langle doc, file\rangle, \langle wps, file\rangle, \dots, \langle ppt, file\rangle\}$. Then, SCA uses the input data of requests to find the mapping in the G. In the process of mapping, we use the methods of location weighting and feature weighting to ensure the request has a precise classification. Feature weighting is that different features have different weighting. For example, the feature "video" has a large weighting than "avi" for indicating video tasks. Location weighting is that we use the location of different features in the URL to weighting. For example, if

a request URL is divided into n segments, then we can use these n segments to form a one-dimensional array $L = \{l_1, l_2, \dots, l_n\}$. We use α_i to denote the location weight of the $i-$th feature:

$$\alpha_i = \frac{(n - n_{\text{loc}})}{n}, \quad (3)$$

where n_{loc} is the location of the $i-$th feature in the order. Moreover, the more forward the location is in the order, the more important the feature is in the description [25]. We use β as the final weight to determine which task line the request should be scheduled. For example, there is a request with some features in G_V, such as fea $= \{k_i \in L \mid i = 1, 2, \dots, m, m < n\}$ and fea$\subseteq G_V$. The weighting of k_i is β_v^i for $i \in \{1, 2, \dots, m\}$. Hence, the total weighting of video features in the request URL is calculated by

$$\beta_v = \sum_{i=1}^{m} \alpha_i \beta_v^i. \quad (4)$$

Similarly, the total weighting of file features in the request URL is calculated by

$$\beta_f = \sum_{i=1}^{m} \alpha_i \beta_f^i. \quad (5)$$

Finally, the attribution of the service request is calculated by

$$\beta = \max\{\beta_v, \beta_f\}. \quad (6)$$

If $\beta_v \geq \beta_f$, this request is transmitted to the video-focus tasks line. Otherwise, this request is transmitted to the file-focus tasks line. If β_v and β_f are equal to zero, this request will be transmitted to the normal tasks line. The process of SCA is shown in Algorithm 1.

4.3. VM Migration. According to Definition 1 and Definition 2, the overall process of SRAF is to find an appropriate location in VM_j and host$_k$ for task$_i$. If we define a location function as $B(t) := [a_{ij}(t), u_{kj}(t)]$, then that task$_i$ is scheduled on VM_j and VM_j located in host$_k$ at time t. In addition, when task$_i$ is classified by classifier traction, it will be scheduled to VM_j and it cannot be transferred. Therefore, when the resource of host$_k$ cannot satisfy the demand of

> Input: request information
> Output: line to which request belongs
> (1) Initialize G_V, G_F
> (2) Divide the URL into L;
> (3) Calculate β_v by circularly comparing the features in L,
> G_V according to (3) and (4)
> (4) Calculate β_v by circularly comparing the features in L,
> G_F following (3) and (5)
> (5) Calculate β by (6);
> (6) Output the attribution of the request.

ALGORITHM 1: The description of SCA.

VM_j, VM_j will be migrated to another host at timeslot t based on practical conditions from the Monitor Manager. Let $D(t) := \{d_{j,kk'}(t) \mid j \in N; k, k' \in Q\}$ represents the set of action, where $d_{j,kk'}(t)$ means VM_j can migrate from $host_k$ to $host_{k'}$ at timeslot t. Correspondingly, each $d_{j,kk'}(t)$ has a migration probability as $p_{j,kk'}(t)$, and all the probabilities make up probability set $\mathcal{P}$, indexed as

$$\mathcal{P}(t) := \{p_{j,kk'}(t) \mid j \in N; k, k' \in Q\}, \quad \sum_{k,k' \in Q} p_{j,kk'}(t) = 1.$$
(7)

The cost function of VM migration is defined as

$$f_j(t) = \sum_{k,k' \in Q} C_{k,k_l} p_{j,kk'}(t),$$
(8)

which means the additional expenditure of VM migration. $C_{k,k'}$ is the migration expenditure of VM_j from $host_k$ to $host_{k'}$. In addition, $C_{k,k'}$ is influenced by the migration distance and the latter operation expenditure. Hence, let $E = \{B(t), D(t), \mathcal{P}(t), f_j(t)\}$ be a basic MDP to represent VM migration because the arrival of service requests is based on the Poisson process [32, 33]. So, if the capacity of VM_j is stationary, the overload and VM migration will be a loop in a long time. Therefore, we can get a stationary policy π to control the overall process of VM migration. Now, we use the Bellman optimality equation and the method of dynamic programming to obtain the optimal control policy π [26, 27]. We introduce the state value function as follows:

$$V_T^\pi(b(t)) = \sum_{d_{j,kk'}(t) \in D(t)} \pi(kk') \cdot \sum_{k,k' \in Q} p_{j,kk'}(t)$$

$$\cdot \left(\frac{1}{T} R(d_{j,kk'}(t)) + \frac{T-1}{T} V_{T-1}^\pi(b(t-1)) \right),$$
(9)

where $R(d_{j,kk'}(t))$ means that the penalty of VM_j operates the action $d_{j,kk'}(t)$; T is the discount factor to determine the importance of history data; $b(t)$ is the location of VM_j at timeslot t in $B(t)$; and $\pi(kk')$ means that VM migrates from $host_k$ to $host_{k'}$ by policy π. We use (9) to select the optimal state at the next timeslot so as to maximize the reward. Then, the action value function is

$$P_T^\pi(b(t), d_{j,kk'}(t)) = \sum_{k,k' \in Q} p_{j,kk'}(t) \left(\frac{1}{T} R(d_{j,kk'}(t)) \right.$$
$$\left. + \frac{T-1}{T} V_{T-1}^\pi(b(t-1)) \right).$$
(10)

We use (10) to determine the action which can satisfy the optimal state at the next timeslot. Finally, we use the value iteration algorithm to handle the control policy π [27–29].

In Algorithm 2, we aim to get and update the control policy π, which can control VM migration based on history data in the History Loads. The control policy π can improve the resource utilization and load balancing in the system. The overall process of Algorithm 2 is to update the state value function by finding the optimal reward path. In other words, we need to traverse all $b(t)$ and choose an optimal location to migrate VM, which can maximize the reward of all the VMs. Then, we find an optimal string of states of $B(t)$ over time. Finally, the algorithm uses the backstepping approach to get control π using the action value function and the optimal string of states in $B(t)$.

4.4. Combination Scheduling Cost Model. We firstly design two types of VMs, the high type and the normal type, which have a different capacity of conducting tasks. The tasks in III-B are divided into three kinds. We use the sharing mode (Figure 1) and Symbiosis in [9] to execute the scheduling according to the tasks' demand for resources.

From Figure 2, we propose a sharing mode for the video-focus tasks and file-focus tasks to jointly share the resources of the high-type VMs. With the sharing mode, one VM can simultaneously execute two tasks, one video-focus task and one file-focus task. Moreover, two tasks share the resources of their owner VM, such as the bandwidth and CPU resources. If the video-focus tasks are less than the file-focus tasks, the VM can have two file-focus tasks. If the video-focus tasks are more than file-focus tasks, the VM will execute one video-focus task and wait for the file-focus tasks. On the contrary, the normal tasks are processed on normal-type VMs by the Symbiosis [9] based on the idea of space sharing in CloudSim [30].

In Figure 3, we use an example to show the process of sharing mode. We design three tasks and give different

Input: $E = \{B(t), D(t), \mathscr{P}(t), f_j(t)\}$, T, θ
Output: π
(1) $\forall b(t) \in B(t), V(b(t)) = 0$.
(2) for $t = 1, 2, 3, \ldots$ do
(3) $\forall b(t) \in B(t), \forall VM_j \in \langle VM_j \rangle : V'(b(t)) = \max_{d_{kk'}(t)} \sum_{b(t) \in B(t)} p_{j,kk'}(t) \cdot (1/TR(d_{j,kk'}(t)) + T - 1/TV_{T-1}^{\pi}(b(t-1)))$
(4) if $\max_{b(t) \in B(t)} |V(b(t)) - V'(b(t))| < \theta$ then
(5) break.
(6) else
(7) $V = V'$.
(8) end if
(9) end for
(10) Output $\pi = \text{argmax}_{b(t) \in B(t)} P_T^{\pi}(b(t), d_{kk'})$.

ALGORITHM 2: Computing π by value iteration.

lengths for each task in transmission by bandwidth and the execution length of CPU. In order to express clearly, we set the execution efficiency of bandwidth as 2 in one interval and that of CPU as 3 in one interval. Due to the sharing mode, one VM can have two tasks, and these tasks share resources based on the percentage of 1:1. So, the process of working is as follows: at time 0, the v-task1 and f-task1 were allocated to the same VM. Firstly, v-task1 and f-task1 begin to transfer their transmission length (T-length) by sharing the bandwidth. At time 2, the T-length of f-task1 is finished. The f-task1 begins to solely execute its execution length (E-length) on CPU. At the same time, v-task1 occupied the bandwidth by itself for transferring its remaining T-length. At time 3, v-task1 finishes its T-length and begins to execute its E-length by sharing with f-task1. At time 5, f-task1 finishes the E-length and leaves the VM. Then, the f-task2 begins to transfer for working. So, the f-task2 uses bandwidth by itself, and v-task1 also uses the CPU by itself. At time 6, the f-task2 finishes its T-length and begins to occupy the CPU with v-task1. At time 12, f-task2 is finished and leaves the VM. V-task1 uses CPU by itself. Finally, v-task1 finishes its work and leaves the VM at time 14. At this point, the overall process is finished. Algorithms 3–5

In the following, we present the overall algorithm process of SRAF. When the system scheduling algorithm (SSA) starts, we will create basic VMs (line 6) for firstly scheduling. The Monitor will constantly monitor $u_{kj}^{bw}(s)$ and $u_{kj}^{cpu}(s)$ of every VM at the beginning of the s–th interval Δt. Then, the algorithm will choose an operation by the control policy π, whether doing VM migration or VM scaling, for minimizing the cost (lines 7–11). When the task$_i$ arrives, SSA will classify task$_i$ into the corresponding task line (lines 1–4 and 13). When the task lines have tasks, we will schedule them according to $u_{kj}^{bw}(s)$ and $u_{kj}^{cpu}(s)$ so that we can make full use of the resources (lines 14–17).

In the process of horizontal-scaling algorithm, we set the CPU maximum utilization threshold and the bandwidth maximum utilization threshold as up_{cpu} and up_{bw}, respectively. Then, we use up_{cpu} and up_{bw} to compare with $u_{kj}^{cpu}(s)$ and $u_{kj}^{bw}(s)$ at the beginning of each interval Δt, respectively. Due to the comparison results, the algorithm chooses to perform different operations of resource-level scaling of every active VM.

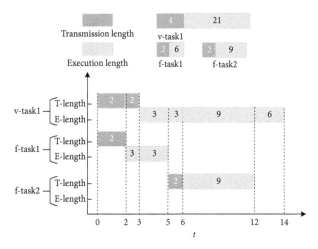

FIGURE 3: An example for the sharing mode.

In Algorithm 5, the arrival ratio and finished ratio represent the quantity of the arriving and finished requests at the beginning of the interval Δt, respectively. Its main duty is to control VMs in the overall framework. According to the situation of resource utilization and the quantity of requests, the algorithm executes the VM-level scaling.

5. Performance Evaluation

To evaluate the proposed framework in this paper, we build the SRAF in CloudSim which is a discrete event simulator [30]. In CloudSim, we can make duplicate and controllable experiments following our idea. CloudSim can support various environments for the resource allocation and scheduling study. We implement all the models and algorithms in CloudSim for comprehensive evaluation and analysis.

In the following, we show the overall calculation process of the execution cost for measuring performance. In the process of scheduling, we can use (11) and (12) to quantify the cost of each VM at the s–th interval Δt:

```
(1) procedure CLASSIFICATION METHOD (task_i)
(2) Uses the SCA in Algorithm 1 to classify task_i;
(3) Puts the task_i into the corresponding line;
(4) End
(5) procedure MAIN
(6) Initializes basic VMs into two types, the high-type VM line and the normal-type line;
(7) while the system is running and in the beginning of an interval do
(8)     Monitors u_{kj}^{bw}(s), u_{kj}^{cpu}(s) of every VM;
(9)     Calls Algorithm 2 to determine the VM migration.
(10)    Calls Horizontal-scaling algorithm; //see Algorithm 4
(11)    Calls Vertical-scaling algorithm; //see Algorithm 5
(12) end while
(13) while (task_i is coming) do
(14)    CLASSIFICATION METHOD (task_i);
(15)    while there have tasks in the three lines do
(16)       Schedules the tasks from the video-focus tasks line and file-focus tasks line on the high-type VM by the sharing mode;
(17)       Schedules the normal task into normal-type VM;
(18)    end while
(19) end while
(18) End
```

ALGORITHM 3: System scheduling algorithm.

```
(1) Begin
(2) if u_{kj}^{cpu}(s) > up_{cpu} and host_k has enough CPU resource then
(3)     Scales up the CPU resource of VM_j;
(4) end if
(5) if u_{kj}^{bw}(s) > up_{bw} and host_k has enough bandwidth resource then
(6)     Scales up the bandwidth resource of VM_j;
(7) end if
(8) if VM_j is idle for a long period then
(9)     Executes the resource scale down;
(10) end if
(11) End
```

ALGORITHM 4: Horizontal-scaling algorithm.

```
(1) Begin
(2) if all the VMs are busy and arrival   ratio > finished   ratio then
(3)     Creates new VM;
(4) else
(5)     Releases the idle VM;
(6) end if
(7) End
```

ALGORITHM 5: Vertical-scaling algorithm.

$$C_{kj}^q(s) = C_{bas}^q + \delta^q \cdot \Delta t \cdot u_{kj}^q(s), \qquad (11)$$

$$g_{kj}^q(s) = C_{kj}^q(s) + C_{sca}^q(s), \qquad (12)$$

where $s \in \{1, 2, \ldots, T_j\}$, in which T_j is the overall execution time of VM_j; $q \in \{cpu, bw\}$, in which cpu represents the CPU resource and bw represents the bandwidth resource; $u_{kj}^q(s)$ contains two aspects, i.e., $u_{kj}^{cpu}(s)$ and $u_{kj}^{bw}(s)$, which represent the utilization of CPU and bandwidth of VM_j at the $s-$th interval Δt, respectively; $C_{kj}^q(s)$ is the ordinary cost of q when the VM_j has tasks and is working; C_{bas}^q is the basic cost of creating the VM on the elements of set q; $C_{sca}^q(s)$ is the scaling cost of q when the VM performs the scaling operation at the $s-$th interval Δt; and δ^q is the execution cost of one interval Δt on the elements of set q. Furthermore, the practical resources cost of all the VMs on $host_k$ in their working period is given by

$$g_{\text{total}} = \sum_{j=1}^{N} \sum_{s=1}^{T_j} g_{kj}^q(s). \tag{13}$$

The total cost of all the resources in VMs is given by

$$g_{\text{all}} = \sum_{j=1}^{N} \sum_{s=1}^{T_j} \left(C_{\text{bas}}^q + \delta^q \cdot \Delta t + C_{\text{sca}}^q(s) \right). \tag{14}$$

Therefore, we get the total resource utilization by using the following equation:

$$U^q = \frac{g_{\text{total}}}{g_{\text{all}}}, \tag{15}$$

where $q \in \{\text{cpu}, \text{bw}\}$. We can get bandwidth utilization and CPU utilization, respectively, by (15). From the analysis above, we get the final optimization cost as follows based on the control policy π and scaling operation:

$$g_{\text{all}} + \sum_{j=1}^{N} \sum_{s=1}^{T_j} f_j^\pi(s) = \sum_{j=1}^{N} \sum_{s=1}^{T_j} \left(C_{\text{bas}}^q + \delta^q \cdot \Delta t + C_{\text{sca}}^q(s) + f_j^\pi(s) \right), \tag{16}$$

where $f_j^\pi(s)$ is the migration cost of VM_j by the control policy π at the s – th timeslot.

5.1. System Configuration. We simulate two physical nodes, and each node has enough resources. VM configuration is shown in Table 1 [9, 31]. Due to the expensive CPU, we use the quantity of CPU to limit the number of VMs. The workload dataset in this paper is from the Laboratory for Web Algorithmics (LAW) (the dataset is named "eu-2015.urls.gz"; see http://law.di.unimi.it/webdata/eu-2015/ for more information). In addition, we use the Poisson process to simulate the arrival process of service requests [32, 33]. To test the performance of frameworks, we try to stabilize the arrival rate. In this paper, we set $\lambda = 8$. In the following, we will set λ from 1 to 10 for testing the robustness of SRAF.

5.2. Performance Analysis. In this section, we compare our framework (SRAF) with the framework (Symbiosis) which is proposed in [9]. We also add the Bejo algorithm [6] into the Symbiosis. In addition, we add the deadline factor into the experiments for clearly showing the difference between SRAF and Symbiosis.

In Figure 4, we make the comparison of SRAF and Symbiosis at different deadlines. All the tasks will be serviced in each framework, and they have three statuses. **Success** means that the task is finished smoothly in its owner VM. **Failed** means that the task is discarded when its service time exceeds the deadline. **Scale** means that the task needs extra resource from the resources pool for its working. In Figure 5, the results represent the utilization of bandwidth and CPU in Symbiosis and SRAF at different deadlines. In Figure 4, with the growing deadline, many failed and scale tasks become success tasks and wait for

processing. The bandwidth and CPU will have more idle time. As a result, the resource utilizations are decreased as shown in Figure 5.

In Figure 4, when deadline is 20 ns and 30 ns, there are still some failed tasks in SRAF, while Symbiosis has none. When the deadline is 40 ns and 50 ns, all tasks in SRAF are success tasks, but there are also some scale tasks in Symbiosis. Therefore, SRAF has a higher absorption rate of requests and higher efficiency than Symbiosis. What caused the status above has two sides. For one thing, the more training the SCA has, the better classification results it has. The better training of SCA can make a better combination of file-focus tasks and video-focus tasks for reducing the waiting time by the sharing mode. Immediately following the operation of SCA, SRAF will make a full use of resources. As a result, SRAF has more time for working and many failed and scale tasks will become success tasks. For another thing, the sharing mode may prolong the execution time of long tasks (video tasks). But this problem can be solved by the method of resetting resource sharing proportion. For example, in Figure 5, the resource utilizations of the Symbiosis are decreasing with the growing deadline. The resource utilizations of SRAF are approximated to 97%. This phenomenon means that there are some long tasks held on CPU resources with the growing deadline. As a result, the bandwidth resources are unoccupied. Finally, the utilization of bandwidth is decreasing and that of CPU is increasing. However, SRAF has a different performance. Because of SCA, the long tasks (video tasks) are executed with short tasks (file tasks) by the sharing mode. In other words, one VM can have two tasks. When the short task has been finished and the long task is still working, a new short task will come for transmitting. As a result, the CPU and bandwidth are occupied by one task. Hence, the resource utilization of CPU and bandwidth decreases, but in a small range. However, the resource utilizations are still higher than those of Symbiosis. The detailed process of the above example is shown in Figure 2. Taking a holistic look of Figure 4, the SRAF also has a shorter dropping rate and scaling rate of all the tasks. Therefore, facing the same tasks, SRAF has a higher resource utilization and task processing rate than Symbiosis.

In order to measure the performance of our control policy π in Section 4.3, we add the operation of VM migration into Symbiosis for comparison, which is named "Symbiosis + vm-mi" (SVM). Additional execution time represents the total execution time of VM migration and scaling operation. Additional cost means the total cost of all the VM migration and scaling operations. For enhancing the veracity, we make ten experiments of SRAF, Symbiosis, and SVM, respectively, for comparison at each deadline. Let us firstly make a comparison with Symbiosis and SVM in Figures 6 and 7 because SVM makes a control policy to measure the VM migration. Therefore, when VM becomes overloaded (resource utilization exceeds thresholds), the overloaded VM will firstly choose to make migration by control policy. If not, SVM will do scaling operation (shown in Algorithms 4 and 5). On the contrary, Symbiosis can only do scaling operation for these overloaded VMs. In theory,

TABLE 1: VM configuration.

VM type	CPU			Bandwidth		
	MIPS (MB/s)	Per cost (dollar/h)	Per scale cost (dollar/h)	Bandwidth (MB/s)	Per cost (dollar/h)	Per scale cost (dollar/h)
High	1000	0.05	0.06	100	0.005	0.006
Normal	600	0.03	0.04	60	0.003	0.004

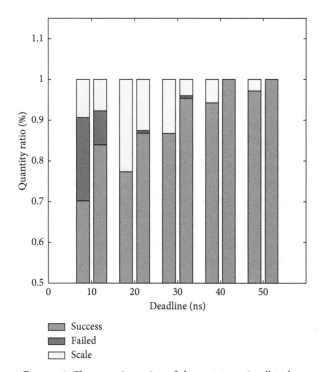

FIGURE 4: The quantity ratios of three statuses in all tasks.

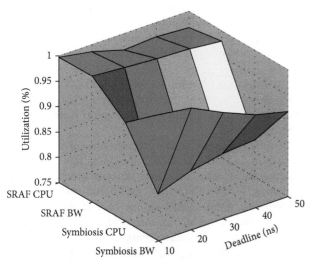

FIGURE 5: The utilization of bandwidth and CPU in SRAF and Symbiosis frameworks at each deadline.

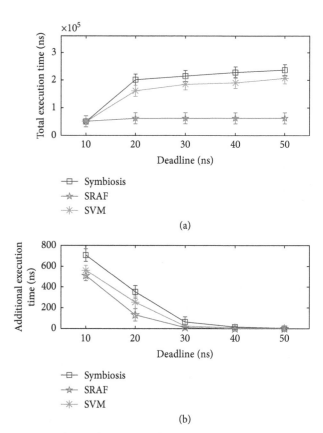

FIGURE 6: The total execution time (a) and additional execution time (b) of SRAF, Symbiosis, and Symbiosis + vm-mi frameworks at each deadline.

VM migration can reduce the cost than scaling operation because scaling operation is easily creating frequent fluctuation. Hence, taking an overall look of Figures 6 and 7, SVM has a shorter execution time and cost than Symbiosis because the control policy can make a long-time prediction to operate VM migration so as to avoid the VM overload and resource lack. In other words, SVM uses the method of VM migration to cut down frequent scaling operation for reducing the additional cost.

From the overall perspective of Figure 6, the total execution time of Symbiosis is much more than that of SRAF. The total execution time of SRAF is almost at the level of 0.6 $\times 10^5$ ns when deadline is 50 ns. Correspondingly, the additional execution time is approximated to zero. But the total execution time of Symbiosis is almost four times higher than that of SRAF when the deadline is 50 ns. The total cost of Symbiosis in Figure 7 is also approximately five times higher than that of SRAF. The analysis above means that the service latency of SRAF is much shorter than that of Symbiosis. For example, in Figure 5, the decrease of resource utilizations on Symbiosis is sharper than that on SRAF with the growing deadline. Therefore, when more failed and scale tasks become success tasks, the total execution time of Symbiosis will continually increase. The total execution time of SRAF is changing in a small range. That is to say, facing more tasks,

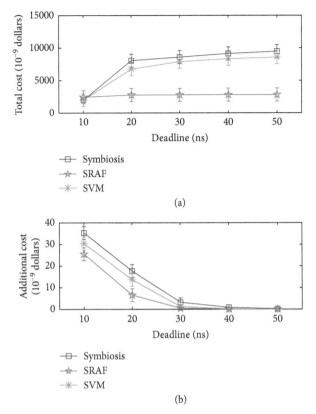

(a)

(b)

FIGURE 7: The total cost (a) and additional cost (b) of SRAF, Symbiosis, and Symbiosis + vm-mi frameworks at each deadline.

SRAF has a stable and better performance on reducing service latency and cost than Symbiosis.

For testing the robustness of SRAF, we set λ from 1 to 10 to simulate different situations of congestion. During experiments, we will provide enough resources. In order to avoid the additional cost made by frequent operation of changing the resources of VMs, we make the operation of resource-scaling can only maximully add twice resources of the original configuration resources on VM. We make ten experiments for each framework at different λ. Then, according to these experiments, we get the mean bandwidth utilization, mean CPU utilization, and total cost of each framework, which are shown in Figures 8–10, respectively. In Figure 8, the bandwidth utilization of SRAF, Symbiosis, and SVM is increasing with the growing value of λ. What caused the increasing phenomenon has two sides. Firstly, because the arrival rate of requests is based on the Poisson distribution, the situation of congestions becomes smooth with the growing value of λ. So, facing the more stable arrival rate, frameworks will have more time for working. As a result, all the bandwidth utilizations have an increasing trend with the growing value of λ. Secondly, if the deadline of requests exceeds the transmission time, they will become the failed tasks. In addition, we do not calculate the time and cost of failed tasks. In other words, failed tasks are the waste of bandwidth, and it is the main factor affecting the bandwidth utilization. Therefore, with the growing value of λ, the more stable arrival rate will make less failed tasks. As a result, the bandwidth utilization of frameworks is

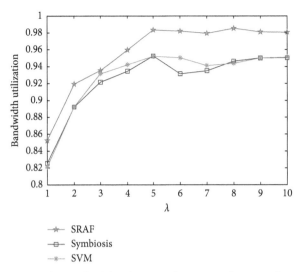

FIGURE 8: The bandwidth utilization of SRAF, Symbiosis, and SVM at different λ.

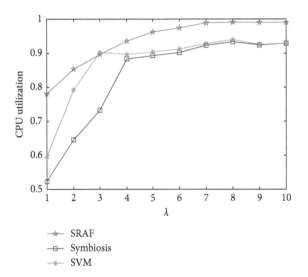

FIGURE 9: The corresponding CPU utilization of SRAF, Symbiosis, and SVM.

increasing. Taking a detailed look of Figure 8, it is observed that the bandwidth utilization of SRAF is higher than that of Symbiosis and SVM. Symbiosis and SVM have an approximate trend. It is because that SRAF can service two tasks simultaneously with the sharing mode (shown in Figure 2). The sharing mode can make full use of bandwidth and CPU resources than Symbiosis and SVM. Therefore, SRAF can effectively reduce the latency and enhance the utilization of free resources. The same trend of SVM and Symbiosis is that they have the same method of bandwidth transmission and do not have the sharing mode. According to these analogies above, for different arrival rates of requests, SRAF has a better performance on bandwidth utilization than Symbiosis and SVM, especially with the stable arrival rate.

Figure 9 presents the CPU utilizations of SRAF, Symbiosis, and SVM at different λ. Taking an overall look of Figures 8 and 9, CPU utilization of different frameworks is

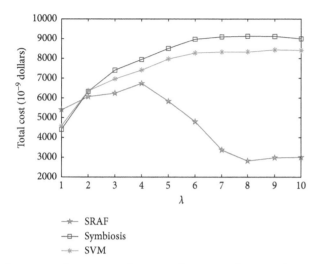

FIGURE 10: The corresponding cost of SRAF, Symbiosis, and SVM at different λ.

also increasing with the growing value of λ as the situation of bandwidth utilization. In addition, the increases of adjacent λ on CPU utilization are slightly higher than those on bandwidth utilization. It is because that the sharp congestion will cause more free time on the CPU resource than the bandwidth resource. The queueing time will exceed the deadline and cause many failed tasks. When the value of λ is growing from 1 to 4, the CPU utilizations of SVM are higher than those of Symbiosis. This phenomenon means that VM migration in SVM cuts down the quantity of scaling tasks and failed tasks. It is because that VM migration can avoid the overload of VMs and reduce the free time of failed tasks. In addition, with the growing value of λ, the arrival rate of requests is becoming stable. The stable arrival rate will reduce the frequent fluctuation of scaling operation. The operation of VM migration can also be reduced. When the value of λ is larger than 4, the CPU utilizations of SVM and Symbiosis are approximately the same. Certainly, CPU utilization of SRAF is stable and higher than that of SVM and Symbiosis because of the sharing mode and VM migration.

Figure 10 represents the total cost of SRAF, Symbiosis, and SVM at different λ. Different to the situation in Figures 8 and 9, the total cost of SVM and Symbiosis is increasing with growing λ. The total cost of SRAF is decreasing. What caused this phenomenon has two sides. Firstly, because of the higher utilizations of bandwidth and CPU in SRAF, SRAF reduces more waiting time and service time for all tasks with the sharing mode, especially for those short tasks behind the long tasks in the queue. In addition, the one-by-one service method of Symbiosis is the main factor which affects the utilizations of CPU and bandwidth. Therefore, facing the same tasks, Symbiosis will waste many resources and prolong the service time than SRAF. Secondly, VM migration can reduce frequent scaling operation. For example, the total cost of SVM is less than that of Symbiosis at each λ in Figure 10. It is because that the control policy can make a trade-off between VM migration cost and scaling operation cost. Control policy has the ability of prediction according to

the history data, which can select the minimal cost of each action for reducing the additional cost. Therefore, taking a holistic look of Figures 8–10, SRAF has a better performance on stability and robustness.

6. Conclusions

In this paper, we have designed, modeled, and evaluated the SRAF, which aims to reduce the latency of service requests in mobile data networks and enhance the utilization of bandwidth and CPU resources. In SRAF, we have proposed the SCA to execute the tasks' classification. We also designed a sharing mode to realize the combination process of two tasks. Sharing mode greatly reduces the waiting time during service. In addition, we also designed an MDP to control VM migration. We use the combination method of VM migration and scaling to enhance resource utilization. Finally, we make many different experiments to show that SRAF has a good performance on resource utilization, stability, and robustness.

Acknowledgments

This work was supported in part by the National Natural Science Foundation of China (NSFC) under Grant nos. U1604155, 61602155, 61871430, and 61370221, in part by Henan Science and Technology Innovation Project under Grant no. 174100510010, in part by the Industry University Research Project of Henan Province under Grant no. 172107000005, and in part by the basic research projects in the University of Henan Province under Grant no. 19zx010.

References

[1] B. Yang, K. Chai, Z. Xu, K. Katsaros, and G. Pavlou, "Cost-efficient NFV-enabled mobile edge-cloud for low latency mobile applications," *IEEE Transactions on Network and Service Management*, vol. 15, no. 1, pp. 475–488, 2018.

[2] A. Mukherjee, D. De, and D. Roy, "A power and latency aware cloudlet selection strategy for multi-cloudlet environment," *IEEE Transactions Cloud Computation*, vol. 99, no. 99, p. 1, 2016.

[3] J. Li, R. Ma, H. Guan, S. David, and L. Wei, "Accurate CPU proportional share and predictable I/O responsiveness for virtual machine monitor: a case study in Xen," *IEEE Transactions Cloud Computation*, vol. 5, no. 4, pp. 604–616, 2017.

[4] B. P. R Killi and S. V. Rao, "Capacitated next controller placement in software defined networks," *IEEE Transactions on Network & Service Management*, vol. 14, no. 3, pp. 514–527, 2017.

[5] C. Birkestrand, J. Heyrman, and C. Prosser, "Virtual machine placement in a cloud computing environment based on factors including optimized processor-memory affinity," US Patent 9600321, 2017.

[6] L. Xu, J. Cao, Y. Wang, L. Yang, and J. Li, "Bejo: behavior based job classification for resource consumption prediction in the cloud," in *Proceedings of IEEE International Conference*

on *Cloud Computing Technology and Science*, pp. 10–17, Singapore, December 2014.

[7] H. J. Kim, H. I. Kim, and J. W. Chang, "A privacy-preserving kNN classification algorithm using Yao's Garbled circuit on cloud computing," in *Proceedings of IEEE International Conference on Cloud Computing*, pp. 766–769, Honolulu, Hawaii, USA, June 2017.

[8] K. Katsalis, G. Papaioannou, N. Nikaein, and T. Leandros, "SLA-driven VM scheduling in mobile edge computing," in *Proceedings of IEEE International Conference on Cloud Computing*, pp. 750–757, San Francisco, CA, USA, June 2016.

[9] J. Jiang, S. Ma, and B. Li, "Symbiosis: network-aware task scheduling in data-parallel frameworks," in *Proceedings of the IEEE International Conference on Computer Communications*, pp. 1–9, San Francisco, CA, USA, April 2016.

[10] N. Armanfard, J. P. Reilly, and M. Komeili, "Local feature selection for data classification," *IEEE Transactions on Pattern Analysis and Machine Intelligence*, vol. 38, no. 6, pp. 1217–1227, 2016.

[11] H. Hajimirsadeghi and G. Mori, "Multi-instance classification by max-margin training of cardinality-based Markov networks," *IEEE Transactions on Pattern Analysis and Machine Intelligence*, vol. 39, no. 9, pp. 1836–1852, 2017.

[12] S. Kumar, G. Kumar, and K. Jain, "an approach to reduce turn around timeand waiting timeby the selection of round robin and shortest job first algorithm," *International Journal of Engineering and Technology*, vol. 7, no. 2, pp. 667–677, 2018.

[13] M. Kumar and S. C. Sharma, "Priority Aware Longest Job First (PA-LJF) algorithm for utilization of the resource in cloud environment," in *Proceedings of International Conference on Computing for Sustainable Global Development*, pp. 16–18, New Delhi, India, March 2016.

[14] N. T. Argon, C. Deng, and V. G. Kulkarni, "Optimal control of a single server in a finite-population queueing network," *Queueing Systems*, vol. 85, no. 1-2, pp. 149–172, 2017.

[15] B. Legros, "M/G/1 queue with event-dependent arrival rates," *Queueing Systems*, vol. 89, no. 3-4, pp. 269–301, 2018.

[16] Z. Han, H. Tan, and G. Chen, "Dynamic virtual machine management via approximate Markov decision process," in *Proceedings of IEEE International Conference on Computer Communications*, pp. 1–9, San Francisco, CA, USA, July 2016.

[17] Z. Li, "An adaptive overload threshold selection process using Markov decision processes of virtual machine in cloud data center," *Cluster Computing*, vol. 2018, pp. 1–13, 2018.

[18] S. Chen, Y. Wang, and M. Pedram, "A semi-Markovian decision process based control method for offloading tasks from mobile devices to the cloud," in *Proceedings of IEEE Global Communications Conference*, pp. 2885–2890, Atlanta, GA, USA, December 2013.

[19] T. Liu and D. Tao, "Classification with noisy labels by importance reweighting," *IEEE Transactions on Pattern Analysis & Machine Intelligence*, vol. 38, no. 3, pp. 447–461, 2016.

[20] A. L. Jin, W. Song, and P. Wang, "Auction mechanisms toward efficient resource sharing for cloudlets in mobile cloud computing," *IEEE Transactions on Services Computing*, vol. 9, no. 6, pp. 895–909, 2016.

[21] T. X. Tran and D. Pompili, "Dynamic radio cooperation for downlink cloud-RANs with computing resource sharing," in *Proceedings of IEEE International Conference on Mobile Ad Hoc and Sensor Systems*, pp. 118–126, Brasília, Brazil, October 2016.

[22] D. Kovachev, Y. Cao, R. Klamma et al., "Learn-as-you-go: new ways of cloud-based micro-learning for the mobile web," in *Proceedings of International Conference on Advances in Web-Based Learning*, pp. 51–61, Hong Kong, China, December 2011.

[23] C. Sing, N. Metz, and S. Dudli, "Machine learning-based classification of 38 Years of spine-related literature into 100 research topics," *Spine*, vol. 42, no. 11, pp. 863–870, 2017.

[24] J. Zhang, Y. Han, and J. Jiang, "Semi-supervised tensor learning for image classification," *Multimedia Systems*, vol. 23, no. 1, pp. 63–73, 2017.

[25] J. Chen, C. Chen, and Y. Liang, "Optimized TF-IDF algorithm with the adaptive weight of position of word," in *Proceedings of International Conference on Artificial Intelligence and Industrial Engineering*, Beijing, China, November 2016.

[26] S. I. Gass and M. C. Fu, *Bellman Optimality Equation*, Springer US, New York, NY, USA, 2013.

[27] D. P. Bertsekas, *Dynamic Programming and Optimal Control*, Athena Scientific, Vol. 2, Nashua, NH, USA, 1995.

[28] A. Heydari, "Stability analysis of optimal adaptive control using value iteration with approximation errors," *IEEE Transactions on Automatic Control*, vol. 63, no. 9, pp. 3119–3126, 2018.

[29] B. Taylor, M. Hendrickx, and F. Glineur, "Smooth strongly convex interpolation and exact worst-case performance of first-order methods," *Mathematical Programming*, vol. 161, no. 1-2, pp. 307–345, 2017.

[30] R. N. Calheiros, R. Ranjan, A. Beloglazov, C. A. De Rose, and R. Buyya, "CloudSim: a toolkit for modeling and simulation of Cloud computing environments and evaluation of resource provisioning algorithms," *Software: Practice and Experience*, vol. 41, no. 1, pp. 23–50, 2011.

[31] Y. Zhao, N. Calheiros, G. Gange, K. Ramamohanarao, and R. Buyya, "SLA-based resource scheduling for big data analytics as a service in cloud computing environments," in *Proceedings of IEEE International Conference on Parallel Processing*, pp. 510–519, Beijing, China, September 2015.

[32] Q. Xie and Y. Lu, "Priority algorithm for near-data scheduling: throughput and heavy-traffic optimality," in *Proceedings of IEEE International Conference Computer Communications*, pp. 963–972, Las Vegas, Nevada, USA, August 2015.

[33] S. Guo, B. Xiao, and Y. Yang, "Energy-efficient dynamic offloading and resource scheduling in mobile cloud computing," in *Proceedings of IEEE International Conference on Computer Communications*, pp. 1–9, San Francisco, CA, USA, July 2016.

14

An Energy-Aware Task Offloading Mechanism in Multiuser Mobile-Edge Cloud Computing

Lan Li, Xiaoyong Zhang⑩, Kaiyang Liu⑩, Fu Jiang⑩, and Jun Peng⑩

School of Information Science and Engineering, China Hunan Engineering Laboratory of Rail Vehicles Braking Technology, Central South University, Changsha, Hunan 410083, China

Correspondence should be addressed to Xiaoyong Zhang; zhangxy@csu.edu.cn

Academic Editor: Bartolomeo Montrucchio

Mobile-edge cloud computing, an emerging and prospective computing paradigm, can facilitate the complex application execution on resource-constrained mobile devices by offloading computation-intensive tasks to the mobile-edge cloud server, which is usually deployed in close proximity to the wireless access point. However, in the multichannel wireless interference environment, the competition of mobile users for communication resources is not conducive to the energy efficiency of task offloading. Therefore, how to make the offloading decision for each mobile user and select its suitable channel become critical issues. In this paper, the problem of the offloading decision is formulated as a 0-1 nonlinear integer programming problem under the constraints of channel interference threshold and the time deadline. Through the classification and priority determination for the mobile devices, a reverse auction-based offloading method is proposed to solve this optimization problem for energy efficiency improvement. The proposed algorithm not only achieves the task offloading decision but also gives the facility of resource allocation. In the energy efficiency performance aspects, simulation results show the superiority of the proposed scheme.

1. Introduction

Internet of things (IoT) devices, such as sensors and wearable devices, are increasingly penetrating into our everyday lives. Gartner forecasted that, by 2020, the Internet of things (IoT) devices will reach 50 billion, representing an almost 30-fold increase from 0.9 billion in 2009 [1]. However, it is well known that mobile devices have their inherent problems, such as finite computing power and particularly limited battery life [2]. These resource-limited mobile devices are difficult to support computation-intensive applications, such as interactive gaming, image/video processing, and online social networking services [3]. Therefore, when dealing with the sophisticated applications on devices, the contradiction between the requirements of network applications and the limited resource of mobile devices poses a significant challenge.

A new architecture and technology known as Mobile Cloud Computing (MCC) brings a new idea, which can augment the processing ability of mobile devices and reduce the energy consumption of mobile devices at the meantime by migrating computational tasks from mobile devices to infrastructure-based cloud servers [4, 5]. Thus, MCC has the potential to address the aforementioned challenge [6].

In recent years, cloud offloading technologies have been widely studied by researchers all over the world. Considering the huge increase of computational demand on the mobile devices in the 5G networking, Chen [7] proposed a game theoretic approach for the distributed task offloading decision to improve the energy efficiency. Barbarossa et al. [8] investigated the problem of multiuser computation offloading in multiradio channel scenarios ignoring the communication interference. In order to save the energy consumption of mobile devices and meet the requirements of the application execution time, Huang et al. [9] proposed a Lyapunov optimization-based algorithm for dynamic offloading, which can reduce the computational complexity at the same time to obtain a near-optimal offloading decision. In large-scale mobile applications, Yang et al. [10] researched the problem of multiuser computing partition for the purpose of minimizing the average task completion of all users and designed an offline heuristic offloading algorithm. Viswanathan et al. [11] proposed a resource provisioning framework for organizing the heterogeneous devices in the vicinity. A joint optimization framework of the wireless resource and computing resource is proposed in [12] for the energy-constrained mobile users in the femtocell.

In order to solve the problem of transmission delay and energy consumption, Satyanarayanan et al. [13] proposed an architecture replacing the remote cloud with nearby cloudlets. Zhang et al. [14] proposed a Markov decision process offloading algorithm for mobile users in an intermittently connected cloudlet system. However, the cloudlet-based mobile cloud computing has some drawbacks. Due to the limited wireless network coverage, the cloudlet cannot guarantee ubiquitous service everywhere for users. Moreover, the computation resource of the cloudlet is insufficient to satisfy the QoS requirements of a large number of users in the future.

These existing task offloading strategies in mobile cloud computing are not sufficient to greatly improve the energy efficiency of the system. The long transmission delay between the mobile device and traditional cloud servers is a critical issue, which is the inherent limitation of mobile cloud computing [15, 16]. The long propagation distance from the mobile device to the remote cloud data center will result in unacceptable long latency for mobile applications. The additional communication transmission delay will decrease the computation offloading efficiency and the QoS of users.

Therefore, a mobile-edge cloud (MEC) computing architecture is adopted. With the development of wireless communication technologies such as Wi-Fi, 4G, and 5G, the MEC is envisioned as a promising and challenging approach to address the abovementioned challenges [17]. In the MEC framework, mobile devices are able to offload their tasks to the MEC clouds through the radio access points nearby rather than the public clouds such as Amazon EC2 and Windows Azure. Thus, this MEC paradigm can provide lower latency and high communication rate and computing agility in the process of computation offloading.

There are a few studies on the efficient computation offloading mechanism of the MEC. For instance, Wang et al. [18] used the Markov decision process to formulate a sequential offloading decision-making problem for dynamic service migration. Considering the finite number of wireless access channels and the interference, Chen et al. [19] presented a distributed offloading decision method based on game theory. Sardellitti et al. [20] investigated the task offloading problem by jointly considering the allocation of radio resources and computational resources and proposed an iterative algorithm to solve the problem. Beck et al. [21] studied the virtual network embedding problems and proposed network virtualization in the context of MEC networks.

However, the abovementioned methods still have some limitations in performance and flexibility. Furthermore, the complexity of the algorithm is not suitable for large-scale network scenarios. Thus, an efficient task offloading mechanism is designed for mobile-edge cloud computing. It is well known that the base stations in most wireless networks are running under multichannel settings. If a considerably large number of mobile device users simultaneously choose the same wireless channel to perform task offloading, interactive communication interference will seriously affect the transmission rate of the data, which further leads to the increase of completion delay of the task and the energy consumption of mobile devices. In this circumstance, the task offloading operation violates the original intention of the mobile users, and these mobile users prefer local processing. The abovementioned effect makes the offloading decision of mobile device users couple with the wireless resource allocation. Thus, to achieve efficient task offloading, two important issues need to be solved: Which tasks are suitable for offloading while others are suitable for local execution? How to choose the appropriate channel performing offloading from a global optimization?

In order to deal with the abovementioned coupling problems and match the mobility of the mobile device, an energy-aware task offloading mechanism is designed to solve the energy-efficient task offloading decision problem for mobile-edge cloud computing in the multichannel wireless environment. The mechanism adopts the auction theory, which is a decentralized market mechanism of resource allocation in economy, which has been widely used in various fields nowadays, such as the management of the spectrum resource in cognitive networks [22, 23] and traffic offloading in the cellular network [24, 25]. Inspired by the methods for resource allocation, the auction is applied into the model of wireless channel access opportunities for mobile devices, namely, task offloading decision. In the auction system, the mobile device user acts as a buyer, and the wireless channels are treated as sellers. Thus, the mobile device broadcasts the offloading request, and wireless channels will send their bid information with their available resources to the mobile device user. Then, the mobile device calculates the cost to determine the winner channel to offload.

The contributions of this paper are listed as follows:

(i) The issue of the task offloading decision is formulated as a constrained 0-1 nonlinear integer programming problem, which minimizes the total energy consumption of the system subjected to the latency and channel communication quality constrains.

(ii) To reduce the complexity of solving the optimization problem, an algorithm at first for classifying mobile devices and priority determination is proposed. The mobile devices are classified into two groups. First category is suitable for local processing, and the remaining mobile devices are classified as the second category, which is determined by the corresponding priority.

(iii) In order to further determine the offloading decision for the mobile device, an auction-based approach is proposed to solve the energy-efficient task offloading decision problem, with high efficiency to approximate the optimal task offloading decisions. Hence, each mobile device can minimize its average energy consumption.

The rest of this paper is organized as follows. In Section 2, the system model is described and the issue of task offloading decision is formulated as a 0-1 nonlinear integer programming problem. The details for solving the constrained optimization problem and the proposed energy-aware task offloading mechanism are provided in Section 3. In Section 4, the simulation results and the analysis of the results are presented. Finally, concluding remarks are drawn in Section 5.

2. System Model and Problem Formulation

In this section, the mobile-edge cloud computing architecture is presented, and the 0-1 nonlinear integer programming problem is formulated for energy saving and delay decreasing. Table 1 lists all the important symbols used in this paper.

2.1. Mobile-Edge Cloud Computing Architecture. As illustrated in Figure 1, the mobile-edge cloud computing system consists of three key components: mobile-edge cloud server, wireless access point, and device users. Mobile device users first connect to Internet through a base station, which is equipped with a mobile-edge cloud server. Then, the users can further access this powerful mobile-edge cloud server through the Internet. After receiving the service request, the cloud control provides the corresponding cloud services to the mobile devices via the base station. In this paper, a set of $\mathcal{N} = \{1, 2, \ldots, N\}$ devices are considered, where each device has a task to be completed. Moreover, the wireless base station runs $\mathcal{M} = \{1, 2, \ldots, M\}$ channel settings.

The types of tasks include interactive gaming, natural language processing, virus scanning, and video transcoding. Each task is set in two terms as $\mathcal{T}_i =^\Delta (O_i, D_i), i \in \mathcal{N} = \{1, 2, \ldots, N\}$, which can be computed either locally on the mobile device or on the cloud side via computation offloading. It is assumed that, in the model, each task is atomic and cannot be further divided. The characteristic information of all tasks can be defined through the total number of $\{O_i\}$ CPU cycles required to accomplish and the amount of $\{D_i\}$ data to be exchanged. The computation overhead is discussed in terms of both energy consumption and processing time for both local and cloud computing approaches.

For the local computing approach, a mobile user i executes the task $\mathcal{T}_i$ locally on the mobile device. Taking into account the different processing capabilities of different mobile devices, as well as a different CPU computing capacity per bit required for different tasks, let C_i^l be the computation capability (i.e., CPU cycles per second) allocated by the mobile device i. Then, the time duration of the local execution of the task $\mathcal{T}_i$ can be obtained as follows:

$$t_i^l = \frac{O_i}{C_i^l}. \tag{1}$$

The classic CPU energy consumption model of mobile devices is applied [26, 27]. Thus, the energy consumption E_i^l of this local execution can be calculated as

TABLE 1: Important notations.

Symbol	Definition
$\mathcal{N}$	Mobile IoT device set
$\mathcal{M}$	Wireless channel set
$\mathcal{T}_i$	The information of task on the device i
O_i	The computing workload of the task i
D_i	The data size for communication of the task i
C_i^l	The computing capacity required for the task i
C_i^c	The computation capacity allocated from the cloud
P_i^{tr}	The data transmission power of the device i
P_i^{id}	The idle power of the device i
G_i	The channel gain
$\psi_{i,j}$	The offloading decision
w_j	The bandwidth of the channel j
$R_{i,j}$	The transmission rate of the mobile device i
t_i^l	Local completion time of the task i
t_i^c	The cloud processing delay of the task i
E_i^l	Local processing energy consumption
$E_{i,j}^{\text{tr}}$	The data transmission power consumption of the task i
E_i^{id}	The idle power consumption of the mobile device i
$E_{i,j}^{\text{Off}}$	The total offloading energy consumption of the mobile device

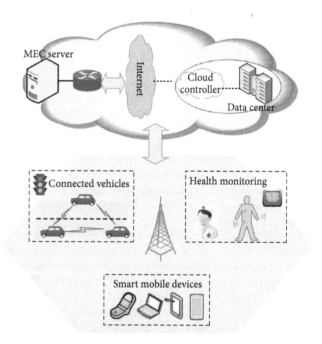

FIGURE 1: The architecture of mobile-edge cloud computing.

$$E_i^l = \left(\alpha_i \left(C_i^l \right)^{\chi_i} + \beta_i \right) t_i^l, \tag{2}$$

where the exponents χ_i, α_i, and β_i are the parameters, which are dependent on the CPU processing model. All the energy consumption parameters are set to conform to the actual mobile scenario in the subsequent simulation experiment.

For the cloud computing approach, once the communication link is established, the task $\mathcal{T}_i$ on the mobile device can be offloaded through the channel j, $j \in \mathcal{M}$, to the mobile-edge cloud server, which will execute the computation task on behalf of the device user. The whole task offloading process can be divided into three phrases and involves corresponding time delay and energy consumption.

To begin with, the mobile device transmits the data load of the task i to the closest base station through the channel j, which involves the data transmission delay $t_{i,j}^{\mathrm{tr}}$ and energy consumption of the data transmission $E_{i,j}^{\mathrm{tr}}$ of the task. The energy consumption of the data transmission is not only related to the data size but also to the uplink data rate. Thus, the binary variable $\psi_{i,j} \in \{0, 1\}$ is denoted as the task offloading decision of the mobile user i, which is given by

$$\psi_{i,j} = \begin{cases} 1, & \text{offload the task of device } i \text{ via channel } j, \\ 0, & \text{process task of device } i \text{ locally,} \end{cases}$$
(3)

where $\psi_{i,j} = 1$ means that the device i chooses to offload the task and transmits the computation data to the cloud through the channel j, $j \in \mathcal{M}$, while $\psi_{i,j} = 0$ denotes that the device executes the task locally. According to the offloading decisions, an appropriate number of VMs are deployed in the data servers for cloud execution. Moreover, given the offloading decisions $\Psi = (\psi_{1,j}, \psi_{2,j}, \ldots, \psi_{N,j})$ of all device users, the uplink data rate $R_{i,j}$ of a device user i who chooses to offload the task to the cloud via the wireless channel j can be calculated as [19]

$$R_{i,j}(\Psi) = w_j \log_2 \left(1 + \frac{P_i^{\mathrm{tr}} G_i}{\sigma^2 + \sum_{r \in \mathcal{N}\{i\}:\psi_{i,j}=1} P_r^{\mathrm{tr}} G_r} \right),$$
(4)

where w_j is the j channel bandwidth and P_i^{tr} represents the data transmission power of mobile devices, which is determined by the wireless base station according to some power control algorithms such as [28] and [29]. Furthermore, $G_i = \ell_i^{-a}$ denotes the channel gain between the device user i and the base station, where ℓ_i indicates the distance between the mobile device i and the wireless base station. Moreover, σ^2 denotes the background noise power, and the parameter a denotes the path loss factor. Therefore, the data transmission delay $t_{i,j}^{\mathrm{tr}}$ can be calculated as

$$t_{i,j}^{\mathrm{tr}} = \frac{D_i}{R_{i,j}}.$$
(5)

Let P_i^{tr} be the data transmission power consumption. The energy consumption of the data transmission can be denoted as

$$E_{i,j}^{\mathrm{tr}} = \frac{D_i}{R_{i,j}} P_i^{\mathrm{tr}}.$$
(6)

In the second phase, the base station transmits the data load to the mobile-edge cloud server through a high-speed wire network [27, 28]. Because of the high-speed link, the time delay of this phase can be ignored. In the final phase,

the mobile-edge cloud server processes the task and returns the results back to the device users. Because the size of results is often considerably smaller than that of the input data load, the time delay from the mobile-edge cloud server to the device is not considered, as in some of the previous studies. Therefore, the delay of this phase is mainly composed of the cloud processing delay t_i^c. Let C_i^c be the computation capacity allocated from the cloud. The cloud processing time t_i^c spent on the cloud side can be formulated by

$$t_i^c = \frac{O_i}{C_i^c}.$$
(7)

When the task is executed on the cloud, the mobile device needs to wait for the return of the response result. Thus, at the period of time t_i^c spent on the task processing on the cloud side, the idle power consumption of the mobile device can be calculated as follows:

$$E_i^{\mathrm{id}} = \frac{O_i P_i^{\mathrm{id}}}{C_i^c}.$$
(8)

Thus, the total offloading energy consumption $E_{i,j}^{\mathrm{Off}}$ is expressed as $E_{i,j}^{\mathrm{Off}} = E_{i,j}^{\mathrm{tr}} + E_i^{\mathrm{id}}$, which is defined as the sum of the energy spent to transmit data to the cloud, $E_{i,j}^{\mathrm{tr}}$, plus the idle power consumption, E_i^{id}, as follows:

$$E_{i,j}^{\mathrm{Off}} = \frac{D_i P_i^{\mathrm{tr}}}{R_{i,j}} + \frac{O_i P_i^{\mathrm{id}}}{C_i^c}.$$
(9)

And the total time delay of offloading is denoted as $T_{i,j}^{\mathrm{Off}} = t_i^c + t_{i,j}^{\mathrm{tr}}$, which is defined as the transmission delay plus the cloud processing delay.

In order to realize energy-efficient task offloading, it is necessary to properly deal with the reasonable allocation of communication resources and computing resources, which are mutually coupled in the case of energy efficiency because of the competition for the resource. However, in this paper, it is assumed that the processing capacity of cloud services is far greater than the processing capacity of each mobile device, and the computing and storage resources are sufficient to satisfy the requirements of all mobile devices. Besides, our research scope of this paper is mainly focused on the mobile devices for the purpose of saving energy by offloading the task onto the cloud side. Therefore, the server overhead and energy consumption of the cloud server are not considered, which does not affect the completeness of the paper.

Among the multiple mobile device users for the mobile-edge cloud computing environment, the mobile device selects the nearest wireless access point in order to get better communication and interaction. Similar to the previous research on mobile-edge cloud computing, from the beginning of the offloading decision until the end of offloading operation, it is reasonable to assume that all mobile devices move very slowly in a quasi-static scenario.

To minimize the total energy consumption of the system, the optimization problem is mathematically modeled as follows:

$$\min_{\{\psi_{i,j}\}} \quad \mathcal{F} = \sum_{i=1}^{N}\left(\left(1-\psi_{i,j}\right)E_i^l + \psi_{i,j}E_{i,j}^{\text{Off}}\right), \tag{10}$$

$$s.t. \quad \sum_{j=1}^{M}\psi_{i,j}E_{i,j}^{\text{Off}} \leq E_i^l, \forall i \in \mathcal{N}, \tag{11}$$

$$\sum_{j=1}^{M}\psi_{i,j}T_{i,j}^{\text{Off}} \leq T_i^l, \forall i \in \mathcal{N}, \tag{12}$$

$$\sum_{i=1}^{N}\psi_{i,j}P_i^{\text{tr}}G_i \leq \mathcal{C}, \forall j \in \mathcal{M}, \tag{13}$$

$$\sum_{j=1}^{M}\psi_{i,j} \leq 1, \forall i \in \mathcal{N}, \tag{14}$$

$$\psi_{i,j} \in \{0,1\}, \forall i \in \mathcal{N}, \forall j \in \mathcal{M}. \tag{15}$$

Constraint (11) ensures that the energy consumption of task offloading is not greater than the local processing energy consumption of the mobile device. Constraint (12) ensures that the total time consumption of mobile devices in the process of task offloading is not greater than the local processing energy consumption of the task. Constraint (13) is to guarantee the communication quality of the wireless channel. The setting of the threshold $\mathcal{C}$ can avoid mobile devices to access the same channel at the same time because the burst data traffic of mobile devices will seriously cause the attenuation of channel quality. Constraint (14) states that the mobile device can only select access to a wireless channel, but the wireless channel can be accepted by a plurality of mobile devices. Constraint (15) states that the cloud offloading decision of the task is a binary variable.

Consider that the task offloading decisions Ψ among the device users are coupled. If too many device users simultaneously choose to offload the task to the cloud via the same wireless channel, they may cause severe interference, which will lead to a low data rate. The two factors related to the energy consumption of data transmission of the mobile device are the inherent transmission power and data transmission time. The transmission energy consumption of mobile devices is proportional to the transmission time. Thus, when the data rate of the mobile device user is low, it would consume high energy and incur long transmission time as well. In this case, more and more device users will avoid offloading and are more willing to choose execution locally. However, this is not our original intention: allowing beneficial cloud computing users to offloading as much as possible. Thus, the $\mathcal{C}$ threshold is set, which can be flexible assignment.

3. Energy-Aware Task Offloading Mechanism

To solve the optimization problem (10), an energy-aware task offloading mechanism is designed in the system of mobile-edge cloud computing. The proposed mechanism mainly includes two aspects:

(1) At the beginning, an algorithm for mobile device user classification and priority determination is designed. The mobile device users can be classified into two types: participation in the auction and not

to participate according to the energy cost features of the task computing process. Namely, the mobile device users who do not participate in the auction choose to process the task locally. Then, the priorities of the first class of users are derived, which represent the intensity of user demand for task offloading.

(2) According to the order of priority, the device users get resource allocation in turn. A reverse auction-based offloading algorithm is proposed to achieve the offloading decision and associate the suitable communication resource with each mobile device who participates in the auction.

3.1. Mobile Device User Classification and Priority Determination. Based on the characteristics of the task and the mobile device, such as the data size of the task, workload density, computing capacity, and energy consumption, the mobile device users are divided into two types.

The first type of users is a group that should compute their task locally. The set of users of this type is denoted as $\mathcal{G}_l$. When the mobile device occupies a channel alone, the data transmission rate of this mobile device can be expressed by

$$R_{i,j}^0 = w_j \log_2\left(1 + \frac{P_i^{\text{tr}}G_i}{\sigma^2}\right). \tag{16}$$

The condition used to determine the devices belonging to this type is given as follows:

Theorem 1. *if* $E_i^l < E_{i,j}^{\text{Off}}$, $i \in \mathcal{N}, j \in \mathcal{M}$, *then the device i belongs to* $\mathcal{G}_l$, *where*

$$E_{i,j}^{\text{Off}} = \frac{D_i P_i^{\text{tr}}}{R_{i,j}^0} + \frac{O_i P_i^{\text{id}}}{C_i^c}. \tag{17}$$

Besides the aforementioned type, the rest of the mobile device users fall into second type $\mathcal{G}_o$. The mobile device users belonging to $\mathcal{G}_o$ can either decide to implement their task locally or to offload the task onto the mobile-edge cloud server. The decision of them depends on the communication quality of the channel. For this type of mobile users, different priorities are set for them in the offloading process, which is defined as

$$\eta_i = \frac{G_i P_i^{\text{tr}}}{\sqrt{E_i^l}}. \tag{18}$$

The complete mobile device user classification and priority determination are illustrated in Algorithm 1.

3.2. Reverse Auction-Based Offloading Algorithm. In this section, a reverse auction-based offloading scheme is proposed for the group $\mathcal{G}_o$ of mobile devices based on the abovementioned analysis of mobile device user classification and priority determination. Our aim is to maximize the energy efficiency of task offloading subjected to the mobile device's minimum energy consumption requirement and

Initialization:
Mobile IoT device set: $\mathcal{N} = \{1, 2, \ldots, N\}$;
Wireless channel set: $\mathcal{M} = \{1, 2, \ldots, M\}$;
The task on mobile IoT device: $\mathcal{T}_i =^{\Delta} (O_i, D_i)$;
Transmission power of mobile IoT device: $\{P_i^{\text{tr}}\}$, $i \in \mathcal{N}$;
Idle power of mobile IoT device: $\{P_i^{\text{id}}\}$, $i \in \mathcal{N}$;
Categorized device sets: $\mathcal{G}_l = \mathcal{G}_o = \varnothing$;
Priority set: $\eta = \varnothing$;
1: **for** mobile device $i = 1$ to N **do**
2:    **for** channel $j = 1$ to M **do**
3:       calculate the exclusive channel data transfer rate $R_{i,j}^0$ of
        each mobile device as in (11), and the energy consumption $E_{i,j}^{\text{Off}}$ as in (12);
4:       **if** $E_i^l \leq E_{i,j}^{\text{Off}}$ **then**
5:          $i \Rightarrow \mathcal{G}_l$;
6:       **else**
7:          $i \Rightarrow \mathcal{G}_o$;
8:          $\eta_i = G_i P_i^{\text{tr}} / \sqrt{E_i^l}$;
9:       **end if**
10:   **end for**
Output:
The categorized device set: $\mathcal{G}_l$, $\mathcal{G}_o$;
The priority set for the devices $\eta = \{\eta_i\}$, $i \in \mathcal{G}_o$.

ALGORITHM 1: The Algorithm for classifying the mobile device and priority determination.

the limited communication resource of channels during the task offloading process.

As illustrated in Figure 2, the mobile device $i, i \in \mathcal{G}_o$, acts as the buyer, who achieves higher system energy efficiency in exchange of transmission power resources provided by the channel. Prior to participation in the auction, namely, deciding whether to offload the task onto the mobile-edge cloud server, mobile users first calculate their cost $\mathcal{P}_i$, which means the reserved prices the mobile device can accept. In this case, the reserve price corresponds to the aforementioned local computing energy consumption of the task in the system of mobile-edge cloud computing. Thus, the reserve price $\mathcal{P}_i$ is expressed as $\mathcal{P}_i = E_i^l$.

On the other hand, the wireless channels are sellers. Each channel can participate in the auction by submitting to the mobile device the bidding information (b_j, s_j), where b_j is the price at which the jth channel agrees to share their available resources to the mobile device. Each seller calculates their s_j and b_j, respectively, by $s_j = P_i^{\text{tr}} G_{i,j}$ and $b_j = E_{i,j}^{\text{Off}}$. The total number of available resources of each seller corresponds to the aforementioned interference threshold of each channel. Then, the mobile device will calculate the energy cost and compare the biding prices provided by the seller to decide whether it could achieve energy saving and choose the target channel or give up task offloading decision. The target of the wireless channel is the winner of the reverse auction process. The mobile device chooses the target channel for task offloading.

In the previous auction researches, they allocate the resources through multiround bidding procedures to determine the final winner. However, this multiround auction method is not suitable for our scenario because mobile device users have to wait for the consequences after multiple rounds of auction, which inevitably generate an intolerable

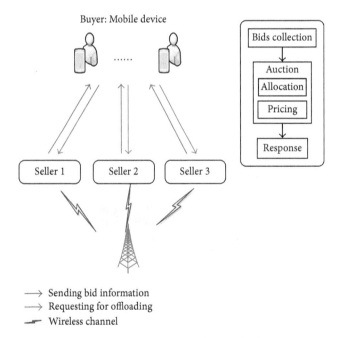

FIGURE 2: The reverse auction system. The mobile device sends the request of transmission with the energy cost, collects the bid information sent by wireless channels, and then chooses the winner channel.

extra delay. In the process of task offloading, mobile devices are sensitive to delay. Therefore, the single-round auction is implemented in this paper in order to improve the energy efficiency and reduce the delay for the offloading users. Moreover, it is assumed that the time delay of the auction process is so small that it can be ignored. The auction is conducted periodically, which means that, after a smaller

---

Input: $\mathscr{G}_l$, $\mathscr{G}_o$, η.
Output: Offloading decision $\Psi = (\psi_{1,j}, \psi_{2,j}, \ldots, \psi_{N,j})$.
1: Set the temporary set $\mathscr{G}'_o = \mathscr{G}_o$;
2: **while** $\mathscr{G}'_o \neq \varnothing$ **do**
3:     Select the device i, where $i = \text{argmax}\{\eta_i\}_i$, $i \in \mathscr{G}_o$;
4:     **for** channel $j = 1$ to M **do**
5:         Update the data transmission rate $R_{i,j}$ and update $E^{\text{Off}}_{i,j}$, as in (4) and (9);
6:         **if** $\mathscr{C}_j > 0$ **then**
7:             Calculate the bid density bd_j of each channel j based on the 2-tuple (b_j, s_j);
8:             Set the bid density $bd = \{bd_j\}$;
9:             **while** $bd \neq \varnothing$ **do**
10:                 Select the channel j, where $j = \text{argmin}\{bd_j\}_j$;
11:                 **if** $E^{\text{Off}}_{i,j} \leq E^l_i$ && $\mathscr{C}_j - \sum_{r=1,r \neq i}^{\mathcal{N}} \psi_{r,j} P^{\text{tr}}_i G_i > 0$ **then**
12:                     Let $\psi_{i,j} = 1$;
13:                     $\mathscr{C}_j \Leftarrow \mathscr{C}_j - \sum_{r=1,r \neq i}^{\mathcal{N}} \psi_{i,j} P^{\text{tr}}_i G_i$;
14:                 **else**
15:                     Let $\psi_{i,j} = 0$;
16:                 **end if**
17:                 $bd = bd\backslash j$;
18:             **end while**
19:         **else**
20:             Let $\psi_{i,j} = 0$;
21:         **end if**
22:     **end for**
23:     $\mathscr{G}'_o = \mathscr{G}'_o \backslash i$
24: **end while**

ALGORITHM 2: Reverse Auction-Based Offloading Algorithm for Offloading Decisions.

auction interval, a new round of auction is started and the relevant information is collected again, which is adapted to the dynamic mobile cloud computing environment. In order to simplify the model, it is assumed that the auction interval is very short and is ignored. The complete reverse auction-based offloading algorithm is illustrated in Algorithm 2.

3.2.1. Allocation. In the allocation steps, the mobile device decides which channel will be the auction winner. In order to avoid the extra delay caused by the multiround auction, the single-round auction is implemented in this paper. Thus, jointly considering the resources and the price that the bidders can provide, the mobile device decides who will win the auction, and b_j is the transaction price. Therefore, given the abovementioned definitions and notation, the optimization problem can be converted into the reverse auction problem. Here, $\psi_{i,j}$ represents the consequence of auction. $\psi_{i,j} = 0$ denotes that there is no winner channel. On the contrary, $\psi_{i,j} = 1$ expresses that the jth channel wins the auction. Our goal is to maximize the utility of the mobile device user, which can be formulated as

$$\max_{\{\psi_{i,j}\}} \mathcal{F} = \sum_{i=1}^{\mathcal{N}} \mathscr{P}_i - \sum_{j=1}^{\mathcal{M}} \sum_{1i=1}^{\mathcal{N}} \left(\left(1 - \psi_{i,j}\right) \mathscr{P}_i + \psi_{i,j} b_j \right). \quad (19)$$

In order to determine the winner and the allocation relationship, the bid densities of the participants are calculated and sorted firstly. In the list of wireless channels, the wireless channels were ranked in ascending order of their bid densities. For mobile users, the lowest call density is the

best communication quality. The bid density of sellers can be calculated by

$$bd_j = \frac{\left(\mathscr{C}_j - \sum_{r=1,r \neq i}^{\mathcal{N}} \psi_{r,j} P^{\text{tr}}_i G_i\right) E^{\text{Off}}_{i,j}}{\sqrt{\mathscr{C}_j - \sum_{r=1,r \neq i}^{\mathcal{N}} \psi_{r,j} P^{\text{tr}}_i G_i}}, \quad (20)$$

where $\mathscr{C}_j - \sum_{r=1,r \neq i}^{\mathcal{N}} \psi_{r,j} P^{\text{tr}}_i G_i > 0$, which is an indispensable condition for the wireless channel to ensure their quality of service. If the value is less than or equal to zero, the channel will give up participating in the auction.

3.2.2. Pricing Model. The final transaction price paid by the mobile device is b_j, which is the bid price submitted by the winner wireless channel. The utility of the mobile device user can be formulated as

$$\mathcal{F} = \sum_{i=1}^{\mathcal{N}} \mathscr{P}_i - \sum_{j=1}^{\mathcal{M}} \sum_{1i=1}^{\mathcal{N}} \left(\left(1 - \psi_{i,j}\right) \mathscr{P}_i + \psi_{i,j} b_j \right). \quad (21)$$

If the mobile user does not participate in the auction, its utility value is equal to 0. In other words, if $\psi_{i,j} = 0$, obviously then $\mathcal{F} = 0$ through the calculation of formula (21). Moreover, the utility of the wireless channel can be formulated as

$$\Theta = \sum_{j=1}^{\mathcal{M}} \sum_{1i=1}^{\mathcal{N}} \psi_{i,j} b_j. \quad (22)$$

If the wireless channel does not win the auction, then $\psi_{i,j} = 0$, obviously the utility of the wireless channel is equal to zero.

3.2.3. Properties. In this section, the properties of the proposed reverse auction model are analyzed. The individual rationality and the truthfulness properties need to be proved.

(1) Individual rationality: when the utility of each participating bidder in the pricing stage is greater than zero, the proposed mechanism is individual rational for each winning bidder. Namely,

$$\mathcal{F}_{B_M \setminus \{b_j\}} - \left(\mathcal{F}_{B_M} - \mathcal{F}_{b_j} \right) \geq 0, \tag{23}$$

where $B_M = \{b_1, b_2, \ldots, b_j, \ldots, b_M\}$ and $\mathcal{F}_{B_M \setminus \{b_j\}}$ denotes the utility of the mobile device under the optimal allocation solution without the presence of the jth channel.

(2) Truthfulness: for each bidder, the truthfulness means that the bid price of each bidder is equal to its private value. If the bidding of channels is untrue, the utility will be unlikely the biggest. In order to get the maximum, the allocation should be formulated as follows:

$$\mathcal{F}_{B_M \setminus \{b_j\}} - \left(\mathcal{F}_{B_M} - \mathcal{F}_{b_j} \right) \geq 0, \tag{24}$$

$$\Omega = \mathcal{F}_{B_M \setminus \{b_j\}} - \left(\mathcal{F}_{B_M} - \mathcal{F}_{b_j} \right) - \left[\mathcal{F}_{B_M \setminus \{b_i\}} - \left(\mathcal{F}_{B_M} - \mathcal{F}_{b_i} \right) \right]$$
$$= \left(\mathcal{F}_{B_M \setminus \{b_j\}} + \mathcal{F}_{b_j} \right) - \left(\mathcal{F}_{B_M \setminus \{b_i\}} + \mathcal{F}_{b_i} \right). \tag{25}$$

Based on the proposed reverse auction mechanism in this paper, because the bid price of the channel is not greater than the reverse price of the mobile device user, $\Omega \leq 0$. Obviously, when $j = i$, the value of Ω is equal to zero. Therefore, each bidder must be truthful to obtain the maximum utility.

4. Simulation and Analysis

In this section, the performance of the proposed mechanism is evaluated through numerical simulations designed by using the MATLAB. The compared algorithms are the competition-based algorithm [30] and the user-satisfaction-based offloading algorithm [31]. Their features are described as follows:

(1) Competition-based algorithm: the system is modeled as a competitive game subjected to the job execution deadlines and user-specific channel bit rates. Each user tries to minimize its own energy consumption when it competes for the shared communication channel. The Gauss–Seidel-like method is executed for achieving the Nash equilibrium to derive the mobile device user's offloading decisions.

(2) User-satisfaction algorithm: a utility function is introduced to choose the best communication resources in terms of user-satisfaction parameters, such as the throughput, used energy, and time spent to execute the application. Based on this, the offloading strategy is obtained by the applications' computation percentage.

Without loss of generality, four performance metrics of the proposed algorithm and the two classical algorithms are compared on the same simulation scenarios fairly. The four metrics are the average energy consumption, delay, energy efficiency factor, and throughput of the mobile device for offloading.

4.1. Simulation Setup. The simulations are deployed based on real-world settings. All the parameters, including the energy consumption rates and computing capacity, are measured from real mobile devices. These real-world datasets, which have been widely used, are measured at various clock speeds and in the cellular network scenarios by using a monsoon power monitor.

At first, a base station is considered that covers a hexagonal cellular network with radius 2 km, and assume that the wireless access point is located at the center. The base station has $M = 4$ channels, and the channels belonging to this base station are orthogonal. The bandwidth capacity of the channels can be different values, but in order to simplify the simulation, four channels of the same bandwidth of the device are set to $w = 1$MHz, which does not affect the effect of the experiment. Besides, the power of the background noise is set to $\sigma^2 = -100$dBm, and the path loss factor is set to $a = 2$ according to the physical interference model. In the system of mobile-edge cloud computing, mobile devices are randomly distributed in the coverage area of the hexagonal cellular network, accessing to this wireless point at any time if needs. And there is a mobile-edge server deployed near the base station, who assigns 5 GHz computation capability for each mobile device, sufficient to satisfy the requirements of all mobile devices.

Conforming to the diversity of the mobile device in the real world, four types of smartphones are considered, namely, Galaxy Note, Galaxy Note 2, Nexus S, and HP iPAQ PDA. Different mobile devices have different CPU computing capacities. The HP iPAQ PDA with a 400 MHz Intel XScale processor [31] has the following parameters: the local processing power $P_i^l = 0.9$W, the standby power $P_i^{id} = 0.3$W, and the transmission power $P_i^{tr} = 1.3$W. In addition, the parameters of the other three mobile devices include CPU processing parameters, such as χ_i, α_i, and β_i. These parameters are adopted as in [30]. In the simulation, the type of the mobile device in the mobile-edge cloud computing scenario is randomly selected among the abovementioned three types, and each mobile device has only one task waiting to be executed. The tasks on mobile devices are set to ten types: face recognition, virus scanning, online gaming, and so on. These ten types of tasks are randomly assigned to each mobile user. Different types of mobile devices have different processing speeds for different task types, whose corresponding parameters are given in Table 2, including workload density, data size, and the allocated computing capacity.

It is clear that, in these tasks, the workload densities of face recognition and virus scanning are larger than those of other types of tasks, and the data size of the two tasks is relatively small, which are computation-intensive tasks. On the contrary, the workload density of video coding is far less than that of the other eight tasks, but the data size is

TABLE 2: Parameters of the system.

Smartphone	χ_i	α_i	β_i	P_i^{tr}	P_i^{id}
Galaxy Note	3.0	0.33	0.10	2605	9.64
Galaxy Note 2	2.7	0.25	0.40	2796	11.70
Nexus S	3.0	0.34	0.35	1217	7.4
Galaxy Nexus	3.0	0.40	0.30	964	22.37
Task	O_i	D_i	C_i^l	—	—
Face recognition	60	31680	1.2	—	—
400-frame game	2048	2640	1.0	—	—
Chess: select	400	1580	0.6	—	—
Chess: move	400	2640	1.0	—	—
Virus scanning	300	32946	1.5	—	—
4-queen puzzle	200	87.8	0.4	—	—
5-queen puzzle	200	263	0.45	—	—
6-queen puzzle	200	1760	0.72	—	—
7-queen puzzle	200	8250	1.04	—	—
Video transcoding	10240	200	0.56	—	—

particularly larger than that of others, which belong to communication-intensive tasks. It is obvious that the parameters given in the table include various types of tasks that satisfy the generality and credibility of the simulation.

In order to accurately evaluate the performance of the algorithm without any loss of generality, a series of simulations are carried out, gradually increasing the number of mobile devices from 50 to 1000. Since the mobile devices are randomly deployed within the coverage of mobile networks, and the type of the mobile device and the task request of mobile users have stochastic features.

4.2. Evaluation Results. Firstly, the energy consumption of the proposed algorithm is evaluated. Figure 3 shows the average energy consumption of the mobile device when the number of mobile devices increases from 50 to 1000 with four different methods. The average energy consumed by one mobile device is approximately 21.2060 J with the local computing approach. Comparing with the local computing approach, both the proposed approach and the other two algorithms achieve the purpose of energy saving through task offloading.

At the beginning, with 50 mobile devices, the three methods exhibit an energy consumption of 6.4219 J, 6.4276 J, and 6.9430 J, respectively. With the gradually increased number of mobile devices, the average energy consumption of the mobile device increases to 11.0077 J, 12.5876 J, and 13.3540 J, respectively. This is because too many mobile devices choose to access the same wireless channel to implement the task offloading simultaneously, which would lead to the augment of mutual interference. According to (4), it is obvious that the severe interference to each other will cause the reduction of the communication quality and the rates for computation offloading. Therefore, with 1000 mobile devices, more and more users tend to choose the local computing method, and the average energy consumption of

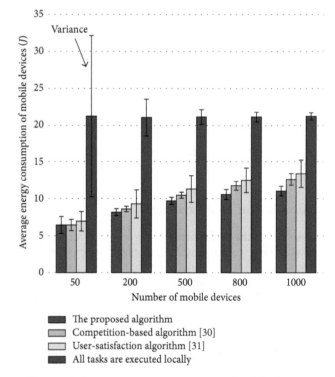

FIGURE 3: Average energy consumption of mobile devices.

mobile devices increases. The proposed mechanism can save at least 56.42% of the energy consumption.

The superiority of the proposed approach is gradually obvious. This is due to the fact that the reverse auction-based offloading mechanism performs task offloading decision in a global long-term perspective, reasonably allocating communication resources for mobile device users to meet the quality of service requirements. It exhibits a relatively lower energy consumption when the number of mobile device users is small. However, with the explosive increase in the

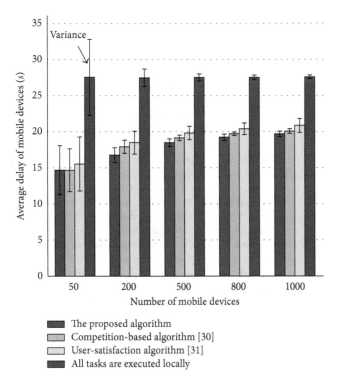

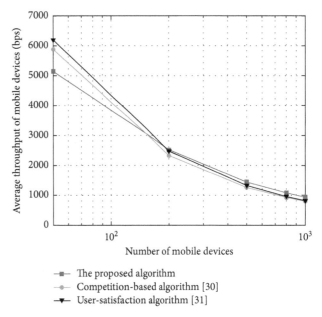

FIGURE 4: Average delay of mobile devices.

FIGURE 5: Average throughput of mobile devices.

number of mobile device users, the performance degrades due to the traffic growth. Obviously, the proposed method can find a better energy-saving solution than other two approaches.

The average task execution delay of mobile devices with the proposed method and the other two schemes is compared. As shown in Figure 4, the average time delay of mobile devices for performing a task is approximately 27.5157 s with the local computing approach. With 50 mobile devices, the time delays of these three methods are 14.6868 s, 14.6605 s, and 15.5306 s, respectively. Compared with the local computing approach, at least 45.69% of the time delay can be saved. When 1000 mobile devices are deployed, the time delays obtained by four methods are 19.6782 s, 20.0610 s, 20.8447 s, and 27.5751 s, respectively. The proposed mechanism can save about 35.43% of the time compared to the local computing approach, which is slightly higher than the performance of other two algorithms.

Third, the throughput of mobile devices is compared in the case of our proposed method and other two methods in addition to the local computing method because the local processing does not need to upload data and the throughput is zero. Figure 5 shows that, at the beginning, with 50 mobile device users, the other three methods exhibit an average throughput of 5144.6 bps, 5883.7 bps, and 6195.7 bps, respectively. Although the throughput of mobile devices in the case of our proposed algorithm is lower at the beginning, when the number of mobile devices is between 50 and 200, the trend of throughput drops more slowly than the other two methods. With the continued growth of mobile devices, the throughput of mobile devices in the case of the proposed method is higher than that in the other methods. At the end, with 1000 mobile device users, the methods exhibit an

average throughput of 932.3 bps, 796.4 bps, and 819.5 bps, respectively. As the number of mobile devices increases logarithmically, the correspondingly mutual interference among the device will grow. Furthermore, the uplink data transmission rate will decrease, which leads the energy consumption of cloud offloading greater than that of local computing. Thus, more and more mobile device users will adopt local computing, substituting for offloading operation. Compared with the competition-based algorithm and the user-satisfaction algorithm, the throughput is higher and the rate of decline is relatively slow when using the proposed method.

Finally, the energy efficiency factor for offloading is evaluated with the proposed method, competition-based algorithm, and user-satisfaction algorithm over 1000 simulation runs. The proposed mechanism is designed to reduce the energy consumption and the response time delay of mobile devices. Thus, a function is proposed representing the QoS degree perceived by the user. The function is modeled as a sigmoid curve, which is widely used to measure user satisfaction and service quality in previous studies [32]. User satisfaction increases as energy consumption and latency decrease, so we use sigmoid functions to represent the relationship between them: $f_1 = 1 - (1/1 + e^{-(E^{aver} - E^l)})$; $f_2 = 1 - (1/1 + e^{-(T^{aver} - t^l)})$.

The function $\mathcal{U} = \omega_1 f_1 + \omega_2 f_2$ is introduced to analyze the energy efficiency factor, where $\omega_1 + \omega_2 = 1$. Moreover, ω_1 and ω_2 represent the weight coefficients of energy consumption and delay, respectively. And E^{aver} and T^{aver}, respectively, denote the average energy consumption and average delay. As shown in Figure 6, with an increased number of mobile devices, the user satisfaction for task offloading gradually reduced. And at the last, with 1000 mobile devices, the values of the two methods of comparison drop sharply. On the contrary, compared with the other two curves, the curve corresponding to the proposed method is relatively stable. Therefore, when there are a large number of

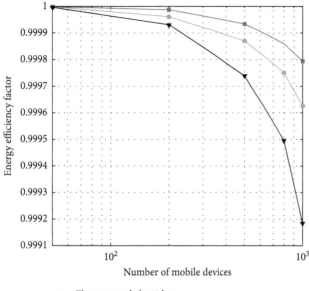

- ■- The proposed algorithm
- ●- Competition-based algorithm [30]
- ▼- User-satisfaction algorithm [31]

FIGURE 6: Energy efficiency factor of mobile devices.

mobile devices in the mobile-edge cloud system, the performance of the proposed method is better to meet the request of mobile device users for cloud offloading.

5. Conclusion

In this paper, an energy-aware task offloading mechanism is designed to perform offloading decisions, with optimization on minimizing the energy consumption of mobile devices. Considering the interference threshold in each channel, the task local execution delay, and the local energy consumption, the task offloading decision problem is formulated as a 0-1 nonlinear integer programming optimization. In order to solve this problem, the algorithm is proposed for classifying the mobile device and priority determination. Furthermore, the reverse auction theory has been implemented with the proposed algorithm to decide the offloading target channel. The individual rationality and truthfulness of the reversed auction model are also discussed in the paper. The performances of the proposed mechanism, comparing with the other two methods, are evaluated with performance metrics of energy consumption, time delay, throughout, and the energy efficiency factor. The simulation results validate that the proposed algorithm can achieve better performances.

Acknowledgments

This work was partially supported by the National Natural Science Foundation of China (Grant nos. 61379111, 61402538, 61403424, 61502055, 61672537, and 61672539).

References

[1] R. Janessa and R. Meulen, *Gartner Says the Internet of Things Installed Base Will Grow to 26 Billion Units by 2020*, Gartner Inc., Stamford, CT, USA, 2013.

[2] CISCO, *The Internet of Things: How the Next Evolution of the Internet is Changing Everything*, CISCO White Paper, 2011.

[3] E. Ahmed, A. Gani, M. K. Khan, R. Buyyac, and S. U. Khan, "Seamless application execution in mobile cloud computing: motivation, taxonomy, and open challenges," *Journal of Network and Computer Applications*, vol. 52, pp. 154–172, 2015.

[4] H. T. Dinh, C. Lee, D. Niyato, and P. Wang, "A survey of mobile cloud computing: architecture, applications, and approaches," *Wireless Communications and Mobile Computing*, vol. 13, no. 18, pp. 1587–1611, 2013.

[5] S. Barbarossa, S. Sardellitti, and P. D. Lorenzo, "Communicating while computing: distributed mobile cloud computing over 5G heterogeneous networks," *IEEE Signal Processing Magazine*, vol. 31, no. 6, pp. 45–55, 2014.

[6] S. Abolfazli, Z. Sanaei, E. Ahmed et al., "Cloud-based augmentation for mobile devices: motivation, taxonomies, and open challenges," *IEEE Communications Surveys and Tutorials*, vol. 16, no. 1, pp. 337–368, 2014.

[7] X. Chen, "Decentralized computation offloading game for mobile cloud computing," *IEEE Transactions on Parallel and Distributed Systems*, vol. 26, no. 4, pp. 974–983, 2014.

[8] S. Barbarossa, S. Sardellitti, and P. D. Lorenzo, "Joint allocation of computation and communication resources in multiuser mobile cloud computing," in *Proceedings of the IEEE 14th Workshop on Signal Processing Advances in Wireless Communications*, pp. 26–30, Darmstadt, Germany, June 2013.

[9] D. Huang, P. Wang, and D. Niyato, "A dynamic offloading algorithm for mobile computing," *IEEE Transaction on Wireless Communications*, vol. 11, no. 6, pp. 1991–1995, 2012.

[10] L. Yang, J. Cao, H. Cheng, and J. Yusheng, "Multi-user computation partitioning for latency sensitive mobile cloud applications," *IEEE Transactions on Computers*, vol. 64, no. 8, pp. 2253–2266, 2015.

[11] H. Viswanathan, E. K. Lee, I. Rodero, and D. Pompili, "Uncertainty-aware autonomic resource provisioning for mobile cloud computing," *IEEE Transactions on Parallel and Distributed Systems*, vol. 26, no. 8, pp. 2363–2372, 2015.

[12] O. Munoz-Medina, A. Pascual-Iserte, and J. Vidal, "Optimization of radio and computational resources for energy efficiency in latency-constrained application offloading," *IEEE Transactions on Vehicular Technology*, vol. 64, no. 10, pp. 4738–4755, 2015.

[13] M. Satyanarayanan, P. Bahl, R. Caceres, and N. Davies, "The case for VM-based cloudlets in mobile computing," *IEEE Pervasive Computing*, vol. 8, no. 4, pp. 14–23, 2009.

[14] Y. Zhang, D. Niyato, and P. Wang, "Offloading in mobile cloudlet systems with intermittent connectivity," *IEEE Transactions on Mobile Computing*, vol. 14, no. 12, pp. 2516–2529, 2015.

[15] W. Li, Y. Zhao, S. Lu, and D. Chen, "Mechanisms and challenges on mobility-augmented service provisioning for mobile cloud computing," *IEEE Communications Magazine*, vol. 53, no. 3, pp. 89–97, 2015.

[16] L. Lei, Z. Zhong, K. Zheng, J. Chen, and H. Meng, "Challenges on wireless heterogeneous networks for mobile cloud computing," *IEEE Wireless Communications*, vol. 20, no. 3, pp. 34–44, 2013.

[17] Y. Mao, C. You, J. Zhang, K. Huang, and K. B. Letaief, "A survey on mobile edge computing: the communication perspective," *IEEE Communications Surveys & Tutorials*, vol. 99, 2017.

[18] S. Wang, R. Urgaonkar, M. Zafer, and T. He, "Dynamic service migration in mobile edge-clouds," in *Proceedings of the IFIP Networking Conference*, pp. 1–9, Toulouse, France, March 2015.

[19] X. Chen, L. Jiao, W. Li, and X. Fu, "Efficient multi-user computation offloading for mobile-edge cloud computing," *IEEE/ACM Transactions on Networking*, vol. 24, no. 4, pp. 974–983, 2015.

[20] S. Sardellitti, G. Scutari, and S. Barbarossa, "Joint optimization of radio and computational resources for multicell mobile-edge computing," *IEEE Transactions on Signal and Information Processing Over Networks*, vol. 1, no. 2, pp. 89–103, 2015.

[21] M. T. Beck and M. Maier, "Mobile Edge Computing: Challenges for Future Virtual Network Embedding Algorithms," *The Eighth International Conference on Advanced Engineering Computing and Applications in Sciences*, pp. 65–70, Rome, Italy, 2014.

[22] Y. Zhang, C. Lee, D. Niyato, and P. Wang, "Auction approaches for resource allocation in wireless systems: a survey," *IEEE Communications Surveys and Tutorials*, vol. 15, no. 3, pp. 1020–1041, 2013.

[23] B. Kollimarla, *Spectrum Sharing in Cognitive Radio*, College of Oklahoma State University, Oklahoma City, OK, USA, 2009.

[24] G. Iosifidis, L. Gao, J. Huang, and L. Tassiulas, "A double-auction mechanism for mobile data-offloading markets," *IEEE/ACM Transactions on Networking*, vol. 23, no. 5, pp. 1634–1647, 2015.

[25] S. Paris, F. Martignon, I. Filippini, and L. Chen, "An efficient auction-based mechanism for mobile data offloading," *IEEE Transactions on Mobile Computing*, vol. 14, no. 8, pp. 1573–1586, 2015.

[26] J. Kwak, Y. Kim, J. Lee, and S. Chong, "DREAM: dynamic resource and task allocation for energy minimization in mobile cloud systems," *IEEE Journal on Selected Areas in Communications*, vol. 33, no. 12, pp. 2510–2523, 2015.

[27] K. Son and B. Krishnamachari, "SpeedBalance: speed-scaling-aware optimal load balancing for green cellular networks," in *Proceedings of the IEEE INFOCOM 2012*, pp. 2816–2820, Orlando, FL, USA, March 2012.

[28] M. Xiao, N. B. Shroff, and E. K. P. Chong, "A utility-based power-control scheme in wireless cellular systems," *IEEE/ACM Transactions on Networking*, vol. 11, no. 2, pp. 210–221, 2003.

[29] M. Chiang, P. Hande, T. Lan, and C. W. Tan, "Power control in wireless cellular networks," *Foundations and Trends in Networking*, vol. 2, no. 4, pp. 381–533, 2008.

[30] E. Meskar, T. Todd, D. Zhao, and G. KarakLondon UKostas, "Energy efficient offloading for competing users on a shared communication channel," in *Proceedings of the IEEE International Conference on Communications (ICC)*, pp. 3192–3197, London UK, June 2015.

[31] D. Mazza, D. Tarchi, and G. E. Corazza, "A user-satisfaction based offloading technique for smart city applications," in *Proceedings of the 2014 IEEE Global Communications Conference*, pp. 2783–2788, Austin, TX, USA, December 2014.

[32] D. H. V. Seggern, *CRC Standard Curves and Surfaces with Mathematica*, CRC Press, Boca Raton, FL, USA, 2015.

15

Interactive and Immersive Learning Using 360° Virtual Reality Contents on Mobile Platforms

Kanghyun Choi ⓘ,[1] Yeo-Jin Yoon ⓘ,[1,2] Oh-Young Song ⓘ,[3] and Soo-Mi Choi ⓘ[1,2]

[1]*Department of Computer Science and Engineering, Sejong University, Seoul, Republic of Korea*
[2]*Mobile Virtual Reality Research Center, Sejong University, Seoul, Republic of Korea*
[3]*Department of Software, Sejong University, Seoul, Republic of Korea*

Correspondence should be addressed to Soo-Mi Choi; smchoi@sejong.ac.kr

Academic Editor: Sang-Youn Kim

Recent advances in mobile virtual reality (VR) devices have paved the way for various VR applications in education. This paper presents a novel authoring framework for mobile VR contents and a play-based learning model for marine biology education. For interactive and immersive mobile VR contents, we develop a multilayer 360° VR representation with image-based interactions such as mesh deformation and water simulation, which enable users to realistically interact with 360° panoramic contents without consuming excessive computational resources. On the basis of this representation, we design and implement play-based learning scenarios to increase the interactivity and immersion of users. Then, we verify the effectiveness of our educational scenarios using a user study in terms of user-created VR contents, interactivity, and immersion. The results show that more experienced elements in VR contents improve the immersion of users and make them more actively involved.

1. Introduction

Following the development of consumer virtual reality (VR) devices using smartphones and tablets, the demand for mobile VR contents in various fields such as games, movies, and education has increased. Recently, user-created VR contents have been rapidly increasing with the advancement in low-cost image or video-capturing equipment for mobile VR. However, VR contents that use images or videos suffer from limitations with the lack of interactivity by only providing passive information; thus, the contents cannot draw active involvement of users. In education, we know that VR encourages learners to become more active through immersive experience [1]. Learning with real-time interactions provides learners instant results and helps in the decision making to reach their goals. To achieve a higher level of immersion and maximize educational effects, more engaging interactions between the learners and mobile VR contents are essential.

We aim to satisfy the interactive and immersive experience using the following criteria in terms of education.

(i) *Environment*. We target mobile VR platforms. Compared with the desktop-based VR, the cost of mobile VR hardware such as smartphones and mobile headsets is affordable, making it suitable for production and distribution of educational contents for a large number of users.

(ii) *Engagement*. We allow users to navigate in the environment and perform appropriate interactions according to the educational purpose within the VR environment, providing natural adaption to the virtual learning environment.

(iii) *Immersion*. We provide a personalized learning environment to maximize the educational effect by increasing the immersion and participation in virtual space compared with traditional teaching methods.

This paper presents a novel authoring framework for mobile VR contents using 360° photographs or videos. To create engaging interactions in the surrounding scenes, we introduce a multilayer representation where each layer can have different interaction types or properties such as deformable objects, moving objects, and water simulation. On the basis of the proposed representation, we develop play-based learning scenarios on a mobile VR platform for education in marine biology and conduct a user study on the interactivity and immersion of the scenarios. The overall pipeline of our play-based learning is shown in Figure 1.

The main contributions of this paper are summarized as follows:

(i) For interactive 360° VR contents, we introduced a multilayer representation using image-based interactions that are effective on mobile devices with limited performance.

(ii) We designed a learning model to increase the interactivity and immersion of users, which includes viewer-directed immersive forms, layers for interactivity, achievements, and engaging contents.

(iii) We verified our learning framework where we employed a user study in terms of user-created VR contents, interactivity, and immersion.

The remainder of this paper is structured as follows: Section 2 provides an overview of related works, and Section 3 describes our multilayer representation using different interactions and immersive experience in mobile VR. We present our active learning model in Section 4 and evaluate the developed learning scenarios through a user study in Section 5. Finally, we discuss our conclusions and future work directions in Section 6.

2. Related Work

The 360° VR contents for mobile platforms are usually created by including panoramic photographs or videos. This image-based content generation can reduce the time and cost of building a virtual space compared with the same process using three-dimensional (3D) graphics in traditional VR such as games. In addition, real-time manipulation is possible on mobile devices because it consumes a small amount of memory. The main advantage of using image-based techniques is that anyone can use their own images; thus, learner-driven personalized education is possible with the use of user-created VR contents.

2.1. Mobile VR in Education. In recent years, mobile VR education has been shown to provide more interactive learning with fully immersive contents than traditional education approaches [1, 2] A study on motivational active learning methods of several subjects that make learners quickly lose interest, such as Science, Technology, Engineering, and Mathematics [3], measured the level of immersion of learners using the Game Engagement Questionnaire (GEQ) to evaluate the ability of these learning methods. To create VR contents using synthetic 3D graphics, special skills, and extensive efforts are required. Furthermore, delivering user personalized experience and recreating virtual spaces according to the change in the learning context are difficult. The study in [4]

provided a conceptual framework for user-generated VR learning across three contexts such as paramedicine, journalism, and new media production. In that study, an active learning environment using student-created 360° videos was proposed where the students could share the videos on social networks to collaborate together. Although 360° contents are more realistic than synthetic 3D VR, limitations on the interactions with the contents have been identified to exist.

2.2. Interactivity in 360° Contents. The 360° contents, which are called 360° photographs or videos, are captured by several wide-angle cameras, which are stitched and mapped in a spherical or hemispherical space to observe the surrounding scenes based on the camera [2]. Most research works for 360° content creation focus on accurate stitching for spherical mapping [5, 6], image-based lighting effects [7–9], resolution improvement, and distortion correction of images from wide-angle cameras [10]. Some researchers investigated methods to improve the interactivity in image-based VR contents by adding URLs to 360° images [11], changing scenes using button clicks [12, 13], and adding special effects to reduce the differences between composite boundaries [14]. The study in [8] recently presented an immersive system using powerful PC-tethered VR headsets that provide interactive mixed reality experiences by compositing 3D virtual objects into a live 360° video. However, there are few studies on interactive 360° contents for mobile VR headsets, such as the Samsung Gear VR and Google Daydream View, where all of the processing is done on mobile phones.

2.3. Immersive Experience in Mobile VR. Immersion comes from the feeling that a user actually exists in a virtual space reproduced by a computer [15, 16]. Unlike the existing computer-based environment, the mobile VR environment can create more immersive feelings owing to the development of VR devices such as head-mounted displays and tracked controllers [17]. The immersion in VR is not only created by elements such as visual, auditory, and haptic information [15] but also by content resolutions, stereo cues, behavioral fidelity of what is being simulated, and system latency [18]. Mobile VR is especially affected by factors such as resolutions and latency because of low-power constraints. Therefore, to increase the immersive experience in mobile VR, visual quality, sound quality, and intuitive interactions can be considered without consuming excessive computational resources [19]. To improve the immersion in mobile VR in terms of user interaction, a few methods to synchronize the virtual and physical spaces have been presented. The methods enable users to navigate the virtual world by mapping the user movement in the physical space to the virtual space [20–22]. In addition, interactive devices that use hand-gesture recognitions have been developed [23, 24]. However, only a few studies have been conducted on the optimization of visual quality or 3D sounds that consider mobile VR platforms.

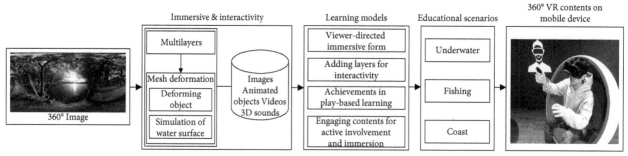

FIGURE 1: Overall pipeline of our play-based learning for education in marine biology.

2.4. Measurement of Immersive Experience. The measurement of immersion in virtual environments has been studied for a long time, mainly in the field of games [25]. GEQ [26] has been widely used as a method for measuring user engagement in games. Recently, a more accurate method to measure immersion in a virtual environment has been proposed in which 10 rating scales and 87 questions are used to show the reliability of the questionnaire [27]. We verify the effectiveness of our framework using the questions corresponding to the flow, immersion, presence, emotion, and judgment among the presented 10 scales.

3. Interactivity and Immersion of 360° VR Contents

In this section, we present methods to improve the interactivity and immersion of mobile VR contents using 360° photographs and videos for education in marine biology.

3.1. Interactive 360° VR Contents with Multiple Layers. Watching 360° contents using a VR headset replaces the user real environment with a simulated one and makes the user feel like he or she is actually there. Thus, 360° panoramic contents create a more immersive experience than regular photographs or videos. However, the viewers can usually look around by moving their head without engaging any interactions because the panoramic contents are generally mapped on a spherical surface as a single layer.

To create more engaging contents, we propose interactive 360° VR contents with multiple layers, as shown in Figure 2. The original input image can be manually segmented by its context such as sky, water, and land by using image tools, and the segmented regions are assigned to different layers. Then, images, animated images, and videos can be added to the original image as separate layers. Images and animated images are usually used for foreground objects, while videos are used for backgrounds. Different interaction properties can be assigned to these multiple layers based on image semantics or user intention. Finally, all the layers are mapped onto each individual sphere. The presented method uses a multilayer representation not only for image composition but also for different types of interactions that enable users to more realistically interact with the 360° panoramic contents.

3.2. Creating Deformable Objects. In our education scenarios for marine biology, interactive image deformation is used to create animation for some objects such as seaweeds and starfish based on transformation of the vertices of a 2D mesh on a particular layer. The resulting visual quality depends on the geometric resolution of the mesh, but mobile devices with limited computational power reduce the maximum achievable quality. Moreover, computationally expensive deformation techniques cause increased latency when the user interacts with the objects within a mobile VR environment. Therefore, we adopt a force-based approach to deform a mesh on a spherical surface, as shown in Figure 3.

The force is attenuated using the inverse proportion of the square of the distance from the center of the force, for example, user picking point p. Attenuated force F_v is defined by dividing original force F by one plus the square of the distance, i.e., $1 + d^2$, to ensure that the force is at full strength when the distance is zero.

$$F_v = \frac{F}{1 + d^2}. \tag{1}$$

The force can be converted into acceleration via $a = F/m$, where m is the mass. To simplify, we set mass $m = 1$ for each vertex. Therefore, m is ignored as $a = F$. Then, the change in velocity Δv is defined by Equation (2), where Δt is the time difference:

$$\Delta v = F\Delta t. \tag{2}$$

Vertex velocity V_v is computed using Equation (3), where the direction is derived by normalizing the vector that points from the center of the force to the vertex position. The vertices are moved to new positions via $\Delta p = v\Delta t$.

$$V_v = \frac{d}{\|d\|}\Delta v. \tag{3}$$

3.3. Realistic Simulation of Water Surface. In our virtual marine contents, realistic simulation of the water surface is an essential method to increase immersion. Real-time performance of the simulation is also required to work with various user interactions in the mobile-based VR environment. To satisfy these requirements, we adopt a physics-based approach to simulate water surface waves by deforming a 2D ocean mesh and implementing it on a graphics processing unit (GPU) hardware. The dynamic

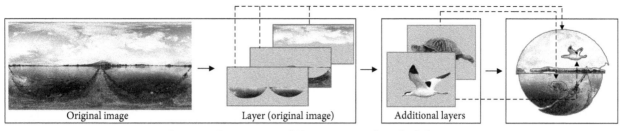

FIGURE 2: Interactive 360° VR contents with multiple layers.

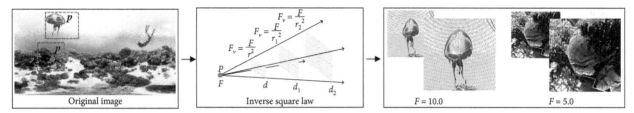

FIGURE 3: Force-based deformation of a mesh on a spherical surface.

behavior of the water surface, such as the ripple effects, is governed by the following 2D wave equation:

$$\frac{\partial^2 h}{\partial t^2} = c^2 \left(\frac{\partial^2 h}{\partial u^2} + \frac{\partial^2 h}{\partial v^2} \right). \tag{4}$$

Equation (4) defines height displacement h on the mesh surface depending on wave speed c. t refers to time, and u and v represent the coordinates of the 2D mesh. To numerically solve this second-order partial differential equation, we discretize it by replacing the partial derivatives with the central differences using a regular spatial grid of spacing ($k = \Delta u = \Delta v$) and a constant time step (Δt), yielding the following equation:

$$\frac{h^{t+1}[i, j] - 2h^t[i, j] + h^{t-1}[i, j]}{\Delta t^2} = c^2 \left(\frac{h^t[i+1, j] - 2h^t[i, j] + h^t[i-1, j]}{k^2} + \frac{h^t[i, j+1] - 2h^t[i, j] + h^t[i, j-1]}{k^2} \right). \tag{5}$$

New height displacement h^{t+1} can be explicitly integrated from old displacements h^t and h^{t-1}.

$$h^{t+1}[i, j] = \frac{c^2 \Delta t^2}{k^2} \left(h^t[i+1, j] + h^t[i-1, j] + h^t[i, j+1] \right.$$
$$\left. + h^t[i, j-1] - 4h^t[i, j] \right) + 2h^t[i, j] - h^{t-1}[i, j]. \tag{6}$$

Although this explicit method is fast and easy to implement as a GPU shader, it is only conditionally stable. If the time step becomes very large, the system can become unstable. To increase the stability of the simulation, an artificial attenuation can be added to the new displacement, i.e.,

$$h^{t+1} = \alpha h^{t+1}, \quad \alpha < 1. \tag{7}$$

In our experiments, we set attenuation coefficient α to 0.985 to generate the ripple effects that naturally disappear. Figure 4 shows the results of water simulation in terms of mouse movement and the attenuation coefficients.

3.4. Immersive Experience in Mobile VR Environments. The immersion experience in a mobile VR environment is mainly influenced by the viewing devices and content types, as shown in Figure 5. Because the VR headsets block out all sights of the outside world, watching 360° images or videos using them makes the users more immersed than watching regular photographs or videos. The wide viewing angle of the headset, such as more than 110°, and the higher quality of the visuals, such as more than 4K resolution, as well as the faster and more stable frame rates would result in deeper immersive experience.

Furthermore, wearing headphones makes the users even more immersed by also blocking outside sound. Realistic 3D audio effects on a mobile platform will increase their perception of immersion and reduce the side effects, including headaches and motion sickness. Therefore, the users can focus on the learning objectives in a virtual environment and become active learners.

4. Active Learning Design: Marine Biology Case Study

The learning style of the new generation tends to be more self-directed, engaged, and flexible. Therefore, interactive engagement and active involvement are key elements in digital learning environments for the new generation. The

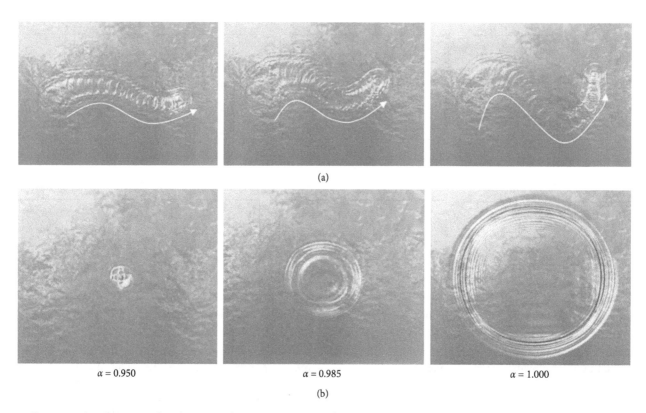

(a)

$\alpha = 0.950$ $\alpha = 0.985$ $\alpha = 1.000$

(b)

FIGURE 4: Simulation results of water surface waves in terms of mouse movement (a) and the attenuation coefficients (b).

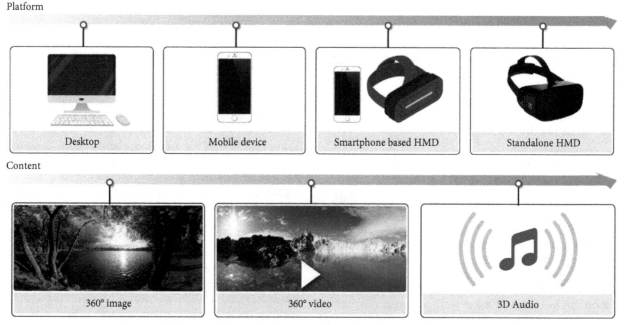

FIGURE 5: Degree of immersion in terms of viewing devices and content types.

simple forms of mobile VR contents are 360° photographs or videos. When we use the content for educational purpose, lack of interactivity makes the learners become passive in the virtual world. Thus, improvement in the interactivity and immersion in mobile VR contents is crucial to engage learners in the contents and keep them focused and motivated.

For active participant learning, we present play-based learning for education in marine biology. The learning scenarios are designed and implemented as mobile VR environment and allow the users to interact with the immersive mobile VR contents. A low-cost approach to user-created VR contents makes the mobile VR environment potentially learner-centered education to enable learners to create active participant VR contents. The main features of our learning scenarios include viewer-directed immersive form, adding layers for interactivity, achieving play-based learning, and engaging contents for active involvement and immersion.

4.1. Viewer-Directed Immersive Form.

The viewer can decide to look at any part of the 360° scenes. The scene can be typically captured from what is around the camera (an outward view) or around an object at different angles (an inward view), as shown in Figure 6. Whereas consumer 360° cameras are mainly focused on outward views, a few apps enable users to capture 360° scenes by having the camera circle around an object. Some cameras simply capture this scene as a hemisphere (a 180° × 360° view), and others can show it as a sphere (a 360° × 360° view). Immersive 360° contents are a valuable tool to create focused learning experience. Figure 7 shows that the user can navigate a 360° marine scene using a VR headset and can control what he or she wishes to view.

4.2. Adding Layers for Interactivity.

Layers are a powerful method of creating interactivity and branched stories. For personalized learning, the user can integrate marine plants and animals on additional layers together with animation effects, as shown in Figure 8, which are designed to more engage the learners to the content.

4.3. Achievements in Play-Based Learning.

Achievements can be a great means of increasing the learner engagement within the learning content and encourage learners to experiment with features that are not commonly used. In addition, achievements can be a fun way for learners to engage in a light-hearted competition. Figure 9 shows that the adapted picture from [28] illustrates the three zones of marine life according to the depth of the sea, and Figure 10 shows the user interface of our virtual fishing-game scenario. A high score is given to the player when he or she catches the more difficult fish in the deep sea. Table 1 shows the different scoring depending on the marine life zones.

4.4. Engaging Contents for Active Involvement and Immersion.

By adding the deformable objects discussed in Section 3.2 on a particular layer or moving objects such as lobsters as shown in Figure 11(a), we can create more interesting contents. During a fishing game, when the player casts a fishing rod, the movement of surface water can be simulated by the method presented in Section 3.3, producing more realistic effects as shown in Figure 11(b). To maximize the immersion, engagement, and learning outcome, we also created marine scenes with head-tracking 3D audio effects and video layers as shown in Figures 11(c) and 11(d).

5. Experiments

We developed the proposed learning framework on a desktop computer using Unity3D engine and then exported the educational scenarios to a mobile VR platform using Samsung Galaxy S7, Gear VR, and a dedicated controller as an interaction tool.

When we created an interactive 360° VR content, the number of vertices in the 3D spherical model for the foreground image layer was very important for realistic mesh deformation. Figure 12 shows that the results using a high-resolution 3D sphere (right) provided smoother interactions than those using a low-resolution 3D sphere (left). Because the realistic interaction methods adopted in this study relatively needed a small amount of computation for mesh processing and used only a simple 3D spherical model compared with the full 3D VR contents, our learning contents were maintained over 60-frame-per-second performance without latency even if we used a 3D spherical model with relatively high resolution.

To verify the effectiveness of the proposed interactive and immersive learning using 360° VR contents, we experimented on three scenarios in marine biology (see scenarios 1–3 in the supplementary materials (available here)), which were created based on the active learning design described in Section 4. In scenario 1, users can explore an underwater environment and interact with diverse marine life on the added deformable image layers. In scenario 2, the users can discover marine life while playing virtual fishing on a boat. In scenario 3, they can learn about the types of fish according to the sea depth while playing a waterside fishing game. When they catch a fish, the detailed information of the fish, such as scientific name and species name, is shown on the billboard, and different scores are given based on the sea depth. Each scenario consisted of three different levels of scenes: (1) simple user-created scenes with multiple layers from different image sources, (2) interactive scenes with animations such as realistic water simulation and deformable objects, and (3) immersive scenes with video layers and 3D audio effects. The participants answered questionnaires on two key VR elements, namely, interactivity and immersion, after finishing their experience in each scenario. Then, we analyzed the response results and evaluated whether the immersion of educational contents increased with the addition of more interactive elements.

For the user study, we used Samsung Galaxy S7, Gear VR, and a dedicated controller. Prior to the user study, the participants were asked about their previous experience in computer usage and VR devices. For a smooth user-study process, the participants used mobile VR devices and controllers for approximately 5 min to adapt to the interaction style. Then, experiments on interactivity and immersion were conducted by randomly choosing two of the three scenarios to keep their concentration on the experiments by limiting the maximum duration to 20 min. Each scenario required approximately 6 min to perform from Levels 1 to 3, and the average time for the experiments took a total of 17 min, including the warm-up time.

In each scenario, Level 1 showed the most basic 360° VR images, whereas Level 2 showed the scene with realistic interactive animation such as moving waves of water and object animation using mesh deformation. In Level 3, the

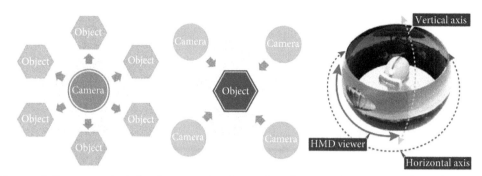

FIGURE 6: Capture methods for the outward and inward 360° views and navigation of a 360° content.

(a)   (b)   (c)

FIGURE 7: The 360° navigation of a boat on the water for virtual fishing.

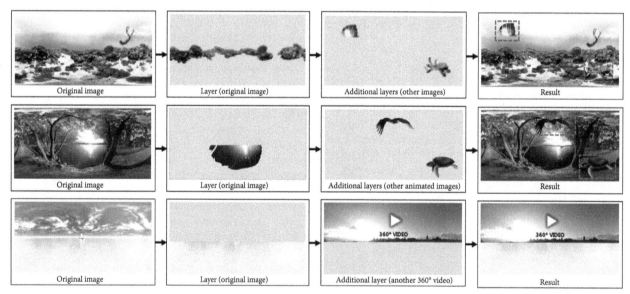

FIGURE 8: Creating multiple layers for interactive engagement.

scenes could have a video layer as a moving object or background, such as sky and water, and 3D audio effects could be added to enhance the immersive feelings. Table 2 lists the added factors to improve the user-created form, interactivity, and immersion in each learning scenario.

The participants were encouraged to experience the scenes using one or more of the techniques included in each experience category, whichever was selected among the given three scenarios. To compare the interactivity and immersion according to the level of experience, the participants provided points from 1 to 10 (10 is the best) for each experience category after finishing all level experience

in each scenario. Then, two different questionnaires in Table 3 were given to measure the interactivity (nine questions) and immersion (17 questions). Twenty-six people who were 14 to 30 years old participated in this user study. We included both, subjects who were familiar with IT and had previous experience in the use of VR and those who were unfamiliar with IT and had no experience of VR, in the user study.

In our experiment, the participants chose Scenario 1 for 20 people, Scenario 2 for 13 people, and Scenario 3 for 19 people. After experiencing the scenarios, they provided score for the interactivity and immersion in each scenario.

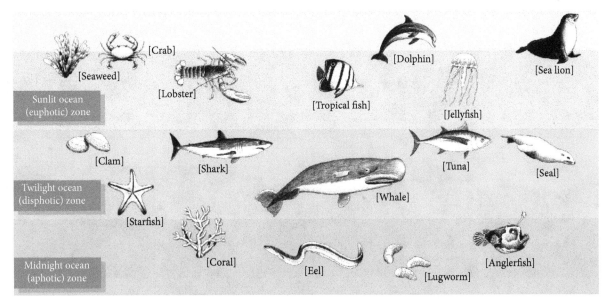

FIGURE 9: Zones of marine life according to the sea depth.

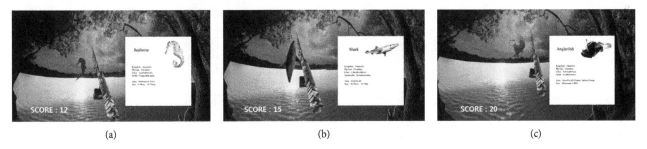

FIGURE 10: Education on marine life through a virtual fishing-game scenario.

TABLE 1: Different scoring depending on the marine life zones.

Marine organism	Depth	Score
Seaweed, crab, lobster, sea horse, tropical fish, dolphin, jellyfish, and sea lion	Sunlit Ocean (Euphotic) Zone	1
Clam, starfish, shark, whale, tuna, and seal	Twilight Ocean (Disphotic) Zone	3
Coral, eel, lugworm, and anglerfish	Midnight Ocean (Aphotic) Zone	5

Figure 13 shows the results of the average scores and standard deviations according to the level of experience in each scenario. In the interactivity case, the number of immersive elements was shown to increase from Levels 1 to 3. The interactivity also increased in all three scenarios. In the immersion case, Scenario 3 showed the highest scores at Level 1. The reason for this result is that the 3D audio effects were added to Scenario 3 to increase the immersion, and it was not reflected in Scenario 1.

Next, we conducted a survey on the interactivity and immersion for each scenario. The questionnaire for the interactivity consisted of nine questions: four for usability, four for interactivity, and one for immersion. The questionnaire for immersion consisted of 17 questions according to the measurement of user experience in the immersive

virtual environment [27]. Our questionnaire consisted of two questions for flow, five for immersion, one for presence, four for emotion, three for judgment, and two for descriptive evaluation of our framework.

We analyzed the response of each participant by calculating the scores. The answers to the positive questions are marked with "P" and the negative questions are marked with "N" in Table 3; we calculated scores according to the 5-point Likert scale (strongly agree = 5, agree = 4, neutral = 3, disagree = 2, and strongly disagree = 1). We also set the scores in the opposite order for the case of negative questions (marked with "N" in Table 3). Thus, we could obtain consistent values from negative and positive responses. The total participant response for each question was calculated as an average score. Then, the average score for each scale was calculated

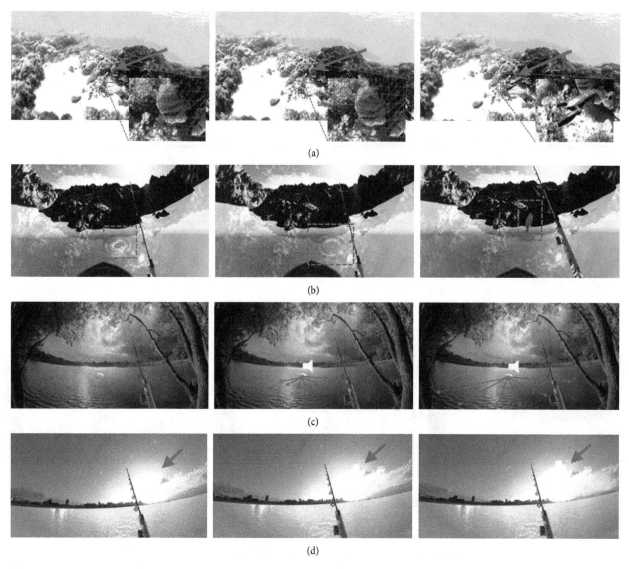

FIGURE 11: Creating engaging contents. (a) Moving objects, (b) water simulation, (c) 3D audio effects, and (d) video layers.

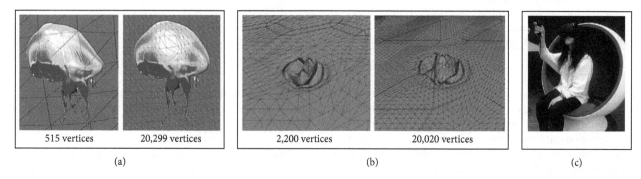

FIGURE 12: Results of the mesh deformation according to the resolution of spherical models. (a) Deforming objects. (b) Simulation of water surface. (c) VR contents.

according to the number of questions in the scale. Figure 14 shows the evaluation scores and standard deviations of the interactivity and immersion for each scenario according to the different scale, which are marked in Table 3. The reason 'immersion' appears in both interactivity and immersion

is that interaction methods or scene elements can affect immersion [27].

When we analyzed the overall answers, no negative values (less than 3) in all scenarios were obtained that appeared to be a negative response, but standard deviations

TABLE 2: Techniques for different levels of experience in three scenarios.

Experience	Technique	Scenario 1 (underwater)	Scenario 2 (fishing)	Scenario 3 (coast)
Level 1: user-created scene	Multiple layers	o	o	o
Level 2: interactive scene	Water simulation		o	o
	Physics-based deformation	o		o
	Image layer (objects)	o	o	o
Level 3: immersive scene	Video layer (backgrounds)		o	
	3D audio effects		o	o

TABLE 3: Questionnaires for interactivity and immersion.

	Scales	
Interactivity		
It was a good time to learn how to control the game	Usability	P
The controller was easy to use	Usability	P
Over time, the controller can be used proficiently	Usability	P
The interaction method was natural	Interactivity	P
The interaction method was clear and easy to understand	Usability	P
It has become possible to interact well over time	Interactivity	P
It was easier to use the controller than traditional input devices (keyboard, mouse, and joystick)	Interactivity	P
Animation (waveform of water, movement of objects) by interaction is reproduced as expected	Interactivity	P
The greater the number of interacting entities, the greater the immersion (depending on the degree of implementation)	Immersion	P
Immersion		
I did not know that the time was running while I was playing the game	Flow	P
After finishing the game play, the time apparently passed faster than expected	Flow	P
I could not figure out what was happening around me while playing the game	Immersion	P
I felt realistic in the configured virtual environment	Presence	P
At some point, I forgot that I had a controller	Immersion	P
I was very immersed in the game	Immersion	P
I needed time to immerse myself in the game	Immersion	N
I tried to get more points (a fishing scenario)	Emotion	P
The game was difficult (a fishing scenario)	Emotion	N
I wanted to stop the game in the middle (a fishing scenario)	Emotion	N
I liked the graphics and images	Emotion	P
Overall, the game was fun	Judgment	P
I want to play the game again	Judgment	P
I would like to recommend it to people	Judgment	P
The sound effect felt like it was real (a scenario with a sound)	Immersion	P
Describe what you liked about the virtual reality content	Description	—
Describe improvements to this content when compared to existing virtual reality content	Description	—

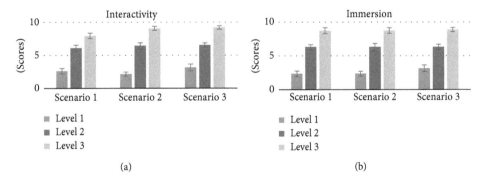

FIGURE 13: Average scoring results of the interactivity and immersion for three scenarios according to the different experience levels.

existed among the scenarios depending on the different scale properties. In the questionnaire on interactivity, Scenario 1 showed the highest values in the interactivity and usability scales. We analyzed that Scenario 1 contained a small number of immersive experience elements; thus, it provided a simple interaction style such as touching marine lives

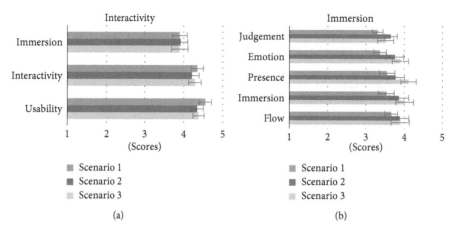

FIGURE 14: Average survey results of the interactivity and immersion of three scenarios according to scales.

underwater. From the viewpoint of usability and interactivity, we found that more mixed immersive experience was included, which showed the lower scores in Scenarios 2 and 3, compared to Scenario 1. In the questionnaire about immersion, each scenario was shown to have different characteristics depending on the scale properties. Scenario 2 showed higher judgment and flow, and Scenario 3 exhibited higher scores in the overall scales, except judgment. It shows that as complex immersive experience elements were added, the addition positively influenced the overall scales.

In addition, in the descriptive evaluation, the participants responded that it was good to create realistic user-oriented content and generate natural and realistic animation using mesh deformation. The participants also suggested that the rendering quality in some scenes should be improved and optimized.

6. Conclusions

We have developed an interactive and immersive 360° content-authoring framework for marine biology education. The developed framework is suitable for many students to study using low-cost mobile VR devices because the developed contents can be installed in smartphones or standalone VR equipment. For active participations, we presented user-created forms using additional layers, realistic image-based interactions utilizing elements in the given images, and more immersive contents using animated objects, videos, and 3D audio effects. We conducted a user study to verify the interactivity and immersion of our education scenarios using different levels of experience. As a result, we found that more experience elements improve the immersion of users and make them become more active.

Because the proposed method is processed based on images, creating contents using various low-cost 360° image-capturing devices is easy; thus, it can be utilized not only for educational contents but also for various image-based applications such as marketing and journalism. In the future, the image-based interactions can be improved by applying adaptive mesh deformation to enable more realistic animation depending on the object properties and by automatically segmenting some elements in the images based on

the contexts. We will also investigate methods to optimize the presented framework on mobile devices to use more layers and various interactions in an effective manner.

Acknowledgments

This research was supported by the MSIT (Ministry of Science and ICT), Korea, under the ITRC (Information Technology Research Center) support program (IITP-2018-2016-0-00312) supervised by the IITP (Institute for Information & communications Technology Promotion).

References

[1] J. Martin-Gutiérrez, C. E. Mora, B. Añorbe-Díaz, and A. González-Marrero, "Virtual technologies in education," *EURASIA Journal of Mathematics, Science and Technology Education*, vol. 13, no. 2, pp. 469–486, 2017.

[2] H. Hartman and A. Gerard, *User-Generated VR and 360° Imaging—An Overview*, Suite48Analytics, Oakland, CA, USA, 2016.

[3] J. Pirker, "Immersive and engaging forms of virtual learning: new and improved approaches towards engaging and immersive digital learning," Doctoral dissertation, Graz University of Technology, Graz, Austria, 2017.

[4] T. Cochrane, "Mobile VR in education: from the fringe to the mainstream," *International Journal of Mobile and Blended Learning*, vol. 8, no. 4, pp. 45–60, 2016.

[5] W. T. Lee, H. I. Chen, M. S. Chen, I. C. Shen, and B. Y. Chen, "High-resolution 360 video foveated stitching for real-time VR," *Computer Graphics Forum*, vol. 36, no. 7, pp. 115–123, 2017.

[6] F. Perazzi, A. Sorkine-Hornung, H. Zimmer et al., "Panoramic video from unstructured camera arrays," *Computer Graphics Forum*, vol. 34, no. 2, pp. 57–68, 2015.

[7] J. Kronander, F. Banterle, A. Gardner, E. Miandji, and J. Unger, "Photorealistic rendering of mixed reality scenes," *Computer Graphics Forum*, vol. 34, no. 2, pp. 643–665, 2015.

[8] T. Rhee, L. Petikam, B. Allen, and A. Chalmers, "MR360: mixed reality rendering for 360 panoramic videos," *IEEE Transactions on Visualization and Computer Graphics*, vol. 23, no. 4, pp. 1302–1311, 2017.

[9] K. Rohmer, J. Jendersie, and T. Grosch, "Natural environment illumination: coherent interactive augmented reality for mobile and non-mobile devices," *IEEE Transactions on Visualization and Computer Graphics*, vol. 23, no. 11, pp. 2474–2484, 2017.

[10] J. Lee, B. Kim, K. Kim, Y. Kim, and J. Noh, "Rich360: optimized spherical representation from structured panoramic camera arrays," *ACM Transactions on Graphics (TOG)*, vol. 35, no. 4, pp. 1–11, 2016.

[11] L. A. Neng and T. Chambel, "Get around 360 hypervideo," in *Proceedings of 14th International Academic MindTrek Conference: Envisioning Future Media Environments*, pp. 119–122, ACM, Tampere, Finland, October 2010.

[12] YouTube: https://www.youtube.com/watch?v=-xNN-bJQ4vI&t=42s.

[13] Facebook: https://www.facebook.com/DiscoveryScienceFrance/videos/1369157583160119/.

[14] A. Chalmers, J. J. Choi, and T. Rhee, "Perceptually optimized illumination for seamless composites," in *Proceedings of 22nd Pacific Conference on Computer Graphics and Applications (Pacific Graphics 2014)*, Seoul, South Korea, October 2014.

[15] D. A. Bowman and R. P. McMahan, "Virtual reality: how much immersion is enough?," *Computer*, vol. 40, no. 7, pp. 36–43, 2007.

[16] C. Dede, "Immersive interfaces for engagement and learning," *Science*, vol. 323, no. 5910, pp. 66–69, 2009.

[17] L. Freina and M. Ott, "A literature review on immersive virtual reality in education: state of the art and perspectives," in *Proceedings of International Scientific Conference eLearning and Software for Education (eLSE)*, vol. 1, Bucharest, Romania, 2015.

[18] M. Slater, "A note on presence terminology," *Presence Connect*, vol. 3, no. 3, pp. 1–5, 2003.

[19] Qualcomm, *Making Immersive Virtual Reality Possible in Mobile*, Qualcomm, San Diego, CA, USA, 2016, https://www.qualcomm.com/documents/making-immersive-virtual-reality-possible-mobile.

[20] Q. Sun, L. Y Wei, and A. Kaufman, "Mapping virtual and physical reality," *ACM Transactions on Graphics (TOG)*, vol. 35, no. 4, pp. 1–12, 2016.

[21] Z. C Dong, X. M Fu, C. Zhang, K. Wu, and L. Liu, "Smooth assembled mappings for large-scale real walking," *ACM Transactions on Graphics (TOG)*, vol. 36, no. 6, pp. 1–13, 2017.

[22] S. Tregillus and E. Folmer, "VR-STEP: walking-in-place using inertial sensing for hands free navigation in mobile VR environments," in *Proceedings of CHI Conference on Human Factors in Computing Systems*, pp. 1250–1255, San Jose, CA, USA, May 2016.

[23] Leap Motion: https://www.leapmotion.com/.

[24] Fingo: https://www.usens.com/fingo.

[25] C. Jennett, A. L. Cox, P. Cairns et al., "Measuring and defining the experience of immersion in games," *International Journal of Human Computer Studies*, vol. 66, no. 9, pp. 641–661, 2008.

[26] J. H. Brockmyer, C. M. Fox, K. A. Curtiss, E. McBroom, K. M. Burkhart, and J. N. Pidruzny, "The development of the game engagement questionnaire: a measure of engagement in video game-playing," *Journal of Experimental Social Psychology*, vol. 45, no. 4, pp. 624–634, 2009.

[27] K. Tcha-Tokey, O. Christmann, E. Loup-Escande, and S. Richir, "Proposition and validation of a questionnaire to measure the user experience in immersive virtual environments," *International Journal of Virtual Reality*, vol. 16, no. 1, pp. 33–48, 2016.

[28] W. Kim, *Marine Life: Easy Written Marine Environment and Creatures*, Daewonsa, Sancheong, South Korea, 2014.

16

CMD: A Multichannel Coordination Scheme for Emergency Message Dissemination in IEEE 1609.4

Odongo Steven Eyobu [iD],[1,2] **Jhihoon Joo** [iD],[1] **and Dong Seog Han** [iD][1]

[1]*School of Electronics Engineering, Kyungpook National University, 80 Daehak-ro, Buk-gu, Daegu 41566, Republic of Korea*
[2]*School of Computing & Informatics Technology, Makerere University, Plot 56, Pool Road, P.O. Box 7062, Kampala, Uganda*

Correspondence should be addressed to Dong Seog Han; dshan@knu.ac.kr

Guest Editor: Mohamed Elhoseny

The IEEE 1609.4 legacy standard for multichannel communications in vehicular ad hoc networks (VANETs), specifies that the control channel (CCH) is dedicated to broadcast safety messages, while the service channels (SCHs) are dedicated to transmit infotainment service content. However, the SCHs can be used as an alternative to transmit high priority safety messages in the event that they are invoked during the service channel interval (SCHI). This implies that there is a need to transmit safety messages across multiple available utilized channels to ensure that all vehicles receive the safety message. Transmission across multiple SCHs using the legacy IEEE 1609.4 requires multiple channel switching and therefore introduces further end-to-end delays. Given that safety messaging is a life critical application, it is important that optimal end-to-end delay performance is derived in multichannel VANET scenarios to ensure reliable safety message dissemination. To tackle this challenge, three primary contributions are in this article: first, a cooperative multichannel coordinator (CMD) selection approach based on the least average separation distance (LAD) to the vehicles that expect to tune to other SCHs and operates during the control channel interval (CCHI) is proposed. Second, a model to determine the optimal time intervals in which CMD operates during the CCHI is proposed. Third, a contention back-off mechanism for safety message transmission during the SCHI is proposed. Computer simulations and mathematical analysis show that CMD performs better than the legacy IEEE 1609.4 and a selected state-of-the-art multichannel message dissemination scheme in terms of end-to-end delay and packet reception ratio.

1. Introduction

Nowadays, intelligent transport systems (ITS) are one of the key drivers for the evolution of smart cities. Among the major enabling technologies to realize this evolution is vehicular communications technology (VCT). VCTs should be able to provide services such as safety on the road and in-vehicle on-demand infotainment content. The IEEE 1609.4 standard [1] is the basic technology designed to achieve and enable the implementation of both cooperative safety message dissemination and provision of infotainment services through multichannel communications. Seven 10 MHz channels have been reserved in the 5.9 GHz frequency band [2] for this purpose.

The multichannels defined therein are the control channel (CCH) and six service channels (SCHs), all operating at fixed intervals. The CCH is dedicated to broadcast safety messages while the SCHs are dedicated to transmit infotainment service content. During the CCH interval (CCHI), all vehicles must tune to the CCH unlike during the SCH interval (SCHI). Furthermore, the standard defines the continuous and alternating channel access modes. In the continuous channel access mode, vehicles tune to the CCH until they demand for a service that has been advertised. The alternating channel access mode allows vehicles to always switch between the CCH and their desired advertised SCH after an interval of 50 ms.

When the different vehicles switch to their desired SCHs during the SCHI, it limits the possibility of transmitting safety broadcast messages to all vehicles in the event of an emergency during the SCHI. This is a threat to the reliability of safety message transmission especially because further end-to-end delays are introduced. Therefore, it is necessary to design interchannel communication mechanisms across service channels which should be able to meet requirements such as minimum end-to-end delay for emergency safety message transmission and delivery.

Various studies [3–11] have proposed approaches on improving end-to-end delay performance for vehicular ad hoc network (VANET) in multichannel conditions. The major considerations in these previous studies include the following: (1) using channel coordination vehicles [6], (2) using road side units (RSUs) as coordinators [10, 11], (3) dynamic variable CCHI and SCHI [5], and (4) time slot utilization based on peer-to-peer negotiation as a multichannel coordination function [4]. A detailed review of studies [3–11] is covered in Section 2. However, for the purpose of this study, the wireless access to vehicular environments—enhanced safety message delivery approach (WSD) [6] is used for comparison with the proposed scheme. During the CCHI, in the WSD approach, each vehicle collects data including the expected SCH that the vehicles in its communication range expect to tune to during the SCHI and computes the delay in each SCH and the number of vehicles expected to tune to a given SCHI. In the event of a high priority message during the SCHI, the invoking vehicle schedules the transmission of the emergency message across all the SCHs based on a schedule determined by the SCH which has the smallest fraction of the delay divided by the number of vehicles in the SCH. This implies that in WSD, the emergency message-invoking vehicle performs the channel coordination function.

In this paper, the information collection routine during the CCHI based on the service advertisements is the same as that of the WSD except that each vehicle only collects the separation distance information between the vehicles in its communication range and the expected SCH they expect to tune to during the SCHI. We consider vehicles expecting to tune to similar specific SCHs as belonging to the same SCH cluster, and for each SCH cluster, a coordinator for each of the other SCH clusters is selected. The cooperative multichannel coordinator (CMD) selection is based on the least average separation distance (LAD). This description of our scheme was first introduced in our paper [12]. Therefore, we extend the concept by (1) detailing the proposed scheme, (2) performing an extensive literature survey of multichannel MAC schemes in VANETs, (3) proposing a Markov chain for the back-off procedure during the SCHI, (4) a mathematical analysis of end-to-end delay which incorporates a proposed model for the optimal slot length when CMD operates during the CCHI, (5) and additional end-to-end delay performance tests in single-hop blind flooding scenarios. The results of the study show that the proposed scheme has a lower end-to-end delay in both non-rebroadcast scenarios and single-hop flooding scenarios when compared to the WSD approach [6]. The original contributions of this article are summarized as follows:

(i) A multichannel coordinator selection approach based on the LAD to vehicles tuned to other SCHs with the purpose of forwarding emergency messages with minimum end-to-end delay

(ii) A Markov chain for the back-off procedure during contention for transmission of safety messages in the SCHI

(iii) A model to determine the optimal slot length in which the proposed CMD operates during the CCHI

(iv) A queueing delay model that depends on the number of vehicles within the carrier sensing range to determine the queue length

(v) A mathematical analysis of the message dissemination end-to-end delay for the proposed CMD scheme and WSD

(vi) A simulation analysis of end-to-end delay while comparing the proposed CMD scheme, WSD, and the legacy IEEE 1609.4

The remainder of this paper is organized as follows. Section 2 discusses the related works. Section 3 describes the proposed CMD system model. Section 4 describes the numerical analysis. Section 5 describes shows the simulation setup and performance analysis. Finally, the conclusion is given in Section 6.

2. Related Work

Various state-of-the-art approaches designed for multichannel VANET scenarios are discussed in this section. The review covers adaptive interval approaches and coordination based approaches used in multichannel VANETs.

Pal et al. [3] proposed to eliminate the fixed CCHI and SCHI intervals by introducing a triggered multichannel medium access control (MAC) scheme where the CCHI is triggered each time there exists an emergency message with the objective of minimizing the end-to-end-delay. Similarly, Chantaraskul et al. [13] and Wang et al. [5] also proposed approaches to dynamically adjust the CCHI based on the channel congestion condition. This approach offers a high trade-off against infotainment content delivery in environments where both safety and content delivery are highly required.

Almohammedi et al. [4] proposed an adaptive multichannel assignment and coordination (AMAC) scheme in VANETs which exploits channel access scheduling and channel switching. The channel access scheduling is done by the RSU based on the traffic conditions to guarantee that all safety messages are disseminated during the CCHI and also achieve higher throughput of the infotainment content. The AMAC scheme also uses a peer-to-peer (PNP) negotiation mechanism between service providers and users for the SCH reservations to adaptively transmit safety messages based on the CCH conditions and the traffic safety state. The PNP negotiation process results into (1) transmission of safety messages over the CCH if the traffic condition is light and (2) transmission over the SCH if the traffic condition is heavy to avoid extended end-to-end delays of safety message delivery. Transmission over the SCH involves negotiating for a time slot during the SCHI. Generally, the PNP negotiation process is an additional process in the synchronization interval (SI) and naturally extends end-to-end delays. Additionally, AMAC uses different adaptive contention windows for safety message and service message transmission in order to minimize on packet collision in the multichannel environment.

Similarly, Wang et al. [5] proposed a variable CCHI (VCI) multichannel MAC which dynamically adjusts the length ratio between the CCH and the SCH mainly for the transmission of safety messages. In the VCI approach, when wireless service advertisements (WSAs) are transmitted during the CCHI, interested nodes request the service provider to reserve a specified content transmission time interval in the SCHI within which they shall receive content. This reservation approach is quite similar to the PNP time slot negotiated for in [4]. The only difference is that in [5], the time slot is used for transmitting infotainment content while in [4], the time slot is used for transmitting safety messages.

The hidden node problem in multichannel VANETS can be minimized using the request to send (RTS)/clear to send (CTS)/data/acknowledgement (ACK) handshake. However, this causes the exposed node problem that hinders concurrent transmissions especially in dynamic environments like VANETs. In particular, SCH selection in multichannel VANETS can result into an exposed node problem hence hindering concurrent transmissions. Lee et al. [8] proposed a scheme based on piggybacking of selected SCHs in the safety message in multichannel VANETs to minimize the exposed node problem. In this case the piggybacked message acts as a coordination agent so that the exposed vehicles do not select a common SCH.

Yao et al. [9] proposed a flexible multichannel MAC (FM-MAC) protocol which allows safety messages to be broadcasted on the service channel and nonsafety messages to be transmitted on the control channel in a flexible way. The SCHI and CCHI are not adjusted dynamically but instead both are utilized for transmitting safety and nonsafety messages. In FM-MAC, finding the optimal bandwidth resource allocation was key in determining the flexibility of using both the SCHI and CCHI. The RSU in [9] performs the major coordination function by the following: (1) setting up a coordination period for the RSU to broadcast frames to all vehicles in range informing them of a contention period to transmit safety messages, (2) safety message broadcasts and SCH service reservation requests are made by vehicles, (3) the RSU as well broadcasts a scheduling period to all vehicles in its range informing them of the schedule assignments and schedule orders, (4) and finally all nonsafety messages are exchanged based on the SCH schedules and assignments which were broadcasted by the RSU. Zhao et al. [10] proposed the demand-aware MAC (DA-MAC) protocol which follows quite a similar criteria like in [9], though DA-MAC does not consider the coordination frames broadcast by the RSU in FM-MAC.

The multichannel coordination schemes in [9–11] seem attractive, but mainly depend on the RSU. It has been reported that RSUs may sometimes face unavailable grid power connection challenges [14], and hence may require being battery powered. The major issue is ensuring that they are power charged. This limitation is the reason for the advocacy of vehicle-to-vehicle (V2V) target multichannel coordination schemes.

The WSD algorithm proposed by Ghandour et al. [6] targets transmitting event driven high priority messages to all service channels with a minimized delay to its neighbours. During the CCHI, WSD operates at each node by gathering information about its neighbours through hello messages thereby forming a database comprising of the available service channels and available vehicles. In case there exists an emergency message event trigger during the SCHI, the SCH with the least average ratio of the channel average delay and the number of nodes is first tuned to by the source vehicle of the emergency message event trigger for message dissemination. SCH switching continues in the order of the least ratio until all SCHs are exhausted. The major point of interest in the WSD protocol is to disseminate information to its neighbours with minimum delay. Due to the multiple switches to different SCHs by the nearest vehicle that acts as a coordinator, WSD logically poses a large total dissemination delay in order to transmit to all the other service channels. The WSD design is based on the argument that nearer vehicles are a greater point of interest for safety.

The scheduling algorithm for high priority message dissemination (SAEMD) proposed by Joo et al. [15] operates by selecting and switching to a SCH belonging to the nearest vehicle. Similar to WSD in [6], SAEMD uses a data collection routine in the CCHI and uses the separation distance data for deciding on the nearest vehicles hence the next SCH to be tuned to for message transmission. Summarily, WSD [6] and SAEMD [15] were designed to work in multichannel WAVE conditions. However, both WSD and SAEMD provide a minimum end-to-end delay benefit in the SCH which the nearest neighbouring vehicles tunes to first. In the case where most SCHs have vehicles tuned to them, the overall total dissemination delay is expected to be larger due to the need to do multiple switching to the different SCHs. Based on WSD and SAEMD, the total end-to-end delay for emergency message dissemination in multichannel WAVE conditions needs to be improved. In our previous work [12], we presented a cooperative multicoordinator scheme (CMD) for multichannel communication in VANETs to take care of the large total dissemination delay in multiple service channels. The proposed CMD addresses multichannel communications in VANETs and uses acquired knowledge from the CCHI.

Like in Dang et al. [16], the proposed CMD advocates for the utilization of the SCH in case an emergency message is invoked towards the time the SCHI takes over in the SI. Utilization of the both the CCHI and SCHI increases the reliability of safety message broadcasting. In the proposed CMD approach, each vehicle maintains a single radio, and the channel coordinator selection approach is distance based. CMD also makes use of multiple coordinators for each SCH cluster based on the available advertised SCHs hereafter referred to as Y. Table 1 shows the comparisons of different state-of-the-art multichannel access schemes used in VANETs.

3. Cooperative Multichannel Emergency Message Dissemination Protocol (CMD)

CMD operates in vehicular multichannel communications with the goal of achieving a low end-to-end delay in the dissemination of messages throughout the entire set of

TABLE 1: Comparison of existing multichannel VANET schemes.

Scheme	Utilizes RSU for coordination?	Nodes hosting the coordination function	Switching times per coordinator
Pal et al. [3]	No	1	Y-1
Chantaraskul et al. [13]	No	1	Y-1
VCI: [5]	No	1	Y-1
AMAC: [4]	Yes	1	Y-1
Lee et al. [8]	No	1	Y-1
FM-MAC: [9]	Yes	1	Y-1
DA-MAC: [10]	Yes	1	Y-1
Li et al. [11]	Yes	1	Y-1
WSD: [6]	No	1	Y-1
SAEMD: [15]	No	1	Y-1
Proposed CMD	No	Y-1	1

Switching times refers to the number of times a coordinator node must switch to different SCHs to transmit a single emergency message until all SCHs receive the message.

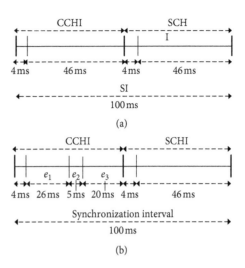

FIGURE 1: (a) Standard channel access in IEEE 1609.4. (b) The SI utilization based on CMD.

vehicles tuned to different SCHs without changing much on the IEEE 1609.4 standard. Like some of the presented multichannel approaches in [6] and [15], the CMD protocol follows the channel coordination principle where the coordinator vehicles are selected using the distance to vehicles tuned to other SCHs. A channel coordinator selection algorithm is presented later in this section.

Figure 1(a) shows the standard IEEE 1609.4 channel access, and Figure 1(b) shows the synchronization interval (SI) utilization based on the proposed CMD which can be described in the following steps:

(1) At the start of the CCHI and after the guard interval, 26 ms are used for broadcasting basic safety messages (BSMs) and advertising available services by service provider nodes. The BSMs broadcasted at this stage includes the vehicle location information and the SCH which a node will use to in order to receive nonsafety data.

(2) In the next 5 ms, using the location information received and piggybacked SCHs from the other nodes, each node calculates the average distance it has from nodes which intend to use each of the different SCHs, respectively.

(3) The calculated average distance to vehicles intending to tune to each SCH is appended to the BSM and broadcasted by each vehicle. In the last 20 ms of the CCHI, the vehicles then broadcast their BSMs. On receipt of each BSM, each vehicle compares its own average separation distances with that in the received BSM if the SCH in the BSM is the same. A node autonomously qualifies itself to be the best fit coordinator if it has the LAD compared to all the other nodes intending to use the same SCH.

(4) During the SCHI, in the event of an emergency event message transmission, the best fit vehicles with the LAD to other SCHs forward the emergency message to the vehicles which tuned to the other SCH by switching to the target SCH.

3.1. Channel Coordinator Selection.

In this subsection, the CMD protocol is described in detail and illustrated by Figure 2. Figure 2 shows three channels SCH1, SCH2, and SCH3 which were advertised during the CCHI and logically clustered to represent the vehicles tuned to the different SCHs during the SCHI. The vehicles selected the advertised SCHs in the CCHI in a random manner. Each vehicle while in the CCHI received and selected an SCH from the WAVE service advertisements (WSAs) and also received and transmitted location information together with their selected SCH. With the received location information and SCH at every instance, each receiving vehicle computes the separation distances in relation to each SCH with the objective of finding the least separation distance to vehicles expecting to tune to a specific SCH.

Considering each SCH as a cluster, the channel coordinator vehicles in each cluster are such that for all vehicles in a given cluster, they have LAD of the connectivity to nodes in another SCH compared to the other vehicles it will share with the same SCH. If Y SCHs were advertised, then there should exist $Y - 1$ SCH coordinators in each cluster. Table 2 describes the notations used in formulating the channel coordinator selection approach. The channel coordinator selection model can be formulated as

$$\exists C_{k\_z} \in SCHk \text{ s.t. } D_{c\_z} \leq d_{i\_z} \quad \forall d_{i\_z}, \quad i = 1, 2, \ldots, m, \\ k = 1, 2, 3, \ldots, 6, \tag{1}$$

where

$$d_{i\_z} = \frac{d_1 + d_2 + \cdots + d_m}{m} \quad \text{for} \quad z = 1, 2, 3, \ldots, 6. \tag{2}$$

Each vehicle keeps the broadcasted SCHz and its associated d_{i_z} in its coordination fitness information base (CFIB) as seen in Table 3. After the d_{i_z} calculation stage by each receiving vehicle, each vehicle again broadcasts its local d_{i_z} and is received through the periodic broadcast BSM. Each incoming d_{i_z}'s is compared with the local d_{i_z}'s as long as the SCHz is the same. The comparison is such that

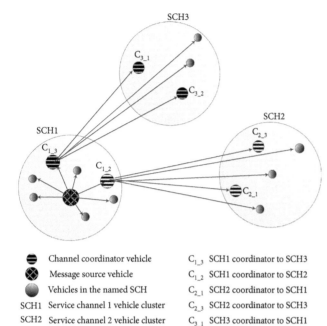

Channel coordinator vehicle	C_{1_3} SCH1 coordinator to SCH3
Message source vehicle	C_{1_2} SCH1 coordinator to SCH2
Vehicles in the named SCH	C_{2_1} SCH2 coordinator to SCH1
SCH1 Service channel 1 vehicle cluster	C_{2_3} SCH2 coordinator to SCH3
SCH2 Service channel 2 vehicle cluster	C_{3_1} SCH3 coordinator to SCH1
SCH3 Service channel 3 vehicle cluster	C_{3_2} SCH3 coordinator to SCH2

FIGURE 2: A logical view of the CMD structure with the emergency message generated from SCH1 and broadcasted to its members and then relayed by the channel coordinators to SCH3 and SCH2.

TABLE 2: Notations used in channel coordinator selection.

Acronym	Description
k	An advertised SCH which a vehicle intends to switch to during the SCHI
z	Any other advertised SCH apart from the one which a given vehicle intends to switch to during the SCHI
c	An SCH coordinator vehicle
c_{k_z}	The channel coordinator for forwarding messages from SCHk to SCHz
Y	The number of advertised SCHs to provide nonsafety services
m	The number of vehicles expecting to switch to SCHz
d_i	The V2V separation distance. $i = 1, \ldots, m$
d_{i_z}	The average d_i for a given vehicle considering the vehicles expecting to switch to SCHz
D_{c_z}	The LAD for the coordinator vehicle in SCHk to SCHz
f_{SCHz}	The coordination fitness value for a given vehicle considering the d_{i_z} to SCHz

TABLE 3: Coordination fitness information base.

Gossiped SCHz	Average d_{i_z}	f_{SCHz}
SCH 1	d_{i_1}	≥ 1
SCH 2	d_{i_2}	≥ 1
SCH 3	d_{i_3}	≥ 1
SCH 4	d_{i_4}	≥ 1
SCH 5	d_{i_5}	≥ 1
SCH 6	d_{i_6}	≥ 1

when d_{i_z} is the least among the incoming d_{i_z}'s for the common SCHz, then the coordination fitness (CF) is 1. Implying that the vehicle i has the least average distance to

SCHz and hence is the service coordinator of its SCH to SCHz. Generally, the value of CF is determined based on the order of greatness of d_{i_z}. That is, the least d_{i_z} has CF equals to 1 and the greatest d_{i_z} has CF equals to m. For clarity, the CF range is $i = 1, 2, 3, \ldots, m$. The least d_{i_z} which represents the coordinators' average distance is then represented as D_{c_z} for purposes of clarity as seen in Equation (1).

Again, as seen in Table 3, the CF value in SCHk is represented as f_{SCHz}. The general representation in Figure 2 shows the vehicle coordinators $C_{k\_1}, C_{k\_2}, C_{k\_3}$ which have the least CF values to the advertised SCHs. It should however be noted that although C_{2_3} and C_{3_2} are represented a SCH coordinators in Figure 2, only C_{1_3} and C_{1_2} are functionally operational as channel coordinators because the emergency message is triggered in SH1. Algorithm 1 elaborates on the CMD channel coordinator selection procedure.

3.2. Challenges in the Proposed CMD. In the CCHI, while transmitting BSMs containing the average separation distance to other vehicles, conditions such as the hidden node problem and shadowing may hinder the BSM delivery to some vehicles. In such a case, more than one vehicle may assume the position of the channel coordinator to a given SCH cluster. During the SCHI, it is also possible that a channel coordinator vehicle may not receive an emergency message from the affected source vehicle due to the hidden node problem.

This is a prominent problem in single-hop broadcast scenarios. To alleviate this reachability problem, the single-hop blind flooding based approach of broadcasting was implemented, and simulation results are shown later in Section 5.4 to describe its impact on delay in each WAVE channel. In single-hop blind flooding, when vehicles receive a message, they rebroadcast it only once. That is, the vehicles receiving the rebroadcasted message do not broadcast the retransmitted message. A comparison of the proposed CMD with WSD is also done for the single-hop flooding scenario.

Again, by applying CMD, it is possible that only one in a given SCH may qualify to be the channel coordinator to all other SCHs by having the LAD to all advertised SCHs. This scenario exists when one node is isolated from its SCH members yet near to all the other SCH cluster members. Another issue about CMD is that when a cluster has less than $Y - 1$ members, then some members will act as coordinators for more than one SCH. These two mentioned scenarios would cause an increase in the total dissemination delay because such coordinators will have to switch between multiple channels.

3.3. Proposed Back-Off Model for Emergency Message Transmission during the SCHI. Figure 3(a) represents the standard back-off process to be adopted in the CCHI and for nonsafety data transmission in the SCHI. The Markov chain proposed and presented in Figure 3(b) operates in the

(1) **while** in CCHI vehicles receive WSA's and broadcast their location information
(2)   Select an SCH to be tuned to
(3)   Append selected SCH and location information to all BSM's and broadcast
(4)   **while** periodic safety messages are received
(4a)   **for** each vehicle
(4b)   **for** each SCH advertised
(4c)   Compute d_{i_z}
(4d)   Append d_{i_z} to the BSM and then broadcast
                end for
(4e)   **if** (BSM is received) **then**
                    for each similar SCHz
(4f)   **if** (all the d_{i_z} values are greater than the local average d_{i_z}) **then**
(4g)   Vehicle is the channel coordinator C_{k_z}
(4h)   **else**
(4i)   Vehicle is just a member of its selected SCHk cluster.
(4j)   **end if**
                    end for
(4k)   **end if**
(4l)   **end for**
(4m)   **end while**
(5) **end while**

ALGORITHM 1: Channel coordinator selection algorithm.

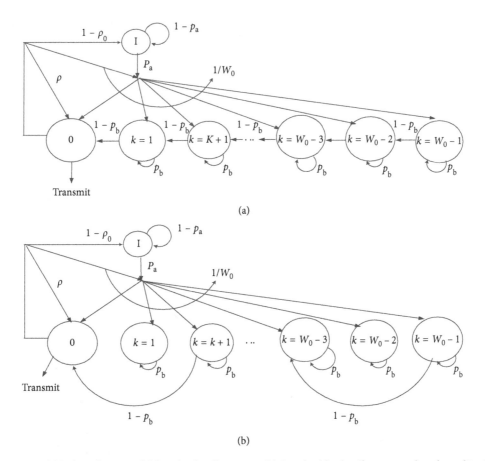

FIGURE 3: One-dimensional Markov chain model for a back-off instance. (a) Standard back-off process to be adopted in the CCHI and for nonsafety data transmission in the SCHI. (b) Proposed back-off process for emergency safety message transmission during the SCHI.

SCHI showing the back-off process when an emergency message is invoked. Safety emergency messages are considered high priority messages during the SCHI; therefore, the model design is tailored to minimize their contention delay. In the standard back-off criteria, waiting state transitions are marked by uniformly reducing contention window sizes.

In the proposed criteria seen in Figure 3(b) the same phenomenon is followed but the size of the reducing contention window (RCW) is two times the size of the RCW compared to when transmitting WSAs, safety messages in the CCHI, and data services during the SCHI.

Let $s_i(t)$ and $b_i(t)$ represent the back-off stage and the back-off counter, respectively at time t. Hence, the state of the Markov chain can be expressed as a two-tuple $\{s_i(t), b_i(t)\}$, and the back-off state of the high priority emergency messages can be simplified as a one-tuple $\{b_i(t)\}$ for $s_0(t) \equiv 0$. Table 4 defines all the probabilities shown in the Markov chains in Figures 3(a) and 3(b). Each of the one-time transition probabilities in Figure 3(b) is described below:

(i) The idle state $\{I\} \longrightarrow$ the back-off state $\{0\}$: node transmits a packet if the channel is sensed as idle: $P\{0|I\} = p_a$

(ii) The idle state $\{I\} \longrightarrow$ the state back-off $\{k\}$: this occurs if a new packet arrives in the queue: $P\{k|I\} = p_a/W_i, k \in (0, W_0 - 1)$

(iii) The back-off state $\{k\} \longrightarrow$ the state back-off $\{k\}$: occurs if the channel is sensed to be busy and in this case the back-off counter freezes: $P\{k|k\} = p_b, k \in (1, W_0 - 1)$

(iv) The back-off state $\{k+2\} \longrightarrow$ the state back-off $\{k\}$: if the channel is sensed to be idle, the back-off counter decrements by two steps: $P\{k|k+2\} = 1 - p_b, k \in (0, W_0 - 2)$

(v) The back-off state $\{0\} \longrightarrow$ the idle state $\{I\}$: node returns to idle state if it has no packet to send: $1 - \rho$.

(vi) The back-off state $\{0\} \longrightarrow$ the idle state $\{k\}$: nodes start back-off procedure if at least one packet is in the queue: $P\{k|0\} = \rho_0/W_0, k \in (0, W_0 - 1)$

In summary, the one-step transition probabilities are as follows:

$$\begin{cases} P\{0 \mid I\} = p_a, \\ P\{k \mid I\} = p_a/W_i, \quad k \in (0, W_0 - 1), \\ P\{k \mid k\} = p_b, \quad k \in (1, W_0 - 1), \\ P\{k \mid k+2\} = 1 - p_b, \quad k \in (0, W_0 - 2), \\ P\{I \mid 0\} = 1 - \rho_0, \\ P\{k \mid 0\} = \rho_0/W_0, \quad k \in (0, W_0 - 1). \end{cases} \tag{3}$$

The stationary distribution of the Markov chain is defined as

$$b_0 = \lim_{t \longrightarrow \infty} P\{b(t) = k\}, \quad k \in (0, W_0 - 1). \tag{4}$$

TABLE 4: Notations.

Acronym	Description
λ	The packet arrival rate
μ	Average service rate of the queue in packets per second
ρ	The probability that at least one packet is in the queue $= \lambda/\mu$
p_b	Is the back-off blocking probability
p_a	The packet arrival probability $= 1 - e^{-\lambda\sigma}$
W_0	Contention window size for back-off
k	Current window size state as an effect of exponential back-off
I	Idle state
β	Traffic density
d_0	The reference distance used in calculating the received signal strength at a particular distance
$P_r(.)$	The received signal strength at specified distance
d_c	The critical distance that refers to the distance where the first Fresnel zone touches the ground and is also referred to as the Fresnel distance
γ_1	Path loss exponent
γ_2	Path loss exponent
h_T	Transmitter height
h_R	Receiver height
ψ	Electromagnetic wavelength fixed at 5.9 GHz
B	The number of vehicles in carrier sensing range
$X_{\sigma 1}$	The zero mean, normally distributed random variables with standard deviation $\sigma 1$
$X_{\sigma 2}$	The zero mean, normally distributed random variables with standard deviation $\sigma 2$
L_{CS}	The carrier sensing range defined as the average distance for a node to detect the other nodes transmissions
c_{th}	The carrier sensing threshold which indicates the receive sensitivity of the radio and is a constant and radio dependent

Given the one-step probabilities, the stationary probabilities can be expressed as

$$b_k = \frac{(W_0 - k)}{W_0(1 - p_b)}b_0,$$
$$b_I = \frac{(1 - \rho)}{p_a}b_0. \tag{5}$$

The sum of the stationary probabilities for the states should be equal to one, therefore,

$$\frac{(W_0 - 1)}{W_0(1 - p_b)}b_0 + \frac{(1 - \rho)}{p_a}b_0 = 1,$$
$$b_0 = \left[\frac{(W_0 + 1)}{2(1 - p_b)} + \frac{1 - \rho}{p_a}\right]. \tag{6}$$

Since transmission occurs when the back-off counter value $k = 0$, the transmission probability τ can be defined as

$$\tau = b_0 = \left[\frac{(W_0 + 1)}{2(1 - p_b)} + \frac{1 - \rho}{p_a} \right], \qquad (7)$$

τ is very important as it is later used in the end-to-end delay analysis seen in the next section.

4. End-to-End Delay Analysis

The key performance indicator in this study is end-to-end delay. The goal of this section is to numerically derive the end-to-end delay while considering the mechanism of the proposed CMD scheme. Generally, the performance of the proposed CMD depends on the communication performance during the 26 ms of transmitting the location information and then the 20 ms of sharing the average separation distances to determine the SCH coordinators. The two decision time slots (26 ms and 20 ms) in this article from now onwards shall be referred to as e_1 and e_3 respectively as shown in Figure 1(b).

Most importantly, all or most of the vehicles should transmit their information within e_1 and e_3 for the channel coordinator selection to be efficient. Therefore, one eminent optimization parameter in this problem is the length of e_1 and e_3 which we believe should depend on the length of an arbitrary time slot T_{slot} existing during the interval e_1 and e_3. And since T_{slot} is one parameter that determines the end-to-end delay of a transmission, and we start by defining the end-to-end delay $E[d]$ model as follows:

$$E[d] = E[q] + E[c] + E[t], \qquad (8)$$

where $E[q]$, $E[c]$, and $E[t]$ represent the average queueing delay, average contention delay, and average transmission delay, respectively.

4.1. Contention Delay Model. The average contention $E[c]$ is defined as

$$E[c] = E[CW] = \left(\frac{CW_{min} - 1}{2} \right) T_{slot}, \qquad (9)$$

where $E[CW]$ is the average contention window size. The size of T_{slot} is relevant for the derivation of the optimal period for e_1 and e_3 for the proposed CMD. Finding T_{slot} requires that (1) we define the stationary probability that a node transmits a BSM in the arbitrary time slot T_{slot} and (2) the time it takes to yield a successful transmission $T_{success}$, collision time T_{coll}, and the idle time σ.

By using the transmission probability τ, the following probabilities can be found:

$$p_{idle} = (1 - \tau)^N,$$
$$p_{busy} = 1 - p_{idle},$$
$$p_{success} = N\tau(1 - \tau)^{N-1}, \qquad (10)$$
$$p_{coll} = 1 - p_{idle} - p_{success},$$

where p_{idle} is the probability that a channel is in an idle state and not being utilized, p_{busy} is the probability that

a transmission is occupying the channel, $p_{success}$ is the probability of having a successful transmission, and p_{coll} is the probability of having a collision in the channel.

The transmission time slot duration T_{slot} is defined as

$$T_{slot} = (1 - p_{busy})\sigma + T_{success} \cdot p_{success} + T_{coll} \cdot p_{coll}, \qquad (11)$$

where σ is the duration of an empty slot. $T_{success}$ is the time required for a successful transmission, and T_{coll} is the average time of a collision event:

$$T_{success} = \text{DIFS} + \sigma + E[t],$$
$$T_{coll} = \text{EIFS} + \sigma + E[t]. \qquad (12)$$

The average transmission delay can be expressed as $E[t] = S/R$, with S representing the message size and R representing the data rate, respectively. DIFS and EIFS are the distributed coordination function interframe space time and extended interframe space time, respectively.

4.2. Optimal Slot Period Allocation Model. At this stage, since T_{slot} has been mathematically defined by Equation (11), the task is now to define the optimal period of that each of e_1 and e_3 slot shall take. In other words, we need to find how many T_{slot}'s should exist in either the 1st or 2nd time slot to enable sufficient coordination selection functionality.

The objective to achieve during e_1 and e_3 is to have most or all of the vehicles to transmit their location, desired SCH and LAD information. In this article, we consider that e_1 and e_3 period should just be long enough to allow all the vehicles denoted by B within the carrier sensing range to transmit their information. The duration V representing either e_1 or e_3 can therefore be defined as

$$V = B \times T_{slot}. \qquad (13)$$

In this article, we define the number of vehicles B in carrier sensing range based on [9] as

$$B = 2\beta L_{cs}. \qquad (14)$$

L_{cs} is given by

$$L_{cs} = \begin{cases} E\left[d_0 10^{(P_r(d_0) - c_{th} + X_{\sigma 1})/10_{\gamma 1}} \right], & d_0 \le L_{cs} \le d_c, \\ E\left[d_0 10^{(P_r(d_0) - 10_{\gamma 1} \log_{10}(d_c/d_0) - c_{th} + X_{\sigma 2})/10_{\gamma 1}} \right], & L_{cs} > d_c, \end{cases} \qquad (15)$$

d_c can be calculated as $d_c = 4h_T h_R / \psi$.

4.3. Queueing Delay Model. In this paper, the queueing delay $E[q]$ is formulated considering that a VANET communication system is best modeled as an M/M/1/B queueing system [17]. In this case, the arrivals are considered to be distributed exponentially through a Poisson process, the service times are exponentially distributed and independent of each other, and a single communication channel acting as a server has a finite queue length B. Where we define B in this article as the number of vehicles within the carrier sensing range. Based on Equation (14), B can be calculated. The expected queue length can therefore be calculated as

$$E[b] = \frac{\rho}{1 - \rho^{B+1}} \cdot \left(\frac{1 - \rho^B}{1 - \rho} - B\rho^B \right). \qquad (16)$$

Using Little's law, the queueing delay can be represented as

$$Q_d = \frac{E[b]}{\lambda(1 - P_B)}, \qquad (17)$$

where P_B is the probability that the queue is full and $\lambda(1 - P_B)$ represents the effective arrival rate which the packets are put into the queue. When $\rho = (\lambda/\mu) \neq 1$, the queueing delay is defined as

$$E[q] = Q_d = \frac{E[b]}{\lambda\left(1 - (1 - \rho/1 - \rho^{B+1}) \cdot \rho^B\right)} = \frac{1}{\mu - \lambda} - \frac{1}{\mu} \cdot \frac{B\rho^B}{1 - \rho^B}, \qquad (18)$$

when

$$\rho = 1,$$

$$Q_d = \frac{E[b]}{\lambda\{1 - [1/(B+1)]\}} = \frac{(B+1)}{2\lambda} = \frac{(B+1)}{2\mu}. \qquad (19)$$

At this stage, all the parameters for numerically finding $E[d]$ using Equation (8) can be computed.

5. Simulation

5.1. Mobility Model and Network Simulator. The Manhattan model is used to emulate the movement pattern of vehicle nodes on streets defined by a map. The map is composed of a number of horizontal and vertical streets. Each street has one lane. The mobile vehicle node moves along the horizontal and vertical grids on the map. At an intersection of horizontal and vertical streets, the mobile node can turn left, right, or goes straight. This choice is probabilistic. The vehicle turn probability is set to 0.5. We consider a two-dimensional 1,500 m by 1,500 m fully connected road network in a Manhattan grid with vehicles moving at a mean speed of 40 km/h. The grid offers a total of 6 km for vehicular motion for the single-lane scenario. Our mobility trace for the vehicles is generated using BonnMotion-2.1.3.

To analyze the performance of CMD, we simulated its system dynamics with the NS-3 simulator, version ns-3-dev. Table 5 summarizes the general simulation parameters, and Table 6 defines the simulation performance metrics.

5.2. End-to-End Delay. In a typical VANET scenario, not all vehicles may demand for the advertised infotainment services. This means that not all SCHs will be utilized during the SCHI. In Figure 4, the total end-to-end dissemination delay is shown for WSD, IEEE 1609.4, and the proposed CMD. Only 5 SCHs were advertised during the CCHI.

Observations show that the proposed CMD maintains lower total end-to-end delays compared to WSD and the legacy IEEE 1609.4 when more than two SCHs are utilized during the SCHI. This observation is true for both the analytical and simulation results. In the legacy IEEE 1609.4,

TABLE 5: Simulation parameters.

Description	Value
Message payload size S	200 bytes
Fading model	Nakagami
Packet interval	100 ms
Data rate R	3 Mbps
Content window size: Min, max	15, 256
Slot time σ	16 μs
Arbitrary interframe space number (AIFSN)	2
Short interframe space (SIFS) time	32 μs
Antenna height	1.5 m
Frequency	5.9 GHz
Transmitter and receiver gain	3 dB
Number of vehicles	50
Vehicle speed	40 m/s
Vehicle mobility model	Manhattan-grid highway

TABLE 6: Simulation performance metrics.

Metric	Description
End-to-end delay	The safety message dissemination single-hop delay
Packet reception ratio (PRR)	The percentage of nodes that successfully receive a packet from a tagged node given that all the receivers are within the transmission range of the sender at the moment that the packet is sent out [18]
Packet transmission ratio (PTR)	The percentage of nodes that successfully transmit a packet given the prevailing contention for channel access

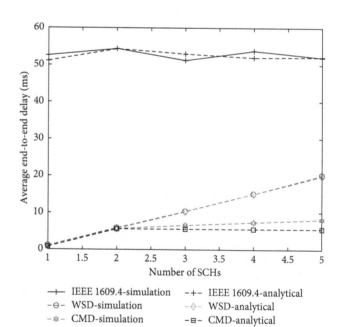

FIGURE 4: Analytical and simulation results of average end-to-end delay versus number of channels.

a vehicle with an emergency message during the SCHI must wait for the CCHI in order to transmit an emergency message. This is the major cause for the much end-to-end delay exhibited by the legacy IEEE 1609.4 system. The better performance realized by the CMD is the effect of using multiple coordinators whereby each coordinator switches to a specific SCH in order to relay a BSM during the SCHI. In WSD, only one channel coordinator is used, hence the need for multiple channel switching in order to relay the BSM to all the SCHs. Therefore, there is an additional delay introduced by the multiple switching and the transmission delays.

The slight differences seen in the theoretical and simulation results are a result of the system dynamics used in generating the results both in theory and in the simulation. In WSD, the theoretical results are generated based on the derivation of a single channel end-to-end delay $E[d]$. We then use the number of SCHs Y as a factor to fix the multichannel condition to find the total message dissemination end-to-end delay T_d as follows:

$$T_d = \begin{cases} E[d], & Y = 1, \\ YE[d], & Y > 1. \end{cases} \quad (20)$$

In the proposed CMD, the theoretical T_d is defined by

$$T_d = \begin{cases} E[d], & Y = 1, \\ 2E[d], & Y > 1. \end{cases} \quad (21)$$

In the simulation, the frequency of each of the SCHs defined by the WAVE standard is different. This has an impact on the end-to-end delay results thus causing the slight differences observed between the theoretical and simulation results. It should be noted that the final T_d represented in the results of Figures 4–6 includes the switching delay where multiple channels are involved. Theoretically, the switching delay was arbitrarily fixed at 2 ms.

5.3. PRR and PTR. The proposed CMD first operates during the CCHI within the time durations, e_1, e_2, and e_3. During the time durations e_1 and e_3, it is important that all or most vehicles transmit and receive the BSMs in order to enable efficient channel coordinator selection. Therefore, Figure 7 is shown to provide an understanding of the PRR and the PTR during the time intervals e_1 and e_3.

It is observed in Figure 7 that as the slot duration of e_1 or e_3 increases, the PRR and PTR also increases. Generally, an increase in the slot duration gives room for more contending nodes to transmit as the available transmission time slots σ would also increase.

Figure 8 represents the PRR and PTR realized when the proposed optimal e_1 model is used. The optimal length in time for e_1 is 8.38 ms given the simulation scenario and settings seen in Table 7. The parameter settings seen in Table 6 are based on realistic channel measurements which were attained in [19].

The key observation in Figures 7 and 8 is that e_1 values greater than 8.38 ms result into relatively the same PRR and PTR values with insignificant differences. This therefore means that lengthening e_1 or e_3 beyond 8.38 ms would simply be a waste in the CCHI.

Figure 5 represents the PRR attained against the total end-to-end delay achieved when transmitting a BSM over single and multiple SCHs. The result shows that the

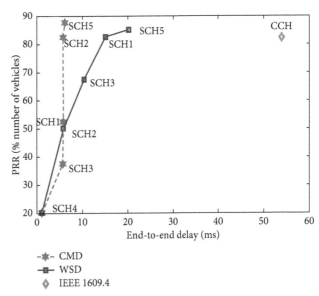

FIGURE 5: PRR versus end-to-end delay: understanding the BSM proliferation rate across various channels.

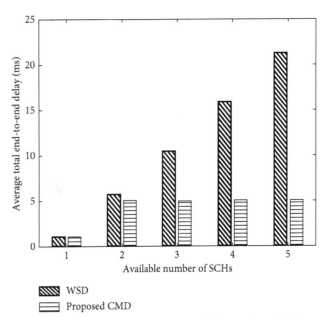

FIGURE 6: Average total dissemination delay in the single-hop flooding scenario given varying numbers of available SCHs.

proposed CMD offers a greater PRR within a shorter end-to-end delay compared to the WSD and IEEE 1609.4 legacy system especially when considering total coverage of all SCHs with the BSM. The order of the SCH switching represented in Figure 5 for each approach depends on the channel switching dynamics of each.

At about 6 ms, CMD covered slightly over 50% of the vehicles and served 3 SCHs while WSD served lesser. The good performance exhibited by CMD is based on the multicoordinator functionality in a scenario where multiple services are demanded and offered by different SCHs. It is important to note again that the IEEE 1609.4 would wait for

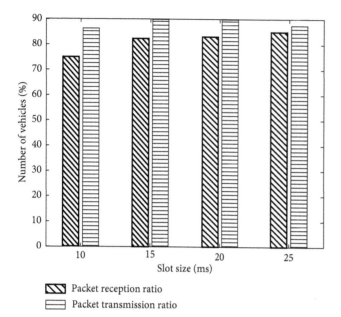

FIGURE 7: PRR and PTR simulation results for various sizes of e_1.

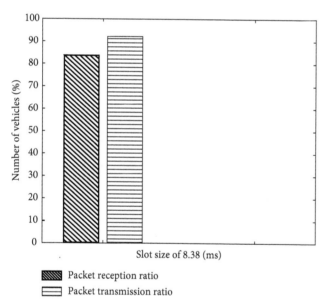

FIGURE 8: PRR and PTR simulation results based on the derived optimal e_1 interval.

TABLE 7: Parameter settings for optimal e_1 determination.

Description	Value
d_0	10 m
$P_r(d_0)$	−60 dB
c_{th}	−85 dB
$X_{\sigma 1}$	5.6 dBm
γ_1	1.9
β	25 vehicles/km

the CCHI to transmit BSMs in case of an emergency during the SCHI. It is for this reason that the end-to-end delay for the legacy system is not better than CMD and WSD.

5.4. Improving Reachability for Reliability by Single-Hop Blind Flooding. In order to provide insights on how to alleviate the hidden node problem which can be a hindrance to the effectiveness of the proposed approach during the channel coordinator selection process, we have implemented the single-hop bind flooding approach well knowing that blind flooding approaches introduce the broadcast storm problem [20] which may affect the end-to-end delay.

The purpose of experimenting the single-hop blind flooding (SHBF) is to provide an understanding that even though using SHBF introduces further end-to-end delays, it can be used as a factor in further determining the optimal size of e_1 and e_3 with the benefit of having a higher reachability during e_1 and e_3.

However, in this study, we have not divulged into further formulating another model for determining the optimal size of e_1 and e_2 based on the SHBF end-to-end delay results. We only present SHBF-based results.

Figure 9 shows the cumulative distribution function of the reachability in both flooding and no flooding conditions in the CCHI given a period of 8 ms. The results captured in Figure 9 are for the first SI in our simulation experiment particularly to understand the influence of the number of vehicles in the simulation playground especially given the fact that the vehicle node generation in the simulation is based on a Poisson process.

Four sections of reachability for analysis can be observed in Figure 9. These are between 0 and 10%, between 10% and 38%, between 38% and 68%, and >68%.

The reachability range between 0% and 10% is realized during the starting period of the SI when few vehicle nodes have been ushered into the simulation environment based on a Poisson process. It can be observed that the no-flooding scenario offers a better reachability compared to the SHBF scenario. This is because at the start, there are few vehicles which are all able to be reached and therefore, introducing the SHBF simply causes unnecessary contention.

As the number of vehicles increases in the simulation environment, the sparsity of the vehicles is larger given the vehicle mobility. This sparsity leads to reduced reachability. This can be observed between 10% and 38% where the SHBF scenario offers a better reachability compared to the no-flooding scenario.

The number of vehicles in the simulation environment increases to a point whereby there is a level of stability in the reachability which can be observed between 38% and 68%. This stability scenario is true for both the SHBF and the nonflooding scenario. This means that SHBF has no effect in the CMD process in dense vehicular scenarios.

After 68% reachability is achieved, using the SHBF scenario does not offer better reachability results because of the broadcast storm. At this moment, all vehicles are in the playground of the simulation environment.

We can generally affirm from the observations that the SHBF is indeed suitable to improve on reachability in sparsely dense vehicular scenarios as seen in the region between 10% and 38%. Therefore, the SHBF is useful in the CMD process in sparsely dense vehicular scenarios.

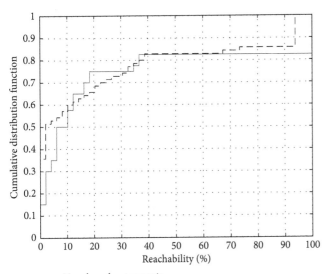

FIGURE 9: Cumulative distribution function of the percentage number of vehicles receiving message transmission during the CCHI.

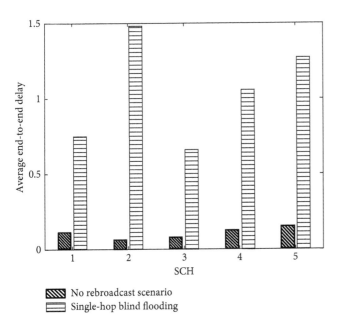

FIGURE 10: Average dissemination delay in each channel while comparing the blind flooding scenario with the non-rebroadcast scenario at each SCH.

To investigate the effect of flooding on delay, the single-hop blind flooding was implemented in five WAVE SCHs with the objective that during the SCHI, there should be a higher guarantee of emergency message delivery to the channel coordinator once invoked by any vehicle.

By observing Figure 10, it is clear that the single-hop blind forwarding introduces a further delay in the message dissemination time compared to when no blind flooding is applied.

Observations in Figure 6 also indicate that as a result of single-hop blind flooding, the average total dissemination end-to-end delay over multiple channels will also increase compared to what was earlier realized in Figure 5 when no flooding was applied. However, it is worth noting that in scenarios of no flooding and single-hop blind flooding, CMD still exhibits a delay lesser than WSD which is desirable for our design goal.

The negative impact of single-hop blind flooding observed in Figures 10 and 6 imply that a good minimum delay flooding mechanism once utilized would further improve the performance of our proposed CMD protocol in the process of disseminating BSMs.

6. Conclusion

In this paper, we proposed a cooperative multichannel message dissemination scheme called CMD for safety message dissemination in the IEEE 1609.4 standard with the goal of improving on the reliability of safety messaging in multichannel scenarios. In order to achieve this, a cooperative SCH coordinator selection approach was developed. The SCH coordinator selection is based on the vehicle which has the LAD to vehicles that expect to tune to other SCHs and operate during the CCHI.

In order to improve on the efficiency of the channel coordinator selection process during the CCHI, a model to determine the optimal slot duration was developed. A channel contention back-off Markov model was developed to operate during the SCHI in order to improve on the transmission of high priority safety messages in the event that they are invoked. Additionally, a queueing delay model that depends on the number of vehicles within the carrier sensing range was proposed and developed to determine the queue length.

Through mathematical and simulation analysis, the proposed CMD achieves lower end-to-end delay and PRR compared to the legacy IEEE 1609.4 system and WSD, which is one of the state-of-the-art multichannel schemes for WAVE.

Acknowledgments

This work was partially supported by Basic Science Research Program through the National Research Foundation of Korea (NRF) funded by the Ministry of Education (NRF-2016R1D1A3B03934420) and the Korea Institute for Advancement of Technology (KIAT) grant funded by the Korean Government (MOTIE) (no. P0000535, multichannel telecommunications control unit and associated software).

References

[1] IEEE, "IEEE standard for wireless access in vehicular environments (WAVE)-multi-channel operation," IEEE 1609.4-2010, 2010.

[2] J. B. Kenney, "Dedicated short-range communications (DSRC) standards in the United States," *Proceedings of the IEEE*, vol. 99, no. 7, pp. 1162–1182, 2011.

[3] R. Pal, A. Prakash, and R. Tripathi, "Triggered CCHI multichannel MAC protocol for vehicular ad hoc networks," *Vehicular Communications*, vol. 12, pp. 14–22, 2018.

[4] A. A. Almohammedi, N. K. Noordin, A. Sali, F. Hashim, and M. Balfaqih, "An adaptive Multi-Channel assignment and coordination scheme for IEEE 802.11 P/1609.4 in vehicular Ad-Hoc networks," *IEEE Access*, vol. 6, pp. 2781–2802, 2018.

[5] Q. Wang, S. Leng, H. Fu, and Y. Zhang, "An IEEE 802.11 p-based multichannel MAC scheme with channel coordination for vehicular ad hoc networks," *IEEE Transactions on Intelligent Transportation Systems*, vol. 13, no. 2, pp. 449–458, 2012.

[6] A. J. Ghandour, M. Di Felice, H. Artail, and L. Bononi, "Dissemination of safety messages in IEEE 802.11 p/WAVE vehicular network: analytical study and protocol enhancements," *Pervasive and Mobile Computing*, vol. 11, pp. 3–18, 2014.

[7] R. Huang, J. Wu, C. Long, Y. Zhu, B. Li, and Y. B. Lin, "SPRCA: distributed multisource information propagation in multichannel VANETs," *IEEE Transactions on Vehicular Technology*, vol. 66, no. 12, pp. 11306–11316, 2017.

[8] D. Lee, S. H. Ahmed, D. Kim, J. Copeland, and Y. Chang, "Distributed SCH selection for concurrent transmissions in IEEE 1609.4 multi-channel VANETs," in *Proceedings of 2017 IEEE International Conference on Communications (ICC)*, pp. 1–6, Paris, France, 2017.

[9] Y. Yao, K. Zhang, and X. Zhou, "A flexible Multi-Channel coordination MAC protocol for vehicular ad hoc networks," *IEEE Communications Letters*, vol. 21, no. 6, pp. 1305–1308, 2017.

[10] H. Zhao, K. Gao, M. Zhang, D. Li, and H. Zhu, "A demand-aware transmission optimization control scheme based on multichannel coordination," in *Proceedings of 2017 International Conference on Cyber-Enabled Distributed Computing and Knowledge Discovery (CyberC)*, pp. 373–377, Nanjing, China, October 2017.

[11] X. Li, B. J. Hu, H. Chen, G. Andrieux, Y. Wang, and Z. H. Wei, "An RSU-coordinated synchronous multi-channel MAC scheme for vehicular ad hoc networks," *IEEE Access*, vol. 3, pp. 2794–2802, 2015.

[12] O. S. Eyobu, J. Joo, and D. S. Han, "Cooperative multi-channel dissemination of safety messages in VANETs," in *Proceedings of 2016 IEEE Region 10 Conference (TENCON)*, pp. 1867–1870, Singapore, 2016.

[13] S. Chantaraskul, K. Chaitien, A. Nirapai, and C. Tanwongvarl, "Safety communication based adaptive multi-channel assignment for VANETs," *Wireless Personal Communications*, vol. 94, no. 1, pp. 83–98, 2017.

[14] K. Tweed, "Why cellular towers in developing nations are making the move to solar power," *Scientific American*, 2013.

[15] J. Joo, H. Lee, and D. S. Han, "SAEMD: a scheduling algorithm for emergency message dissemination in vehicular ad hoc networks," in *Proceedings of 2014 Sixth International Conference on Ubiquitous and Future Networks (ICUFN)*, pp. 501–504, Shanghai, China, 2014.

[16] D. N. M. Dang, C. S. Hong, S. Lee, and E. N. Huh, "An efficient and reliable MAC in VANETs," *IEEE Communications Letters*, vol. 18, no. 4, pp. 616–619, 2014.

[17] J. Li and C. Chiga, "Delay-aware transmission range control for VANETs," in *Proceedings of 2010 IEEE Global Telecommunications Conference GLOBECOM 2010*, pp. 1–6, Houghton, MI, USA, December 2010.

[18] X. Ma, X. Chen, and H. H. Refai, "On the broadcast packet reception rates in one-dimensional MANETs," in *Proceedings of IEEE GLOBECOM 2008-2008 IEEE Global Telecommunications Conference*, pp. 1–5, New Orleans, LA, USA, November 2008.

[19] L. Cheng, B. Henty, D. Stancil, F. Bai, and P. Mudalige, "Mobile vehicle-to-vehicle narrow-band channel measurement and characterization of the 5.9 GHz dedicated short range communication (DSRC) frequency band," *IEEE Journal on Selected Areas in Communications*, vol. 25, no. 8, 2007.

[20] Y. Yi, M. Gerla, and T. J. Kwon, "Efficient flooding in ad hoc networks using on-demand (passive) cluster formation," *Contract*, vol. 14, p. 0016, 2002.

Permissions

All chapters in this book were first published in MIS, by Hindawi Publishing Corporation; hereby published with permission under the Creative Commons Attribution License or equivalent. Every chapter published in this book has been scrutinized by our experts. Their significance has been extensively debated. The topics covered herein carry significant findings which will fuel the growth of the discipline. They may even be implemented as practical applications or may be referred to as a beginning point for another development.

The contributors of this book come from diverse backgrounds, making this book a truly international effort. This book will bring forth new frontiers with its revolutionizing research information and detailed analysis of the nascent developments around the world.

We would like to thank all the contributing authors for lending their expertise to make the book truly unique. They have played a crucial role in the development of this book. Without their invaluable contributions this book wouldn't have been possible. They have made vital efforts to compile up to date information on the varied aspects of this subject to make this book a valuable addition to the collection of many professionals and students.

This book was conceptualized with the vision of imparting up-to-date information and advanced data in this field. To ensure the same, a matchless editorial board was set up. Every individual on the board went through rigorous rounds of assessment to prove their worth. After which they invested a large part of their time researching and compiling the most relevant data for our readers.

The editorial board has been involved in producing this book since its inception. They have spent rigorous hours researching and exploring the diverse topics which have resulted in the successful publishing of this book. They have passed on their knowledge of decades through this book. To expedite this challenging task, the publisher supported the team at every step. A small team of assistant editors was also appointed to further simplify the editing procedure and attain best results for the readers.

Apart from the editorial board, the designing team has also invested a significant amount of their time in understanding the subject and creating the most relevant covers. They scrutinized every image to scout for the most suitable representation of the subject and create an appropriate cover for the book.

The publishing team has been an ardent support to the editorial, designing and production team. Their endless efforts to recruit the best for this project, has resulted in the accomplishment of this book. They are a veteran in the field of academics and their pool of knowledge is as vast as their experience in printing. Their expertise and guidance has proved useful at every step. Their uncompromising quality standards have made this book an exceptional effort. Their encouragement from time to time has been an inspiration for everyone.

The publisher and the editorial board hope that this book will prove to be a valuable piece of knowledge for researchers, students, practitioners and scholars across the globe.

List of Contributors

Abdallah A. Z. A. Ibrahim, Muhammad Umer Wasim, Sebastien Varrette and Pascal Bouvry
FSTC-CSC/ILIAS–Parallel Computing and Optimization Group (PCOG), University of Luxembourg, Avenue de l'Université, L-4365 Esch-sur-Alzette, Luxembourg

Muhammad Umer Wasim and Pascal Bouvry
Interdisciplinary Centre for Security, Reliability and Trust (SnT), Luxembourg City, Luxembourg

Ying Yuan
Department of Clothing & Textiles, Hanyang University, Seoul, Republic of Korea

Jun-Ho Huh
Department of Software, Catholic University of Pusan, Busan, Republic of Korea

Michał Sybis, Paweł Kryszkiewicz and Paweł Sroka
Chair of Wireless Communications, Poznan University of Technology, Poznan, Poland

Ho-Kyeong Ra, Hee Jung Yoon and Sang Hyuk Son
Information and Communication Engineering, Daegu Gyeongbuk Institute of Science and Technology (DGIST), Dalseong-Gun, Daegu, Republic of Korea

John A. Stankovic
Computer Science, University of Virginia, Charlottesville, VA, USA

Jeong GilKo
Software and Computer Engineering, Ajou University, Yeongtong-gu, Suwon, Republic of Korea

J. Madhusudanan
Department of Computer Science and Engineering, Sri Manakula Vinayagar Engineering College, Puducherry, India

S. Geetha and V. Prasanna Venkatesan
Department of Banking Technology, Pondicherry University, Puducherry, India

U. Vignesh
Department of Computer Science and Engineering, K L University, Guntur, Andhra Pradesh, India

P. Iyappan
Department of Computer Science and Engineering, Manonmaniam Sundaranar University, Thirunelveli, Tamil Nadu, India

Xin Li and Yan Wang
School of Computer Science and Technology, China University of Mining and Technology, Xuzhou 221116, China

Dawei Liu
Air Force Logistics College, Xuzhou 221008, China

Xiying Fan, Chuanhe Huang, Shaojie Wen and Xi Chen
School of Computer Science, Wuhan University, Wuhan, China
Collaborative Innovation Center of Geospatial Technology, Wuhan, China

Bin Fu
Department of Computer Science, The University of Texas Rio Grande Valley, Edinburg, TX, USA

Seokhee Jeon, Hongchae Lee and Jiyoung Jung
Department of Computer Engineering, Kyung Hee University, Yongin 17104, Republic of Korea

Jin Ryong Kim
Smart UI/UX Device Laboratory, ETRI, Daejeon 34129, Republic of Korea

Jose Maria Garcia-Garcia, Víctor M. R. Penichet and María Dolores Lozano
Research Institute of Informatics, University of Castilla-La Mancha, Albacete, Spain

Juan Enrique Garrido
Escuela Politécnica Superior, University of Lleida, Lleida, Spain

Effie Lai-Chong Law
Department of Informatics, University of Leicester, Leicester, UK

Weibei Fan
School of Computer Science and Technology, Soochow University, Suzhou, Jiangsu 215006, China

Weibei Fan, Zhijie Han and Ruchuan Wang
Jiangsu High Technology Research Key Laboratory for Wireless Sensor Networks, Nanjing, Jiangsu 210003, China

Zhijie Han
College of Computer & Information Engineering, Henan University, Kaifeng, Henan 475001, China

Nekane Larburu, Arkaitz Artetxe and Jon Kerexeta
Vicomtech, Paseo Mikeletegi 57, 20009 Donostia/San Sebastian, Spain

Nekane Larburu and Arkaitz Artetxe
Biodonostia Health Research Institute, P. Doctor Begiristain s/n, 20014 San Sebastian, Spain

Vanessa Escolar and Ainara Lozano
Hospital Universitario de Basurto (Osakidetza Health Care System), Avda. Montevideo 18, 48013 Bilbao, Spain

Ching-Hsue Cheng and Chung-Hsi Chen
Department of Information Management, National Yunlin University of Science and Technology, 123, University Road, Section 3, Douliou, Yunlin 64002, Taiwan

Md Abdullah Al Hafiz Khan, Nirmalya Roy and H. M. Sajjad Hossain
Department of Information Systems, University of Maryland Baltimore County, Baltimore, MD, USA

Kang Liu, Ruijuan Zheng, Mingchuan Zhang, Junlong Zhu and Qingtao Wu
College of Information Engineering, Henan University of Science and Technology, Luoyang 471000, China

Chao Han
Institute of High Energy Physics, Chinese Academy of Sciences, Beijing 100049, China

Lan Li, Xiaoyong Zhang, Kaiyang Liu, Fu Jiang and Jun Peng
School of Information Science and Engineering, China Hunan Engineering Laboratory of Rail Vehicles Braking Technology, Central South University, Changsha, Hunan 410083, China

Kanghyun Choi, Yeo-Jin Yoon and Soo-Mi Choi
Department of Computer Science and Engineering, Sejong University, Seoul, Republic of Korea

Yeo-Jin Yoon and Soo-Mi Choi
Mobile Virtual Reality Research Center, Sejong University, Seoul, Republic of Korea

Oh-Young Song
Department of Software, Sejong University, Seoul, Republic of Korea

Odongo Steven Eyobu, Jhihoon Joo and Dong Seog Han
School of Electronics Engineering, Kyungpook National University, 80 Daehak-ro, Buk-gu, Daegu 41566, Republic of Korea

Odongo Steven Eyobu
School of Computing & Informatics Technology, Makerere University, Plot 56, Pool Road, Kampala, Uganda

Index

Printed in the USA
CPSIA information can be obtained
at www.ICGtesting.com
JSHW051433221024
72173JS00006B/1459

9 781632 409218